T0328639

Microgrid Protection and Control

Microgrid Protection and Control

Dehua Zheng
Goldwind Sc. & Tec. Co. Ltd., Economic & Technological Development
Zone, Beijing, China

Wei Zhang
Beijing Etechwin Elec. Co., Ltd., Goldwind Sc. & Tec. Co. Ltd.,
Economic & Technological Development Zone, Beijing, China

Solomon Netsanet Alemu
Goldwind Sc. & Tec. Co. Ltd., Economic & Technological Development
Zone, Beijing, China

Ping Wang
Beijing Etechwin Elec. Co., Ltd., Goldwind Sc. & Tec. Co. Ltd.,
Economic & Technological Development Zone, Beijing, China

Girmaw Teshager Bitew
Goldwind Sc. & Tec. Co. Ltd., Economic & Technological Development
Zone, Beijing, China

Dan Wei
Goldwind Sc. & Tec. Co. Ltd., Economic & Technological Development
Zone, Beijing, China

Jun Yue
Shenyang Institute of Engineering, Shenyang City, Liaoning province,
China

ACADEMIC PRESS
An imprint of Elsevier

ELSEVIER

Academic Press is an imprint of Elsevier
125 London Wall, London EC2Y 5AS, United Kingdom
525 B Street, Suite 1650, San Diego, CA 92101, United States
50 Hampshire Street, 5th Floor, Cambridge, MA 02139, United States
The Boulevard, Langford Lane, Kidlington, Oxford OX5 1GB, United Kingdom

British Library Cataloguing-in-Publication Data
A catalogue record for this book is available from the British Library

Library of Congress Cataloging-in-Publication Data
A catalog record for this book is available from the Library of Congress

ISBN: 978-0-12-821189-2

For Information on all Academic Press publications
visit our website at https://www.elsevier.com/books-and-journals

Publisher: Brian Romer
Acquisitions Editor: Lisa Reading
Editorial Project Manager: Chris Hockaday
Production Project Manager: Prem Kumar Kaliamoorthi
Cover Designer: Miles Hitchen

Typeset by MPS Limited, Chennai, India

Working together
to grow libraries in
developing countries

www.elsevier.com • www.bookaid.org

Contents

Preface

Microgrid is a concept relative to the conventional large power grid that constitutes interconnected loads and distributed energy resources (DER) on a small scale and in a geographic area. Microgrids are becoming a common fixture in power systems across the world for different reasons. While the environmental incentive due to convenience for deployment of renewable resources is the key factor policy wise, there are technical advantages such as improved power supply reliability and optimal utilization of DER attached to the growing interest in microgrids. However, due to the size, architecture, operating modes, and incorporation of DER, microgrids face several challenges, including fault protection and transient and dynamic control issues. Some of the issues exhibited in microgrids are bidirectional flow of power, fault current being supplied from multiple directions, loss of synchronism among DER units after fault incidents, small fault current magnitude in island mode, and lower system inertia. Those and other technical issues are even more pronounced when the microgrid is dominated by converter-based generators and operating in island mode. Thus conventional protection and control techniques and equipment may not be adequate for microgrids, which leads to the need for the special protection and control methods that make up the core of this book. The book is especially appealing in the fact that it has modular content that makes it convenient for readers and includes practical application and experimental examples.

There are 11 chapters contained in this book. The concept of the microgrid and related terminologies are presented in Chapter 1, The Concepts of the Microgrid and Related Terminologies. Chapter 2, Current Industrial Practice and Research Trends in Microgrids, discusses the practices and experiences in relation to the microgrid in the power system industry. The current and forecasted market trends of microgrids around the world, as well as the different regions are reviewed. Chapter 3, Key Technical Challenges in Protection and Control of Microgrid, introduces the main technical challenges related to the protection and control of microgrids. Short-term forecasting of generation and demand is very important for the dynamic control and protection and it is explained in Chapter 4, Short-term Renewable Generation and Load Forecasting in Microgrids. The short-term forecasts which target forecasting the generation or load for the next few hours to few days range are the focus of this chapter. The chapter presents basic concepts, classification, different techniques, and practical application examples. In Chapter 5, Fault and Disturbance Analysis in Microgrid, a discussion on how to identify faults from other types of disturbances in microgrids is provided. In addition, there are farther discussions on the basic concepts of power system fault analysis and special features of fault analysis of microgrids. Chapter 6, Protection of Microgrids, addresses the protection system which is one of the most challenging issues in microgrid operation. This chapter first introduces what a protection system in general is and what the main requirements for the protection of microgrids are. Discussions of advanced approaches such as adaptive protection and machine learning–based methods make up the subsequent part of the chapter. The widely accepted hierarchical structure of microgrid control includes primary, secondary, and tertiary control levels. The three chapters from Chapter 7 onward address those control levels. Causes, characteristics, and ways of handling dynamic disturbances in microgrids are discussed in Chapter 7, Dynamic Control of Microgrids. The chapter starts with an introductory description of dynamic disturbance and control. Possible causes of dynamic disturbance and the resulting changes in waveforms and system parameters are described. Some state-of-the-art techniques are presented with detailed formulations. Control strategies which can fit the special requirements of microgrids are presented with simulation and experimental results. Causes, characteristics, and ways of handling transient disturbances in microgrids are discussed in Chapter 8, Transient Control of Microgrids. Descriptions of transient disturbance and control are detailed in the earlier part of the chapter, followed by theoretical and mathematical explanations of how to design control systems to handle transient disturbances. Chapter 9, Tertiary Control of Microgrid, is composed of three critical aspects of tertiary control in microgrids: optimal energy dispatching, demand side management, and energy efficiency. Algorithms, mathematical modeling, and techniques presented in the literature and practically implemented in the industry are elaborated. Chapter 10, Communication Requirements of Microgrids, discusses the communication requirements and the available communication media and protocols for application in microgrids. Chapter 11, Application Cases of Industrial Park Microgrids Protection and Control, presents application cases of two microgrid projects. Detailed elaboration of the topology, constituting elements, and the tested techniques of protection and control systems in the respective microgrid projects are provided. The operational results from the protection and control systems deployed in the microgrids are also presented.

Acknowledgments

In the process of writing this book, we have obtained the strong support of Goldwind Science & Technology Co., Ltd., Zhixue Shi, Yanpeng Xue, and Xun Zhang. Professor Boming Zhang of Tsinghua University and Professor Zhijie Wang of Shanghai Dianji University have also contributed as editors. The publishing team was also very supportive and devoted. We would like to express our deep gratitude to all of them.

Authors
February 2021

Chapter 1

The concept of microgrid and related terminologies

1.1 Introduction

The existing large-scale electrical grids which have been operational for many the years are based on generation of a massive amount of power from conventional generating plants, such as coal, hydro, nuclear, and others; transmitting the power over long distances and serving large communities through distribution networks. However, the idea of having generation units closer to the point of demand and coordinated operation of the distributed units has gained a lot of attention in recent years. One aspect contributing to such a paradigm change is the growing concern with regard to the depletion of the traditional energy resources, the environmental issues related to them, and the new renewable energy sources becoming more affordable. As such, we observe a large number of buildings being equipped with small local solar systems that can serve one property.

The idea of a microgrid is changing the energy infrastructure from how it was perceived for long time. The large grids were observed, on multiple occasions, to fail to ensure a reliable and secure power supply. Simple events such as a tree falling are enough to cause power outages to hundreds of properties. Localized grids, in the form of microgrids, would have the advantage of staying operational during such events in the utility grid by disconnecting themselves from the network and making use of the local generation and stored energy. The maintenance of such systems is also much easier as it would mostly involve fixes that are much closer to the property and easier to troubleshoot.

Although there is a little variation on the way microgrids are defined in the literature, the more widely acknowledged definition from International Electrotechnical Commission (IEC) is:

> Microgrid is the group of interconnected loads and distributed energy resources with defined electrical boundaries forming a local electric power system at distribution voltage levels, that acts as a single controllable entity and is able to operate in either grid-connected or island mode [1].

Microgrid Protection and Control. DOI: https://doi.org/10.1016/B978-0-12-821189-2.00008-5

1

Microgrids have recently been deployed widely in many countries. They have been applied in different parts of the world for different reasons. They are a better alternative for rural electrification in areas without access for electricity while they are also being widely implemented in electrified areas as well. The basic reason behind their deployment in the developed areas is the growing need for reliable and secure power supply. With repeated severe weather events, equipment failures, and sudden spikes in demand causing blackouts and brownouts in may developed countries, the idea of converting the power grid to an aggregate of smaller microgrids is appearing as a viable and appreciated option. The basic structure and common components of microgrid are presented in Fig. 1.1. Some of the key features of a microgrid are [2,3]:

- ability to operate in grid-connected and island modes;
- having one or more points of connection (POC) to the utility grid;
- acting as a single controllable entity to the bigger utility grid;
- incorporation of interconnected loads and local generation sources;
- providing improved power quality and reliability for customers;
- accommodating total system energy requirements; and
- ability to interact with the utility grid and optimize performance and savings.

1.2 Related concepts

It is important to note that microgrids are different from other concepts such as virtual power plants (VPPs), active distribution networks (ADNs), or energy internets. There seems to be some confusion on the understanding and presentation of those different concepts in both academia and the

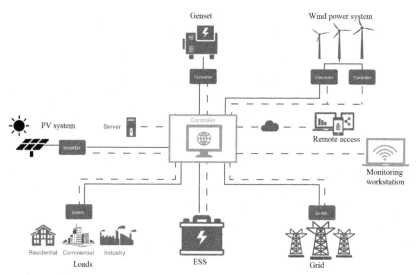

FIGURE 1.1 Schematic of typical microgrid.

industry. In order to give an insight into the major differences between these concepts, we shall try to briefly outline what those terms actually represent.

1.2.1   Active distribution network

ADNs represent distribution networks with systems in place to control a combination of distributed energy resources (DERs) and permit the distribution network operator to manage the energy exchange through a flexible network topology. Central systems that are used to realize ADN are Supervisory Control and Data Acquisition systems and Distribution Management System. The distribution network automation includes control center information systems, substation automation, and customer interfaces such as smart meters [4]. This description of ADN indicates that an ADN is a larger and broader concept than a microgrid. The supervision and control of network state, network breakers and switches, DERs, and customers are included in ADN operation. A wide area of information of all actors in the distribution system, including transmission system operator, energy retailers, and local communities, is collected and utilized [4].

1.2.2   Energy internet

Energy internet is one of the technologies or concepts that have been introduced to take advantage of advanced information communication technologies (ICT) to address the limitations of the electrical grid. The concept of "energy internet" was introduced by Huang et al. [5] as an alternative architecture for a future electric power distribution system that is suitable for plug-and-play of distributed generations (DGs) and energy storage systems (ESS) and which allows automated and flexible electric power distribution. There are two different views of energy internet; one is that it refers to a physically localized system which is analogous to microgrids, and another which prefers to represent the energy internet as a wider and higher level system, mostly referring to the communication and controllability functions. The energy internet is characterized by the integration of diverse energy generation and consumption, interconnection of multiple DERs, and extensive application of ICT [6]. One way of identifying the two concepts suggested by Cao and Yang [7] is to see the energy internet as an internet-based wide area network for information and energy fusion that takes the electrical grid as the backbone network and the microgrid or DGs as the local area network.

1.2.3   Virtual power plant

The term VPP is a relatively new one compared to those discussed above. As is the case with most of these new terminologies, there are differences in defining VPP. However, the IEC's online vocabulary tool, International Electrotechnical Vocabulary, defines VPP as a group of distributed energy resources which combine to function as a dispatchable unit [1]. The definition further adds a note that a VPP may be used for the purpose of participating in the electricity market or aggregating ancillary services. Some definitions focus on the special need of VPPs for software systems to enable remote automatic dispatching and optimization of generation or the management of demand side or storage units in a single web-connected system [8]. Though microgrid and VPPs share some critical features, there are major differences between them. The major difference with microgrids is that generating units of a VPP do not necessarily need to be physically located in a localized area with close proximity to each other. Other differences between the two [8] are microgrids being dependent upon hardware innovations like inverters and smart switches, while VPPs are heavily reliant upon smart meters and IT. It is also stated that storage units are mostly needed in microgrids while they may not feature in VPPs.

1.3   Misconceptions about microgrid

One very common misconception regarding microgrids is the trend to consider residential grid-tied conventional photovoltaic (PV) systems as microgrids while they actually are just DGs systems. Those systems are expected to operate only in parallel with the grid with either the generation surplus to the local demand or the whole generated power is fed into the grid. However, they cannot operate isolated from the grid and in events of grid outage the load cannot be powered even when there is available generation.

Some of the other typical misconceptions regarding microgrids are:

- *Microgrids are unreliable and highly vulnerable to failures and black-outs because they are composed of intermittent and fluctuating renewable energy resources:* This is a wrong perception which actually contradicts the main plus from the use of microgrids. A considerably large number of microgrids have been successfully installed across the spectrum of critical facilities, from airports to hospitals and utility providers with the intended and proved target of ensuring backup to those critical facilities during possible outages due to storms and other events. It is also important to remind ourselves that conventional sources such as diesel gensets or microturbines can be elements of microgrids. Microgrids are furnished with a central controller or energy management system (EMS) that ensures the stable and secure operation.
- *Microgrids are exclusively independent and isolated systems:* This is addressed better in the next section where the classification of microgrids is discussed. However, it is worth noting that even the definition of microgrids incorporates the option of grid-connected operation and microgrids are not necessarily isolated. There is a group of microgrids called nonisolated microgrids which function in parallel with the grid.
- *Customers who own microsources make up a microgrid:* This again is similar to counting roof-mounted solar systems in individual homes as

microgrids. It is not only PV systems but also others such as small wind systems or other microsources that are simply DGs systems, and not microgrids, because of the fact that they do not involve a central microgrid controller and lack coordination between units.

- *Microgrids are too expensive to construct and their applicability is limited to field tests or feasible only for application in remote locations:* However, it is possible to build cost-effective microgrids without subsidies or incentives and microgrids and components are becoming more and more affordable. There are also various available business models which allow the utilization of microgrids without the need to invest the whole upfront cost.

- *Microgrid operators force end-users to shift their loads based on availability of renewable generation:* Though demand response techniques are becoming popular not only in microgrids but generally in the modern power system, it should not be interpreted as some forceful action applied on customers. It rather is an option which allows for greater involvement of consumers in decision-making with some incentive mechanisms. It should also be clear that demand response mechanisms are not a necessity in microgrids.

- *Construction of microgrid requires rebuilding the entire power network:* This is related to an understanding that the microgrid is a completely new concept which cannot be incorporated in the existing power system. However, the simple fact is that although smart features and today's modern equipment are part of microgrids, the concept of the microgrid is not significantly different from the early time grids of Thomas Edison. So, it doesn't require getting rid of the existing infrastructure and rebuilding. It is rather a matter of adding some components and controllability (as a single unit) features to an existing distribution system.

- *There will not be any power interruption in microgrid loads:* Similar to labeling microgrids as specially vulnerable to failure, it is absolutely unrealistic to expect them to be completely shielded from possible failures. It is part of human nature that you cannot completely avoid failure with 100% certainty in any man-made device or system. Microgrids are not any different. The redundancy and incorporation of smart technologies in microgrids allows for improved power supply availability though there may still be some extreme events and internal faults that could lead to power interruptions within microgrids.

1.4   Types of microgrid

The most commonly known way of classifying microgrids is based on the availability of a physical link with the larger grid. As such, we can have two types of microgrids:

1. *Isolated microgrids* are microgrids that cannot be connected to the wider electric power system, which means they are physically isolated from the

utility grid. Such microgrids are usually built and operated for geographical islands or for rural electrification. Isolated microgrids are also referred to as stand-alone microgrids.

2. *Nonisolated microgrids* are, on the other hand, microgrids that can be connected to the wider grid and hence have two possible modes of operation: grid-connected and island modes. Island refers to a portion of a power system which is disconnected from the remainder of the system while still remaining energized [1]. This island can be either intentional or unintentional.

Microgrids can also be of different types based on their applications. The classification varies from literature to literature. The most common types of microgrids are [2,9]:

- Campus/institutional microgrids
- Remote off-grid/stand-alone microgrids
- Commercial microgrids
- Industrial microgrids
- Military base microgrids
- Community microgrids
- Utility microgrids

There also are "nanogrids," which are mentioned in some of the literature, to represent the smallest discrete network units (to the level that it can be a single building) that are capable of operating independently.

The other way to classify microgrids is as Alternating Current (AC) microgrids and Direct Current (DC) microgrids, based on the system architecture or the voltages and currents adopted in the microgrids [10]. The basic arrangements of both types of microgrids is shown in Fig. 1.2. In AC

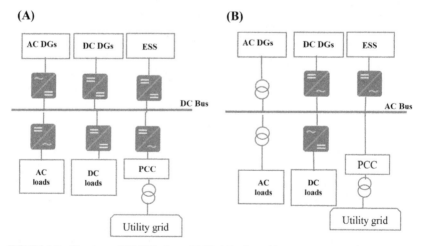

FIGURE 1.2 Structure of (A) DC microgrid (B) AC microgrid.

microgrids, all DERs and loads are connected to a common AC bus. DGs whose output are DC and the ESS are connected to the AC bus through converters while DC loads are also supplied through rectifiers. In DC microgrids, on the other hand, the common bus is DC and this time it is the ac output DGs and AC loads that would need the converter interface. There could also be hybrid microgrids, which are a combination of AC and DC microgrids and consist of both types of buses.

1.5   Components of microgrids

A typical microgrid consists of the following basic components [3,11,12]:

- DERs—includes DGs units and distributed ESS
- Power conversion system (PCS)
- Controllers and EMS
- Communication system
- Loads (uninterruptable, critical, and noncritical)
- Protection system

1.5.1   Distributed Generation

DGs can be of conventional type or electronically coupled type. The conventional type DGs are those such as synchronous generators or induction generators. Electronically coupled type DGs are those whose output does not fit with the system frequency and hence need solid-state converters to connect to the microgrid. In terms of controllability of the generated power, DG units can also be classified as dispatchable or nondispatchable.

Some of the most common DGs in microgrids are photovoltaic arrays, wind turbines, and combined heat and power microturbines. The conventional sources such as synchronous generators driven by internal combustion engines or small-scale hydro can also serve as DGs in a microgrid.

1.5.2   Energy storage systems

ESS are a very common constituent of microgrids. ESSs can be of two major types: energy intensive and power intensive. Some of the common energy intensive ESSs are batteries of different types, fuel cells, and thermal energy storages. They are particularly important to achieve the demand side management and other EMS applications. Though the initial cost of ESS is relatively high, they contribute a lot to the economic operation of a microgrid. They also contribute to the reliability of the microgrid system ensuring power availability in the case of interruptions.

The power intensive ESSs are those such as supercapacitors, flywheels, superconducting magnetic energy storage, and compressed air energy storage. Power intensive ESSs can be used in microgrids to stabilize the system during sudden changes and fluctuations in generation and/or loading conditions. Among the available energy storage technologies, the ones particularly mentioned as suitable in microgrid applications are batteries, flywheels, and supercapacitors [9].

1.5.3 Power conversion system

The PCS is the component of a microgrid through which electronically coupled type DERs are connected to the microgrid. They convert AC input with a frequency different from the system frequency or DC input to an AC output of standard frequency (50 or 60 Hz). DGs such as variable speed wind turbines (e.g., direct driven turbines), microturbines, PV arrays, batteries, and supercapacitors need a PCS to connect to the microgrid. The existence of a PCS in a microgrid offers the possibility for control of the active and reactive power, voltage, and frequency.

1.5.4 Controllers and energy management system

Microgrids are equipped with different levels of controllers. The most commonly known way to classify the control applications and respective controllers in a microgrid is as:

- Primary control
- Secondary control
- Tertiary control

The primary control of microgrids is targeted at providing voltage stability, plug-and-play capability of DERs, preserving frequency stability, and avoiding circulating current among DERs [10]. The secondary control on the other hand compensates for the voltage and frequency deviation caused by the primary control. The tertiary control is responsible for the achievement of optimal operation and energy exchange with the utility grid in grid-connected operation.

One of the requirements in a microgrid is an economical and smart way of energy management and this can achieved through EMS. In the case of a microgrid comprising more than two DERs, the EMS is needed to impose the power allocation among the DERs, the cost of energy production, and emissions [10]. The EMS takes inputs such as forecasts of generation from renewable DGs, load demand, and the power tariff to make an optimal decision on exporting or importing power to/from the grid, optimal dispatching of the dispatchable DGs, and charging or discharging of ESS (Fig. 1.3). In doing so, the operational objectives of achieving lower cost or lower pollution are the targets, while the constraints, in form of capacity limits of the DGs, charging/discharging power limits of the ESS, generation-demand matching and other technical limitations, are satisfied.

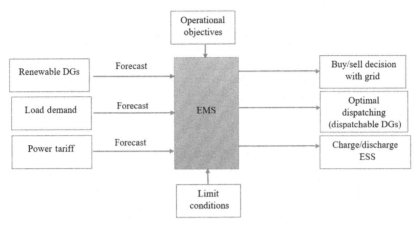

FIGURE 1.3   Energy management system of microgrid.

1.5.5   Communication system

The communication system is an integral part of a microgrid and it integrates smart grid features in terms of control and monitoring features of the microgrid. The type and role of the communication system in a microgrid is dependent on the design of the control system and the composition of the constituting components [13]. The deterministic nature of real-time control demanded in microgrid operation is related to a communication system where signals can be delivered without delay. The use of an appropriate communication system in a microgrid guaranties safe, secure, reliable, sustainable, and economic operation and control. This can be achieved by utilizing an internet communications protocol suite which is made up of a layered architecture with each layer employing one or more protocols. The most widely used and available suite in a microgrid is the one known as Transmission Control Protocol/Internet Protocol (TCP/IP) which consists of four layers [13]. The four layers in a TCP/IP protocol are the application layer, the transport layer, the network layer, and the link layer.

Several protocols are employed in microgrid control systems to enable communication between the system components and intelligent electronic devices. The microgrid controller may communicate with intelligent electronic devices and other components using the standard IEC 61850 via an ethernet using TCP/IP. The architecture, shown in Fig. 1.4, also suggests the presence of human−machine interfaces in order to provide monitoring and controlling functions.

1.5.6   Loads

Microgrids can have electrical, thermal, or other types of loads. It is common to have more of controlled and smart loads in microgrids as that

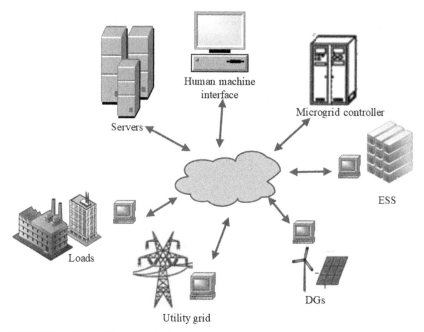

FIGURE 1.4 Microgrid communication systems.

would allow the microgrids to implement demand response actions and allow the customer to have greater control over their consumption and energy bill.

Loads can also be classified as critical and noncritical based on whether they can be controlled or not and the level of criticality. Uninterruptable loads are critical loads which are usually supplied from an uninterruptible power supply to avoid supply interruption even for a fraction of seconds [14]. Critical loads are loads which must be served all the time and cannot be shed regardless of the amount and cost of generation [15]. Noncritical loads, on the other hand, can be scheduled to achieve economic operation of the microgrid [14]. There is also another way of distinguishing loads in a microgrid, that is, as forecastable and nonforecastable loads [16]. Forecasting the load demand for a short-term future helps to realize economic demand and response management.

1.5.7 Protection system

Microgrids must be equipped with a protection system that shall respond when faults occur both in the utility grid and within the microgrid itself. When faults appear in the utility grid, the protection system should trip the circuit breaker at the POC with the grid and disconnect the microgrid from the utility grid as rapidly as possible. When internal faults happen, the protection system should be able to isolate the smallest possible section of the microgrid to eliminate the fault [17].

The conventional protection system in the traditional distribution system is usually based on short-circuit current sensing. The special case in microgrids is the large-scale presence of converter-coupled DGs whose short-circuit currents are limited by the ratings of the switching devices to not more than twice the rated current. Fault currents in microgrids dominated by converter-interfaced DGs operating in island mode may not have adequate magnitudes to use the conventional overcurrent protection techniques. This results in requirements for an advanced protection strategy [18].

References

[1] IEC, International Electrotechnical Vocabulary (IEV), IEV Definitions. <http://www.electropedia.org/>.

[2] E. Hayden, Introduction to Microgrids, Securicon, Alexandria, VA, 2013.

[3] D.N.V. Kema, Microgrids – Benefits, Models, Barriers and Suggested Policy Initiatives for the Commonwealth of Massachusetts, Burlington, MA, 2014.

[4] S. Repo, et al., Active distribution network concept for distributed management of low voltage network, 2013 4th IEEE/PES Innovative Smart Grid Technologies Europe, ISGT Europe 2013, 2013. Available from: http://doi.org/10.1109/ISGTEurope.2013.6695428.

[5] A.Q. Huang, M.L. Crow, G.T. Heydt, J.P. Zheng, S.J. Dale, The future renewable electric energy delivery and management (FREEDM) system: the energy internet, Proc. IEEE (2011). Available from: https://doi.org/10.1109/JPROC.2010.2081330.

[6] K. Zhou, S. Yang, Z. Shao, Energy internet: the business perspective, Appl. Energy (2016). Available from: https://doi.org/10.1016/j.apenergy.2016.06.052.

[7] J. Cao, M. Yang, Energy internet - towards smart grid 2.0, in: Proceedings of the International Conference on Networking and Distributed Computing, ICNDC, 2014, doi: 10.1109/ICNDC.2013.10.

[8] P. Asmus, Microgrids, virtual power plants and our distributed energy future, Electr. J. (2010). Available from: https://doi.org/10.1016/j.tej.2010.11.001.

[9] C.S. Peter Asmus, Utility Distribution Microgrids - Research Report, 2012.

[10] H. Lotfi, A. Khodaei, AC versus DC microgrid planning, IEEE Trans. Smart Grid (2017). Available from: https://doi.org/10.1109/TSG.2015.2457910.

[11] Z. Sun, X. Zhang, Advances on distributed generation technology, Energy Procedia (2012). Available from: https://doi.org/10.1016/j.egypro.2012.02.058.

[12] S. Mu, M. Huang, J. Yang, J. Yu, T. Li, J. Hu, Overview of communication and control techniques in the microgrid, Appl. Mech. Mater. (2011). Available from: https://doi.org/10.4028/www.scientific.net/AMM.71-78.2382.

[13] A. Bani-Ahmed, L. Weber, A. Nasiri, H. Hosseini, Microgrid communications: state of the art and future trends, in: 3rd International Conference on Renewable Energy Research and Applications, ICRERA 2014, 2014, doi: 10.1109/ICRERA.2014.7016491.

[14] S. Ganesan, S. Padmanaban, R. Varadarajan, U. Subramaniam, L. Mihet-Popa, Study and analysis of an intelligent microgrid energy management solution with distributed energy sources, Energies (2017). Available from: https://doi.org/10.3390/en10091419.

[15] Z. Guo, D. Sha, X. Liao, Energy management by using point of common coupling frequency as an agent for islanded microgrids, IET Power Electron. (2014). Available from: https://doi.org/10.1049/iet-pel.2013.0564.

[16] C. Chen, S. Duan, T. Cai, B. Liu, G. Hu, Smart energy management system for optimal microgrid economic operation, IET Renew. Power Gener. (2011). Available from: https://doi.org/10.1049/iet-rpg.2010.0052.

[17] N. Hatziargyriou, Microgrids: Architectures and Control, Wiley-IEEE Press. (2013).

[18] H. Nikkhajoei, R.H. Lasseter, Microgrid protection, in: 2007 IEEE Power Engineering Society General Meeting, PES, 2007. Available from: https://doi.org/10.1109/PES.2007.385805.

Chapter 2

Current industrial practice and research trends in microgrids

2.1 Introduction

Microgrids facilitate optimal utilization of distributed renewable energy, as they provide a better local energy supply and reduce losses in energy transmission while ensuring environmental friendliness through reducing the emission of greenhouse gases. They have capabilities such as self-repair, allowing participation of the consumer in the network operation, and, above all, efficient operation of the electrical network. For those and other reasons, microgrids are being widely applied in different parts of the world.

It has been stated in the previous chapter that one of the misconceptions regarding microgrids is the perception that they are a thing of the future rather than today. In this chapter, we briefly review the current trends in the application of microgrids so as so to demonstrate the applicability and current status of the technology. The technical challenges and the research works being carried out to address those challenges are also discussed.

In recent years, the electrical distribution network has been facing continual changes. A vital part of the evolution of the power system is the manifestation of microgrids as a basic element of the new smart grid. Microgrids allow distributed energy resources to be integrated with the main network. Thus we are witnessing the take-off of small-scale renewable generation, distributed and incorporated into the traditional electricity distribution networks.

One of the reasons for the recent evolution is the technological changes and the reduction in costs of the technologies that followed. In the last 20 years, the cost of installing renewable generation has dropped significantly. According to the International Renewable Energy Agency, the cost of energy has dropped since 2010 by 82% for photovoltaic (PV) solar, by 47% for concentrated solar energy, by 39% for onshore wind, and by 29% for offshore wind [1]. Electrical storage technology, especially in the form of batteries, has also undergone improvements both in terms of performance and cost. Those technological advancements and the reduction in cost have allowed for the possibility of providing electricity to a small grid disconnected from the main grid, generating the electricity with renewables and completing the

Microgrid Protection and Control. DOI: https://doi.org/10.1016/B978-0-12-821189-2.00009-7

system with batteries and minimal use of conventional sources like diesel gensets.

Despite the advances in distributed generation and energy storage technologies, the business case of a renewable system disconnected from the grid is not always justified, unless the broader vision of its advantages, such as the carbon footprint, is taken into account. Many small grids around the world are isolated from the main grid and a lot of them are powered by diesel generators. It is in these remote diesel-powered systems that renewable energy can have the greatest impact on profitability. The recent technological and economic changes in renewable systems and batteries make it possible to power a system more economical than just using diesel generators, even on small projects. It is important to note that any renewable technology system is effective only when it is located in an area that has the natural resources such as the sun or the wind. In an area with a good wind resource, wind turbines could be more suitable, whereas in areas with strong solar irradiation solar panels could be more suitable.

Microgrids are being applied in different sectors at different sizes. The earlier microgrids were mostly demonstrational and installed in universities and research centers. A typical microgrid and usually considered as the first microgrid is the CERTS (The Consortium for Electric Reliability Technology Solutions) microgrid. The CERTS microgrid was initiated in 1999 and led by Lawrence Berkeley National Laboratory under the US Department of Energy. Universities, electric power companies, manufacturers, public research institutes, etc. have participated in the microgrid project.

There are different viewpoints regarding the motivation and significance of microgrids in different parts of the world according to the characteristics of the power system in each region. For example, in the United States, the reliability of the electric power system is an important issue, as symbolized by the power outages of 1996 and 2003. Thus the microgrid is regarded as a means to ensure reliable power to customers. In Europe, the main objectives are to expand the introduction of renewable energy and to build an independent energy system on remote islands. In Japan, the latest natural disasters causing catastrophic impacts on large nuclear power plants has resulted in greater attention toward the application of localized (specially renewable) energy systems. In developing countries, on the other hand, microgrids serve the very basic purpose of providing a modern power source to the nonelectrified rural areas. A report by the International Energy Agency states that 675 million people, 90% of them living in sub-Saharan Africa, will not have access to electricity until 2030 [2]. Renewable energy-based microgrids are expected to contribute to a great extent in improving this situation.

As we move toward the power grid of the future, microgrids take on greater importance, as they are able to unify small-scale and flexible generation to clean energy and smart controls. In the domestic sector, microgrids are implemented to supply electricity to residential buildings. Near Zero Energy Buildings, better known as houses with almost zero energy consumption, form exemplary practice for the application of microgrids in the sector. Financial models, such as Microgrids-as-a-Service, which allow the deployment of microgrids without upfront investment, are also becoming important tools for the deployment of microgrids, especially in the less developed areas.

In general, although the objectives and financial models differ depending on the region, the application and research of microgrids are being promoted almost in every country.

2.2 The current industrial trends in microgrids

2.2.1 Microgrid global market trends

Although the microgrid is still an emerging industry in many regions, its market has been growing in recent years and the market trend is expected to continue with an even steeper trajectory in the coming decades. Microgrids play great significance in supplying power to rural communities, enterprises, and close to a billion people without electricity worldwide. A report by International Market Analysis Research and Consulting Group (IMARK) estimates the global market of microgrids to be worth US$ 21.6 billion in 2019 [3]. Over the period of 2020−25 this market is expected to grow at a compound annual growth rate of more than 13% [4]. Some of the factors contributing to the growth of the microgrid market are:

- increasing demand for reliable power supply;
- growing interest and commitment to reduce the overdependence on large grids;
- avoiding risks of power blackouts related to natural calamities;
- improved economic efficiency of electrification through renewable energy-based island systems for remote locations;
- growing concerns and pledges to address environmental impact;
- cyber security issues in large power grids and the need for secure and reliable power supply options in sectors such as military and research institutions; and
- low transmission losses in microgrids (especially those located in far places) compared to transmitting generated power over long distances.

The unanticipated and unfortunate pandemic of COVID-19 is expected to have either a positive or negative influence in almost every sector. What it will bring to the microgrid market has not been studied in detail yet. However, there are initial reports suggesting the sector will benefit in the long term from the pandemic, against the evident near-term negative impacts including project delays in 2020 or even 2021 [5].

Navigant Research's *Microgrid Deployment Tracker* is considered to be one of the most comprehensive databases published periodically for tracking the deployment of microgrids in the world. The 16th edition of the report identified a total of 4475 microgrid projects up to the second quarter of 2019, representing 26,965 MW (including planned and installed) power capacity [6]. The latest edition (3Q 2020) of this tracker takes those figures to a total of 7968 projects and 34.6 GW capacity [7]. The year 2019 saw a change of leader in the regional share of the microgrid market with Asia-Pacific overtaking North America both in terms of capacity and number of projects. In terms of installed and planned capacity, 37% (i.e., 9935.4 MW) of the global microgrid market is reported to belong to the Asia-Pacific region compared to 33% (8878.6 MW) of North America, according to Navigant Research report [6,8]. According to the same report, the Middle East and Africa contributes to 14% of the global microgrid market which translates to 3627.7 MW power capacity. Europe's growing inclination toward virtual power plants (VPP) and other options means the regional market assumes a lower rank with an 11% share, only bettering the 6% of Latin America [6].

North America is considered to be the pioneer for microgrid deployment and has been the leader in terms of the market share until 2019 [3]. The 17th edition of Navigant Research's tracker reports North America coming back to the leading position in the first quarter of 2020 with 36% of the global market share [9]. The United States microgrid market is dominated by a grid connected system and Alaska is the state with the largest share (50%) followed California and New York contributing 13% and 10%, respectively, up to the second half of 2019 [6]. There is an increasing use of microgrids in the defense sector, especially in remote military bases; the growing demand for security against cyber attacks is one of the major contributing factors for the wider application of microgrids in this area.

After overtaking North America as a leader in 2019, the *Asia-Pacific zone (APAC)* is expected to continue a similar trend as the fastest-growing market during the forecast period of 2020–25, according to Mordor Intelligence [4]. The major factors that favor a growing market and deployment of microgrids in the Asia-Pacific region are:

- rural electrification programs in developing countries;
- development of commercial microgrids in advanced nations of the region;
- Japan's policies to boost renewable energy utilization and energy efficiency supported by sustained efforts toward smart grid deployment; and
- affirmative government policies such as feed-in-tariffs.

The fact that *Japan* is intending to shift away from large-scale power grids toward microgrids for the sake of increasing resiliency against natural disasters after the latest unfortunate events is a major specific point related to the recent and forecasted progress in microgrid deployment in the region. One typical example of such a move is the launching of a microgrid project (in late 2017) to electrify 117 homes in Shioashiya Solar-Shima of Japan to be developed by PanaHome Corporation, ENERES Co. Ltd and IBJ Leasing Co. Ltd. [10]. The growth in rate of rural electrification in developing economies such as India, Malaysia, and Philippines, especially for the nonelectrified islands and remote areas in those countries, is expected to contribute to the market growth in APAC.

The microgrid market in *China* is also worth a special mention. The announcement by the National Energy Administration and the National Development and Reform Commission in October 2017 for an initiative of market-oriented distributed power generation was one important milestone contributing to the microgrid sector development in China [11]. In a more recent event on January 20, 2020, the Electronic Information Department of the Ministry of Industry and Information Technology released a list of smart PV pilot demonstration projects [12]. Five of the 22 demonstration projects directly include microgrids, namely the 13 MWh microgrid demonstration project of Suixi Bus Station, and the Tebian Electric Apparatus (TBEA) Xi'an Industrial Park microgrid; the Beijing-Tianjin-Tang Expressway 5 km of PV-energy storage microgrid project; Triad Smart PV Microgrid System Pilot Demonstration; and Qingdao Neusoft Carrier Park Microgrid Project [12]. This move has sent two positive signals to the promotion of the country's microgrid. One is that there are a number of national-level microgrid projects while the second and most important message is that that in addition to the previous efforts by the National Development and Reform Commission and the National Energy Administration [13], another heavyweight department, the Ministry of Industry and Information Technology, has joined the promotion of microgrid development. This has ushered in new development opportunities for the microgrid, which has always been in the air but never heated up. There also are more commercial microgrids such as Hebi, Dafeng, and Jiaze microgrids which are discussed in more detail in Chapter 11, Application Case of Industrial Park Microgrids Protection and Control, of this book.

Europe has often been touted as a global leader in the utilization of renewable energy and low-carbon energy. There was also the expectation that the integration of microgrids with the wholesale power market would be a big boost for the microgrid market in Europe. However, the regional market constitutes a tiny 9% of the global market in early 2020 [9]. One of the factors for the relatively lower deployment of the microgrid in the region is the policy inclination toward the development of large-scale renewable energy, including offshore wind farms, that favors VPP rather than microgrids. It was also not possible to consider the large number of feed-in tariff-based rooftop solar PV systems as they do not satisfy the key defining requirements of a microgrid [9].

With increasing investment in electrification projects by governments as well as national and international companies, the *African market* is also expected to present a big opportunity for microgrid market players, especially in remote off-grid systems. Africa is one of the regions in the world with majority of the population being unable to access basic electricity services despite the availability of abundant renewable energy resources (RES) [14]. Recent times have seen an increasing number of microgrid projects implemented in Africa with the intention to promote the electrification of rural and remote areas. Reports show that microgrids have proved the most suitable means to integrate renewable resources into the electricity generation and access to clean energy for the African continent. Liu [14] did a review of literature on the practice of microgrids implementation in Africa with a special focus on sub-Saharan Africa, which is one of the less developed regions of the continent. They indicated that the region has been using the microgrid as a tool for the transition toward clean energy. They stated that the lack of adequate policy support and proper project planning were the major barriers and suggested the sufficient investigation and integration of technological and socioeconomic aspects in the project planning and designs [14].

In terms of the companies involved in the microgrid market, it wasn't until recent years that large international companies started to be engaged in microgrid development; most microgrid developers used to be small businesses or startups. We are, nowadays, observing some big companies participating in the sector in different forms such as in acquisition of energy storage systems (ESS), uninterruptible power supply (UPS) systems and control, energy management (software) technologies, and the like. Utilities and oil companies have also started to invest in the installation of microgrids by themselves or cooperate with developers.

2.2.2 Microgrid application trends

Based on the applications that microgrids serve or where they are installed, there can be different types of microgrids such as:

- remote/off-grid;
- campus/institutional/facility;
- community/utility;
- military base; and
- commercial and industrial (C&I) microgrids.

Remote off-grid application is the most mature segment which leads the global microgrid market with a total of 7604.4 MW microgrid capacity (nearly 40% of the global capacity) at the end of 2018 [15]. The next largest application segment by capacity share is the C&I sector with 5542.9 MW installed and planned capacity in the same year [15]. The two sectors remain the dominant ones in 2019 as well, collectively covering up to 77% of the global microgrid capacity [8]. C&I microgrids make up 36% of the 77% while they are expected to be the fastest growing market segment in the coming years [6]. The Navigant Research microgrid deployment tracker for the second quarter of 2019 reports the share of utility and community microgrids individually as 8% and 4%, respectively, while campus and military microgrids make up a respective 6% and 5% [6]. Microgrids in North America have almost uniform distribution of customer applications (among military, commercial/industrial, etc.) with neither of the segments accounting for more than 25% of the microgrid market in the region [16]. The other big market, the Asia-Pacific region, presenters an opposite pattern, where 91% of the microgrids are remote off-grid type [16].

Microgrids are currently majorly used in remote areas which offer localized power to islands and industrial and military installations. Grid expansion has always been the main way to provide power supply services to remote areas. However, compared with independent power sources, the "blue ocean" of main grid expansion is gradually shrinking. With the cost of acquiring renewable energy systems getting lower and lower, off-grid microgrids (especially renewable energy-based ones) have become a popular option in recent years. There are two basic challenges to be overcome for the utilization of microgrids to their full potential. The first one is the fact that rural users usually have limited electricity demand and financial capacity to pay for electricity. Secondly, there is a general lack of microgrid support policies and regulations.

The commercial and industrial facilities are known to be more sensitive to power outages than other sectors, and a short period of an outage can lead to substantial losses. The C&I sector is increasingly moving toward locally built independent power systems through microgrid networks. Savings on energy bills and increasing security and reliability demands are assumed to be the main factors leading to the growing market share of the segment.

2.2.3 Microgrid business models

One of the most critical challenges facing the deployment of microgrids is the large initial investment and shortage of effective financial mechanisms for the development and operation of microgrids. However, recent years saw a variety of innovative financing schemes being devised and implemented for microgrids that allow the financial burden and risk to be distributed among stakeholders. Some of the models allow companies to harvest the benefit from microgrids without a significant upfront investment. The most popular currently available microgrid business models are:

- Power Purchase Agreement (PPA)
- Energy Management Services Agreement (EMSA)

- Pay-as-you-go (PAYG)
- Microgrid-as-a-Service (MaaS)
- Energy savings performance contracts (ESPC)
- Enhanced use leases (EUL)

The *Power Purchase Agreement (PPA)* business model is based on the provision of energy services under a medium−long-term contract, typically 20 years. A starting price for energy services is agreed with the future variation restricted to be between 1% and 5% throughout the contracted period. In this way, the client transfers the cost of the investment and the management of the assets to the seller of the PPA reducing the risk. A third entity may also intervene to take care of the financing. The model allows for Independent Power Producers or investment funds to finance a microgrid by signing a PPA with a client (most commonly C&I institutions) that will be committed to purchase of a certain amount of power with no need to raise the upfront capital for the installation [17]. The 3.2 MW solar−wind-storage microgrid to be developed by Pash Global on the island of Saint Helena is a recent example of a PPA model with the contract lasting for 25 years [18].

An *Energy Management Services Agreement (EMSA)* is a business model that involves a time period similar to that of PPA, where the final end user pays fixed monthly equipment fees and performance-based incentive awards. The calculation of performance is based on the net energy cost savings from the operation of the microgrid. An example for such a modality is the Inland Empire Utilities Agency (IEUA) microgrid in San Bernardino County, California. In this microgrid, IEUA pays a fixed-fee of US$65/kW per year to the developer Advanced Microgrid Solutions based on the 10 years EMSA that includes a performance-based incentive award (50/50 split of profit) triggered when the net energy cost savings exceed US$100/kW [19].

Pay-as-you-go (PAYGO) allows end-users to pay for the energy from the microgrids in periodical (daily or weekly) installments or whenever the projects are financially liquid. Under this model, the companies may provide finance to customers in addition to the products and services. The end-users are expected to pay 10%−20% of the cost upfront while the remaining cost is treated as a loan over a short period (1−2 years). Smart meters that allow providers to remotely control and connect/disconnect the power supply to end users are central to the PAYGO model. PAYGO is prominently oriented to small and remote microgrids in developing regions, with electricity demand limited to lighting and household devices. As a critical tool for universal access to electricity, it has been supported by international development organizations. Between 2015 and 2020, around 8 million people gained energy access with PAYGO models [20]. A typical example of a PAYGO model is the experience in Kenya by companies such as D.light and Mobisol who are offering mobile "pay as you go" home solar products and services packages to rural Kenya. The D30 off-grid home solar system by D.light includes an initial downpayment (US$25 in 2017) and succeeding payments (40 cents per day for a year in 2017) are made through mobile money services (interest-free) with the option that the systems will be unlocked in perpetuity when the cost is paid in full [21].

Microgrid-as-a-Service (MaaS) is a new, industry-leading financing mechanism that enables organizations to implement microgrids without any initial investment. This financing model enables municipal, district, institutional commercial precincts and large buildings to stabilize long-term energy costs and upgrade critical energy infrastructure without capital outlay. In many higher cost areas, microgrids with solar microgrid power or medium distributed cogeneration can be less expensive than grid power. Examples for MaaS business model are the two microgrids developed collaboratively by Schneider Electric and Duke Energy Renewables for Montgomery County Public Safety Building and Correctional Facility [22].

Energy savings performance contracts (ESPC) are financing mechanisms that seek to update energy facilities (including microgrids) through future energy savings. With this objective, an agreement is established between the owner of the installation and an energy services company, so that the latter audits the installation, negotiates and selects the measures to be applied with the owner, and guarantees energy savings. Therefore the reduction of energy costs (or renewable generation) is used to cover the costs of the project, although this implies that essentially the energy services company will not receive payments until the installation demonstrates that the agreed energy savings have been guaranteed. The model may also involve a third party to take care of the financing, in which case the earnings from the savings would go first go to cover the financing agreement. In the United States, ESPCs, in combination with utility service contracts, have led to infrastructure improvements worth US$613 million, financed through efficiency savings [23]. An example for this business model is the ESPC for 24.9-year agreement (worth US$170 million) between the US Naval Facilities Engineering Command and Constellation (a subsidiary of Exelon Corporation) to implement an ESPC with the objective of increasing energy efficiency and installation of distributed generation (DG) for the Marine Corps Logistics Base in Albany, Georgia [24].

Enhanced use lease (EUL) is a financing instrument for clients in the public sector (usually military organizations) with underutilized assets and authorized agencies that allows the underutilized or deteriorated property to be leased for the development of a microgrid in exchange for payment or in-kind consideration. The microgrid, in this system, normally supplies power to the utility grid and to the federal facility when there is an utility grid outage. The developer takes all the risk by covering the full or a portion of installation and O&M costs with the expectation of exploiting the assets in the long term (generally 25 years or more) in exchange for returning to the military body a value (in capital or services) not less than the market value of the assets. The EUL signed between Naval Facilities Engineering Command (NAVFAC) and the developer Bright Canyon Energy (BCE) for the construction and operation of a 2.5 MW PV with 2.5 MW battery energy storage on approximately 8.3 acres of land at Naval Weapons Station Seal Beach in Detachment Norco is an example [25]. The EUL allows BCE to sell the generated energy to the local grid while serving a backup function to the Navy installation during grid outage.

There are also other modalities such as *utility services contracts* (USC) and *utility privatization* (UP) [26]. An utility service contract is a bilateral agreement between a federal agency and a distribution utility for electric connection, demand side management services, or special facilities services. A USC is an ideal option when there is an agency interested in microgrid functionality as a service, and when energy savings are not large enough to cover the microgrid cost. *UP* involves the transfer of ownership of the entire microgrid facility and responsibility of operating, maintaining, and upgrading the system to a third-party utility provider. Where a microgrid may be necessary to meet reliability and resilience requirements, the private owner of the system may be able to install the microgrid with the agency repaying the costs through UP payments over the term of the contract. Other financing and contracting concepts such as Equipment Lease (EL), Efficiency Savings Agreement (ESA), Shared Savings Agreement (SSA), and Energy Asset Concession Agreement (EACA), are also reported in some publications [27]. Owner Financing (OF) and Government Funding (GF) are also very commonly utilized models that do not include a third party involvement.

Navigant Research uses a different approach of categorizing the microgrid business models where most of the above listed concepts, namely EL, PPA, ESA, ESPC, PAYG, SSA, and EACA are grouped into an Energy-as-a-Service (EaaS) business model [27]. Another major finding from the 16th edition of Navigant Research's microgrid development tracker, next to changes in the geographic disparity, is the growing rate of adoption of the EaaS business model, which constitutes a stunning 81% of microgrids deployed globally until 2Q 2019 [6]. The proportion is rather fair on a capacity basis, however, with EaaS and utility rate base models featuring a similar 22% market share while OF, GF, and other models make up 21%, 20%, and 15%, respectively [6].

2.2.4 Technological trends

2.2.4.1 Distributed energy resources technologies

The distributed energy resources (DER) technologies applied in microgrids include the generation and energy storage units. The generation technologies can be generally classified as RES that include solar PV, solar thermal, wind, small-scale hydro, biomass, biogas, etc., and nonrenewable or fuel-based DG that comprise diesel genset, stream turbine, natural gas generators, etc. [28]. A recent report by Guidehouse Insights estimates the revenue generated for the deployment of DER technologies in microgrid projects to increase from US $6.3 billion in 2020 to US$27.7 billion in 2029 [29].

Most of the recent microgrids are composed of PV and battery energy storage with a considerable part of them incorporating diesel generators. The sharp decline in PV module price and the technological advances in site identification, development, construction, and operations and maintenance of PV systems are some of the factors that make PV systems a popular element of microgrids. According to a report by Bloomberg New Energy Finance (NEF), as of March 2020, there are 7181 microgrid projects in sub-Saharan Africa, Asia, and island countries (some of them in Latin America). There are currently 5544 microgrids in operation, of which 63% are PV microgrids or PV hybrid microgrids, 21% are connected to hydropower, and 11% are connected to diesel/heavy fuel oil generators [30]. It should be noted that this reported is focused on customer segments in rural communities in sub-Saharan Africa, developing Asian countries, and small island nations. As the proportion in the report testifies, the fastest growing part of the global microgrid market is the PV hybrid microgrid.

Combined heat and power (CHP) is one of the generation technologies traditionally dominant in the earlier microgrids. CHP has the advantage that it can provide year-round base-load, which the renewable resources cannot guarantee. Unlike the early dominance, CHP is becoming a less favored option than PV and non-CHP natural gas generators (which do not incorporate the heat recovery option) in recent and planned microgrids, especially those in the United States [31]. However, CHP is expected to continue having a significant share in future microgrids. According to a blog by ICF, a renowned consulting firm that runs databases for microgrids and CHP installations in the United States, close to 20% of the microgrids planned to be operational up to 2023 in the United States will incorporate CHP [32].

Gas *microturbines* that can operate on various types of fuel, including those low-percentage methane fuels like biogas, flare gas, landfill gas, etc., are the other type of generation getting wider deployment in microgrids for their superior efficiency, controllability, and stability [33]. Energy conversion efficiency is the most important feature in this type of generation technology. The total CHP efficiency of gas microturbines has improved from 61% to 68% in 2000 [34] to a remarkable 80%−90% currently [33]. A total efficiency of 93% is also reported at 70,000 rpm with suitable testing conditions [35]. Thus the efficiency from gas microturbines is clearly much higher than that of the conventional cogeneration systems (typically 65%). A report by VERGENT power solutions states that there were a total of 3000 microgrid projects (isolated and nonisolated) that incorporated microturbines in 2017 [36].

Fuel cells (FCs) are an environment-friendly, efficient, and flexible generation technology with promise in microgrid projects. The technology of FCs evolved through years from the gaseous voltaic battery of 1839 (when the concept of the reaction between H_2 and O_2 using platinum electrodes and tetraoxosulphate (VI) acid electrolyte was introduced by William Grove) to molten carbonate FCs used in stationary applications in the 2010s [37]. Fuel cells are expected to be applied widely in power systems (including microgrids) in the coming 5−10 years, especially after the molten carbonate and solid oxide FCs pass the developmental phase and are commercialized [37]. The global stationary fuel cell installation capacity is expected to increase from 262 MW in 2016 to more than 3000 MW in 2025, according to estimates by Navigant Research [38]. Japan is one of the countries that has paid due attention to the deployment of fuel cells technology, especially for decentralized application, with 1.4 million and 5.3 million residential fuel cells-based systems planned to be installed in Japan by 2020 and 2030, respectively, according to Bloomberg NEF 2017 [39].

Though *wind* energy is one of the RES with growing deployment in the world, it is much less common in microgrid projects, except for some that combine it with diesel or PV. Though small-size wind turbines are showing a reduction in price, it is not at the same level as the reduction for PV panels [30]. However, there still is a considerable volume of distributed wind generation systems. A typical example of small-scale wind systems are the iSmall-scale wind systems with power capacities enough for a single home, farm, or small business. The boundary limit for size of the so called "small-scale" systems is approximately 50 kW as per International Electrotechnical Commission (IEC), while the World Wind Energy Association assumes a limit of 100 kW. The cumulative distributed wind capacity installed in the United States during the period 2003−18 is 1127 MW from over 83,000 wind turbines [40]. This number includes all wind DER units connected at the distribution level.

The future microgrids are expected to incorporate multiple DER technologies such as PV, energy storage, and others into new roles and use cases. Such a strategy allows harvesting of particular benefits from each technology while guaranteeing the additional advantages of improved reliability, resilience, power quality, and renewables integration from their strategic combination [32].

The most common microgrids are composed of RES and ESS. The *ESS* in microgrids serve the basic purpose of ensuring generation−load balance, which is especially essential in microgrids operating in island mode. The ESS can also provide more functions such as arbitrage, voltage and frequency support, operating reserve, peak shaving, and others. The total annual capacity addition of ESS in microgrid projects was estimated to be around 650.4 MW in 2020 and is projected to increase to 8633.4 MW in 2029 [41]. Lead−acid batteries are the cheapest and widely utilized energy storage in microgrids with the Bloomberg NEF report estimating their contribution to 66% of the microgrids with storage installed in 2019; and lithium-ion (Li-ion) batteries make the second largest share [30]. The limited cycling capability and environmental impact of lead—acid batteries are aspects that require improvements and, as a result, paved the way for other types of battery such as Li-ion, nickel—cadmium (NiCd), nickel metal hydride (NiMH), and sodium-sulfur (NaS) batteries [42]. According to the Guidehouse Insight report, the technological and price changes in the last 10 years mean that Li-ion batteries are becoming the dominant technology for microgrid applications, replacing lead—acid batteries while flow batteries appear as a better choice for longer-term storage needs by microgrids [41]. The intended targets of the advancements in battery energy storage are high energy density, long cycle life, low maintenance requirements, and environmental friendliness.

In addition to the above-discussed electrical ESSs, there are mechanical ESSs, such the *flywheel* that stores electrical energy in the form of rotational kinetic energy. Flywheels and supercapacitors are usually regarded as power-intensive ESSs due to their high power density and application for short-term energy boosts. Flywheels can provide an effective spinning reserve for about 2—10 minutes. *Supercapacitors* are high-capacity capacitors with their characteristics lying somewhere between conventional electrolytic capacitors and batteries, typically known for their larger energy density than capacitors and larger power density compared to batteries. They are suited for transient disturbance control applications, such as frequency regulation and voltage regulation by quickly absorbing or releasing a large volume energy within a very short time during power quality events [42]. Battery—supercapacitor hybrid ESS are also recommended for application in microgrids in order to mitigate the impact of dynamic power exchanges on a battery's life span [43]. The application of a flywheel was demonstrated at an off-the-grid port in Kodiak, Alaska where the 2 MW flywheel storage handles the special needs of deep and frequent charges and discharges [44]. Supercapacitor storage is also used for the purpose of transient stability control in the demonstrational microgrid discussed in Chapter 11.

2.2.4.2 *Microgrid control and monitoring technologies*

Given the smaller size, lower inertia, involvement of the power electronic interfaced DGs, and the limited smoothing of demand spikes, microgrid applications may face significant control, stability, power quality, and security challenges. Hence, it is important to quantify and analyze the challenges and implement the appropriate control systems to manage them.

Digital control platforms are what coordinate the diverse DER units in microgrids so that the system can satisfy the basic element of definition of microgrid, that is, acting as a single controllable entity. Thus advances in the control technology are critical in dictating the performance of the microgrid projects, as a result, influencing the success of the microgrid market. However, unlike their vital role, the control systems of microgrids often represent a low percentage (5%−10%) of the project cost [45]. As the microgrid deployment is on the rise globally, the market for microgrid control systems is also growing with the projected market by 2025 reaching US$3.9 billion [46]. The United States, China, and Europe constitute a combined 63.7% of the global market while China's market is the fastest growing with a Compound annual growth rate (CAGR) of 15.3% over the forecast period of 2020−25 [46].

Control systems are a combination of hardware and software while the recent trend is more focused on the software part than the hardware. The older "local gateway"-based systems are evolving toward cloud computing and edge computing-based architectures, especially for larger microgrids, such as community microgrids. Artificial intelligence (AI) is an important tool in this advancement. Hardware-in-the-Loop is also one of the key innovations in the microgrid controls system technology. The other aspect of technological advance in control platforms is the incorporation and impact of Internet of Things in microgrid management.

The Institute of Electrical and Electronics Engineers (IEEE) p2030.7 classifies functions of a microgrid control into three categories: device-level control (primary control), local area control and supervisory control (secondary control), and grid-interactive control (tertiary control) [47].

The *primary control* functions include active and reactive power control and voltage frequency control, which are usually performed in the microsecond to second range. This control is based on local measurements of electrical parameters. The most common primary control strategies are droop control and virtual impedance-based methods [48]. Primary controls that employ ESS and controllable loads are also applied to improve the inertia of microgrids so that a better level of stability can be ensured [49]. The Distributed Energy Communications & Controls Laboratory of Oak Ridge National Laboratory develops controls for inverter-based DER to provide local voltage, power, and power quality support for the campus distribution system [50].

Secondary control performs functions such as energy management, economic dispatch, automatic generation control, spinning reserve, resynchronization, and fast load shedding, which usually take a time span longer than a second and involve communication [47]. Secondary control can be implemented through either centralized or decentralized control. Model predictive control, multiagent-based control, droop control, probabilistic analytical approaches, and gossip-based techniques are some of the secondary control strategies [51].

Microgrid Central Controller (MGCC) is a typical example for centralized secondary control that utilizes a communication medium to collect the information of the constituting components of the microgrid and provides reference values for primary or local controllers. Application examples of MGCC are the Santa Rita Jail microgrid [52], microgrid lab of Aalborg University [53], and a microgrid testbed of the University of Texas at Arlington [54].

MGCC may also include some tertiary control functions. *Tertiary control* includes functions related to market participation, energy exchange with the utility grid, and coordination of multiple microgrids. *Supervisory Control and Data Acquisition (SCADA)* is considered to constitute the tertiary control of microgrids. SCADA and *Energy Management System (EMS)* are usually considered as interchangeable terms, while sometimes SCADA is seen as a component of the EMS. EMS along with SCADA have been integral elements of the control and monitoring of a power system. The earlier and conventional EMS/SCADA systems were based on the collection of information using RTUs (remote terminal unit) regarding some electrical parameters (such as power and voltage) and the operational status of the system, which usually involves a certain time delay.

With the requirements and expectations reformed in today's power system and due to the special needs of microgrids, such a type of monitoring system may not be sufficient. Thus EMS/SCADA of today's microgrids usually comprise faster, more accurate, and smarter components and functions. The RTU, synchronized phasor measurements, customized human—machine interface, wide area monitoring and control, programmable logic controller (PLC), phasor data concentrator (PDC), super PDC, and telecommunications infrastructure are some the components that make up EMS/SCADA of modern microgrids [55,56]. Synchronization of measured values with global position system on a microsecond level accuracy is also being applied [57].

The latest microgrid EMS/SCADA systems have features such as remote monitoring of functions and system operation, alarms, PLC set points, and alarm history with custom-designed screens and the possibility of incorporating a simulation system. Such incorporation of simulation models in the physical system was tested in the experimental microgrid of the Centre for Renewable Energy Sources and Saving in Pikermi, Greece with Matlab/SIMULINK interconnected to the existing SCADA [58]. This scheme allows for off-line operator training without affecting the operating system parameters. A central EMS based on a real-time SCADA system is reported to be applied in a LAMBDA nearly zero-energy microgrid [59].

2.2.4.3 Microgrid protection technologies

Protection of microgrids is one of the most challenging aspects of microgrid deployment and the technologies are disparate and still evolving. Dominance of power electronic interfaced DER and island mode of operation where there is no utility grid are the critical issues resulting in challenges in microgrid protection design. Devising a protection architecture applicable for both grid-connected and island modes is a real challenge. A broader analysis of the challenges and special requirements of microgrid protection is provided in Chapter 6, Protection of Microgrids, of this book.

Electromagnetic relays are the most widely applied relays since the introduction of protective relaying in the 1830s and still applicable. However, more flexible relays such as solid-state and microprocessor relays are featuring most in the contemporary protection technologies. Microprocessor relays are preferred for their higher precision, flexibility, and speed. They also allow more than one protection function, which operate independently or coordinately, to be encompassed in a single relay. Such advantages make microprocessor-based or digital relays the choice of microgrid applications. *Multifunctional microprocessor relays*, which are Intelligent Electronic Devices, are the primary microgrid protection devices applied in some of the most successful microgrids [60]. The digital relays provide functions such as digital communications, adaptive protection, very fast (subcycle) protection, harmonic restraints, etc., while they also allow data collection which is very important for the monitoring and troubleshooting of fault events. Digital relays also allow for implementation of an *adaptive protection* scheme which is mostly under research.

Smart switches are the other major development in protection technologies applied for microgrids. The intelligence embedded in those switches provides benefits that wouldn't be provided by conventional ones. Smart switches can have the capability to capture waveforms, event logs with very short (as small as a millisecond) time-stamping, and oscillographic data [61]. Smart switches include smart fuses, smart reclosers, and smart circuit breakers. For example, smart fuses have more functionalities than the conventional fuses such as current limiting characteristics, inherent self-monitoring ability, and capability to be triggered from an external source [60]. Smart reclosers with PulseClosing technology are also reported to have the advantage over conventional reclosers in terms of reducing the deterioration of power system equipments by avoiding the stress of fault current as a result of repeated unsuccessful reclosing attempts [62]. They also reduce the risk of fire by cutting the let-through energy. Smart circuit breakers are transforming ordinary wiring into a smart system with automation. Internet-connected or Wi-Fi-enabled circuit breakers such as those are reported to be already available in the market. Those smart circuit breakers have multiple firmware embedded in them to enable metering and communication capabilities that allow provision of functions such as real-time monitoring and tracking, alarming, remote tripping, and system updates.

2.2.4.4 Instrumentation and communication technologies

Instrumentation devices, including sensing and measuring devices, are an important part of the control and protection of microgrids. Current transformer and voltage/potential transformer are the traditionally popular devices in the power system. They are still applicable for microgrids though efforts are being applied toward harvesting the benefits of the growing digital technologies for development smarter instrumentation devices with real-time sensing and monitoring capabilities.

Remote monitoring systems are increasing in popularity in microgrids. They are being applied to an even larger extent in off-grid microgrids in the less developed countries. Experiences and challenges in application of such monitoring systems for off-grid microgrid systems in Kenya and Zambia are investigated and reported by Louie et al. [63].

Wired and wireless technology-based robust and reliable communication networks are important for real-time monitoring and control of microgrids. Wired communication is the simplest mode of communication, and the popular examples are optical fiber, power lines, and twisted-pair cables [64]. Wireless communication technologies are preferred under some circumstances such as congested areas, harsh environments, installations involving large number of sensors, when equipment is located in close proximity, etc. Wireless communication has the advantages of ease of installing remote terminals, low cost, modularity, and the ease of operation. Wireless communications in microgrid applications may be short range or long range (e.g., cellular). The cellular communication network is the most commonly applied communication in the off-grid microgrids in the rural and less developed areas mentioned above. There are different communication technologies reported to be applied in microgrids. Some of them are listed in Fig. 2.1.

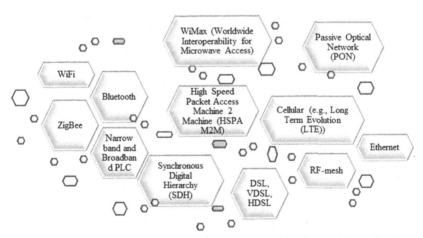

FIGURE 2.1 Communication technologies used in microgrids. *Compiled from S. Kumar, S. Islam, A. Jolfaei, Microgrid communications - protocols and standards, in: Variability, Scalability and Stability of Microgrids, 2019; J. Cao, M. Ma, H. Li, Y. Zhang, Z. Luo, A survey on security aspects for LTE and LTE-A networks, IEEE Commun. Surv. Tutor. (2014), https:// doi.org/10.1109/SURV.2013.041513.00174; S. Safdar, B. Hamdaoui, E. Cotilla-Sanchez, M. Guizani, A survey on communication infrastructure for micro-grids, in: 2013 9th International Wireless Communications and Mobile Computing Conference, IWCMC 2013, 2013, https://doi. org/10.1109/IWCMC.2013.6583616 [64–66].*

The most important issues regarding communication technologies in microgrid applications are speed, reliability, and cyber security. The integration of communication technologies may make microgrids vulnerable to cyber-attacks. Thus robustness against security attacks is an important feature for the selection of a communication technology.

2.2.4.5 Microgrid planning, modeling, and simulation tools

Modeling and analysis tools are important for development of microgrids. Some of the available tools are summarized in Table 2.1.

TABLE 2.1 Software tools for microgrid planning, modeling, and simulation.

Software	Developer	Description
HOMER *Hybrid Optimization Model for Multiple Energy Resources*	US National Renewable Energy Laboratory (NREL)	A simulation model that simulates, optimizes, and analyzes electrical system (microgrid) design using resources, load profiles, and components to deliver the least-cost solution [67]
DER-CAM *Distributed Energy Resources Customer Adoption Model*	Lawrence Berkeley National Laboratory	A technoeconomic model of customer DER adoption that uses input of market information, system load, and DER technology information to provide the optimal selection of DER technologies [68]
MDT *Microgrid Design Toolkit*	Sandia National Laboratories	A decision support software tool that employs search algorithms to identify and characterize trade space of alternative microgrid design decisions based on user defined objectives [69]
GridLAB-D	U.S. DOE at Pacific Northwest National Laboratory (PNNL)	A distribution system simulation and analysis tool that examines the interplay of every part of a distribution system with every other in detail [70]

(Continued)

TABLE 2.1 (Continued)

Software	Developer	Description
REopt *Renewable Energy Optimization*	NREL	An energy planning platform that models the interactions of multiple thermal and electrical technologies operating concurrently and provides the optimal, least-cost solutions based on a mixed integer linear program solver [71]
RETScreen	Natural Resources Canada	A clean energy management software for energy efficiency, feasibility analysis, and energy performance analysis of renewable energy and cogeneration projects [72]
PV SOL*	Valentin Software	A 2D solar software design tool for simulating detailed configuration and shade analysis for PV systems and calculating solar output, panel size and economic forecast [73]
RAPSim *Renewable Alternative Power systems Simulation*	The University of Klagenfurt	An open source microgrid simulation framework to simulate microgrids with a renewable energy source that can conduct a power flow analysis and optimal placement of DG units in a microgrid [74]
ETAP μGrid *Electrical Transient Analysis Program—Microgrid*	Operation Technology, Inc (OTI)	An electrical digital twin model combined with automation and protection to optimize and control microgrids with electric and thermal components; includes ETAP Microgrid Control and ETAP Microgrid Management System [75]

(Continued)

TABLE 2.1 (Continued)

Software	Developer	Description
Matlab/Simulink	MathWorks	A block diagram environment for multidomain simulation and model-based design that supports simulation, system-level design, with additional features such as automatic code generation and testing and verification of embedded systems [76] can be used to simulate, design, and test operation, control, and protection of microgrids
PSCAD/EMTDS *Power System Computer Aided Design (PSCAD)/Electro-Magnetic Transient Design and Control (EMTDC)*	Manitoba Hydro International Ltd. (MHI)	An electromagnetic transients (EMT) simulation tool for building, simulating, and modeling power systems including microgrids with comprehensive library of models for passive elements, control functions, electric machines, and other more complex devices [77]
OPAL-RT	OPAL-RT Technologies. Inc.	PC/FPGA-based real-time simulator, Hardware-in-the-Loop testing equipment, and Rapid Control Prototyping systems for designing, testing, and optimizing control and protection systems in power grids (including microgrids) [78]; a special platform for microgrids is available based on Simscape Power Systems of MathWorks
RTDS Simulator *Real-time digital simulator*	RTDS Technologies Inc.	A simulator for hardware in the loop testing of protection and control equipment which is capable of physically connecting the simulated network with the control and protection system hardware [79]

(Continued)

TABLE 2.1 (Continued)

Software	Developer	Description
Digital Twins		Refers to the digital representation of a real-world entity or system in the cloud of the product, process, or service which also allows running digital simulations a new concept with possibility of a wide range of industrial applications including microgrids

2.3   Current research trends of microgrid

2.3.1   Microgrids research issues

With growing interest in the deployment of microgrids, suitable implementation techniques of microgrids are being actively researched on a global scale. Some of the barriers to implementation of microgrids, for which solutions are still evolving, are [80]:

- Lack of adaptable microgrid performance metrics due to variety in architecture
- Limited financing models that ensure bankability of the microgrid projects
- Cyber security concerns due to communication networks in control and monitoring systems
- Limited standards and grid interconnection protocols
- Technical challenges
  - power quality (e.g., harmonics, voltage, and/or frequency control);
  - grid synchronization;
  - control strategies and architecture;
  - energy management and optimization;
  - stability;
  - protection systems; and
  - initial design process of microgrids.

Microgrids are characterized by their small size compared to the larger utility grid that causes challenges in the energy management. The involvement of different generation technologies and power electronic interfaces also causes some technical challenges in the control, protection, and safety of microgrids. There also is more frequent incidence of relatively large imbalances between generation and load which leads to the need for significant load participation and hence for new technologies. More detailed technical issues regarding the control and protection of microgrids are presented in Chapter 3, Key Technical Challenges in Protection and Control of Microgrid, of this book.

Microgrid research and development is one of the core R&D areas identified by the US Department of Energy. The bulk of the effort of this department has been channeled into the development of demonstration projects focused on functions of microgrids, such as reducing peak demand, integration of renewables, and energy security and reliability [81]. Robust control system is vital for stable and efficient operation of microgrids. The US Department of Energy (DOE) has identified several core areas for microgrid control, including frequency control, VAr control, mode transfer between grid-connected and islanding modes, energy management, protection, ancillary service, black starting, and user interface and data management [82].

Protection of microgrids is one of the most researched topics. The growth of digital protection relays has been a motivation for investigating different smart protection schemes. Adaptive protection is one of those schemes that have drawn a significant amount of interest from researchers and developers. The possibility of applying AI tools in protection relays is also being studied. Adaptive and AI-based protection schemes are specially proposed to achieve effective protection during both modes of operation and the continuous changes in configuration. Islanding detection is an issue which is critical for the protection and control of microgrids, which is also a topic of interest in recent research works.

One of the other areas related to microgrids which is addressed in research works is the optimal sizing of microgrids and their components. Another aspect that is related to implementation of microgrids, and which is a key input for the design, management, and upscaling of microgrid solutions, is the accurate forecasting of electricity demand and generation.

2.3.2 Selected microgrid R&D projects

2.3.2.1 Consortium for electric reliability technology solutions

As expressed earlier, CERTS is usually regarded as the first microgrid. That is excluding the very early small power grids of Edison's era, which may comply with the basic requirements of a microgrid, although they do not render the functionalities and features of the modern-day microgrids. The CERTS project was devised with the basic intention of conducting research on improving power system reliability, testing performance of emerging technologies, and understanding economic, regulatory, and environmental impacts of the microgrid [83]. As specifically stated in the definition of the microgrid, adopted from the definition by the US DoE, what is taken as the defining factor of the CERTS microgrid is the requirement that a microgrid shall operate as a single self-controlled entity [84].

CERTS has made major contributions to the microgrid industry through the development of advanced control and integration techniques. Those techniques are collectively referred to as the CERTS Microgrid Concept. Field testing of the concept is carried out based on a full-scale microgrid test bed whose construction began in 2006 in partnership with the American Electric Power (AEP). The CERTS/AEP microgrid test bed is located near Columbus, Ohio. It initially consisted of three generators with cumulative capacity of 60 kW + 60 kVAr and four load banks (100 kW + 20kVAr each) [85]. Concepts such as peer-to-peer control and plug-and-play were introduced and effectively applied in the CERTS/AEP microgrid. The peer-to-peer control allows microgrid stable operation with no single unit acting as a master controller while the plug-and-play feature ensures the possibility of connecting DER units at any point in the microgrid without modifying the controls [86]. The CERTS/AEP microgrid is characterized by [87]:

- autonomous load following;
- seamless separation and automatic resynchronization;
- autonomous load transfer from overloaded source to other sources;
- UPS level power quality; and
- stable operation with multiple sources.

The microgrid at the Santa Rita Jail in Alameda County, California, which consists of a 1.2 MW PV system, five wind turbines with total capacity of 11.2 kW, a 1.0 MW fuel cell, and 2.0 MW battery, is another of the early field demonstrations of the CERTS Microgrid Concept.

2.3.2.2   *Microgrids, infrastructure resilience, and advanced controls launchpad*

Microgrids, infrastructure resilience, and advanced controls launchpad (MIRACL) is a collaborative research project funded by the Office of Energy Efficiency and Renewable Energy of the US DOE that focuses on enabling distributed wind technology to be an affordable, accessible, and compatible DER. There are multiple institutions involved in the project including four DOE national laboratories, together with electric utility, wind, microgrid, and DER industries [88]. The collaborating laboratories are the National Renewable Energy Laboratory, Pacific Northwest National Laboratory, Sandia National Laboratories, and Idaho National Laboratory. The project is involves research capabilities and DER infrastructure worth around a billion dollars across the four laboratories. The project has three primary R&D priorities [89]:

- valuation and modeling;
- advance controls; and
- resiliency and cyber security.

The "Advanced controls" topic aims at developing and demonstrating grid support functions of wind turbines in both grid-connected and isolated microgrids and increasing the control and communications compatibility of wind turbines with other DERs. Some of the research grid services from advanced controls are [89]:

- active and reactive power control;
- inertial response;
- primary and secondary frequency response;
- voltage support;
- blackstarting; and
- mode transfer between island and grid-connected modes.

The microgrid test bed of the MIRACL project integrates simulation, ESS, solar panels with smart inverters, load banks with control capabilities, and switchgear sets. Issues such as demand response, peak shaving, and other ancillary services are researched based on this platform.

2.3.2.3 Renewable energy integration demonstrator singapore

Renewable energy integration demonstrator Singapore (REIDS) is a project commenced by Nanyang Technological University, conceived in 2015 with the support of the Singapore Economic Development Board and National Environment Agency. It is comprised of eight microgrids built on Semakau Island, 8 km south of Singapore. Different types of DG units, such as wind, PV, tidal, diesel, power-to-gas, and battery energy storage, are integrated in those microgrids. The project is generally aimed at developing technological solutions that can be commercialized and applied for electrification, especially for islands with limited or no access to electricity in Southeast Asia [90]. International energy companies (Accenture, Alstom, Engie, Schneider Electric, and Metron) are involved in cooperation with local companies in the development of the microgrids in the REIDS project.

The REIDS microgrid is used for research and development activities on issues such as [91,92]:

- Interoperability and energy trading of multimicrogrid networks
- Cybersecurity of distribution networks
- Power grid dynamics
- DER management systems
- Resiliency and reliability
- Power and energy management
- Supply-side and demand-side requirements reconciliation
- Plug-and-play capability
- Centralized versus local monitoring and control
- Economic performance evaluation
- Standardization

Automation in microgrids though the use of AI is one of the interests of the project with a smart energy management platform for ensuring interoperability between the constituting microgrids being an important part of the project [90]. The platform makes use of weather, energy market, and other types of data for real-time optimization of energy utilization so as to provide energy efficiency and savings.

2.3.2.4   Other research projects

There are also other microgrid research projects in different parts of the world. *China* is one of the countries who paid attention toward the deployment of microgrids supported with research and development works. Xi'an Jiaotong University, Tianjin University, Hefei University of Technology, The Chinese Academy of Science's Institute of Electrical Engineering, etc., are some of the research institutions in China which are playing leading roles in microgrid research and development [93]. The utility companies, State Grid Corporation of China, China Southern Power Grid, etc., have also undertaken several microgrid demonstration projects to give a further boost to application of microgrids [93]. One of the early microgrid research projects in China was by State Grid Electric Power Research Institute, and it was approved for implementation in August 2011. The project proposed a microgrid system intended to address issues such as strategies for grid response to microgrids, technical standards and policies, etc. [94]. In August 2015 the Yantai Changdao microgrid project passed the acceptance of the National Development and Reform Commission and was officially completed and put into operation. It can realize isolated grid operation to ensure continuous power supply to important users when the external large power grid collapses, greatly improving the power supply capacity and reliability of the Long Island power grid [94]. On July 22, 2015, the National Energy Administration issued guidance on advancing the construction of new energy microgrid demonstration projects. The National Energy Administration encourages the use of advanced information and communication technology in the construction of new energy microgrids in accordance with the concept of the Internet of Energy, to achieve intelligent matching and coordinated operation of energy production and use, and to participate in the power market in a new format to form an efficient and clean new era of energy utilization [94]. The demonstrational smart microgrid in China is another good example of the R&D microgrid projects. The microgrid is used for the developing and testing of different state-of-the-art products for control and protection of microgrid. It also served as a testing platform in the development of two IEC standards, IEC-62898-3-1 and IEC-62898-2.

There are also some projects that address very important issues related to microgrids, although they are not ultimately limited to microgrids. One such project is the *INDORSE* (International Cluster on Inverter Dominated Power

Systems) implemented by various stakeholders worldwide, including the European Distributed Energy Resources Laboratories (DERlab). Microgrids make the most typical case of an inverter dominated power system, and the dominant existence of converter-based generators is one of the most critical aspect of microgrids. The main aim of the INDORSE project is to contribute to the harmonization of grid codes and testing procedures regarding inverter dominated power systems through knowledge exchange and cooperation. The project was endorsed in a workshop at the 2018 in Vienna with the aim of addressing the particular research topics of power system stability, power quality, grid connection requirements, protection, testing procedures, and power system modeling [58].

2.3.3 International standards related to microgrids

One of the factors which has been mentioned as a reason holding back the deployment of microgrids is the lack or shortage of standardization. However, there are significant number of standards on global as well as regional and national levels that have been developed or are being developed in recent years on different aspects of microgrids. IEEE and IEC are two of the most popular institutions for the development of standards in the electrical and electronic engineering sector. Table 2.2 summarizes the available standards on different aspects of microgrids at the international level.

TABLE 2.2 IEEE and IEC standards related to microgrid.

Area	Name of standard
Microgrid control	IEEE 2030.7-2017—IEEE Standard for the Specification of Microgrid Controllers
	IEEE Std 2030.8-2018: IEEE Standard for the Testing of Microgrid Controllers
	IEEE P2030.11—IEEE Draft Guide for Distributed Energy Resources Management Systems (DERMS) Functional Specification
	IEC TS 62898-3-1:2020 Standard | Microgrids—Part 3–1: Technical requirements—Protection and dynamic control
Microgrid protection	IEEE P2030.12: Guide for design of Microgrid Protection Systems
	IEC TS 62898-3-1:2020 Standard | Microgrids—Part 3–1: Technical requirements—Protection and dynamic control

(Continued)

TABLE 2.2 (Continued)

Area	Name of standard
Microgrid communication	IEC 61850:2020 SER Series Communication networks and systems for power utility automation—All parts
	IEEE 1646–2004—IEEE Standard Communication Delivery Time Performance Requirements for Electric Power Substation Automation
	IEEE 802.15.4-2020—IEEE Standard for Low-Rate Wireless Networks
	IEC 61968 Application integration at electric utilities—System interfaces for distribution management—All parts
	IEEE 1646-2004—IEEE Standard Communication Delivery Time Performance Requirements for Electric Power Substation Automation
Design and other issues in microgrids	IEEE 2030.9-2019—IEEE Recommended Practice for the Planning and Design of the Microgrid
	IEEE—P2030.10/D07 Draft Standard for DC Microgrids for Rural and Remote Electricity Access Applications
	IEEE 1547-2018—IEEE Standard for Interconnection and Interoperability of Distributed Energy Resources with Associated Electric Power Systems Interfaces
	IEEE 1547.4-2011—IEEE Guide for Design, Operation, and Integration of Distributed Resource Island Systems with Electric Power Systems
	IEC TS 62898-2:2018 Microgrids—Part 2: Guidelines for operation
	IEC TS 62898-1:2017 Microgrids—Part 1: Guidelines for microgrid projects planning and specification
	IEC TS 62257-9-3:2016 Recommendations for renewable energy and hybrid systems for rural electrification—Part 9–3: Integrated systems—User interface
	IEC TS 62257-9-4:2016 Recommendations for renewable energy and hybrid systems for rural electrification—Part 9–4: Integrated systems—User installation

References

[1] International Renewable Energy Agency (IRENA), Renewable Power Generation Costs in 2019, 2020.

[2] International Energy Agency, World Energy Outlook 2017 - Chapter 1: Introduction and scope, World Energy Outlook 2017, 2017.

[3] IMARC Group, Microgrid Market: Global Industry Trends, Share, Size, Growth, Opportunity and Forecast 2020–2025, 2020.

[4] Mordor Intelligence, Microgrid Market - Growth, Trends and Forecast (2020–2025), Gachibowli, Hyderabad, 2020.

[5] N. Nhede, Microgrids market to record double-digit growth, thanks to COVID-19, Smart Energy International. <https://www.smart-energy.com/industry-sectors/distributed-generation/microgrids-market-to-record-double-digit-growth-thanks-to-covid-19/>.

[6] Navigant Research, Update on World Global Markets, in: Fort Collins 2019 Symposium on Microgrids, 2019.

[7] Navigant Research, Microgrid Deployment Tracker 3Q20: Projects and Trends in the Global Microgrid Market by Region, Segment, Business Model, and Top States and Countries, 2020.

[8] N. Nhede, Interesting statistics on global microgrid projects, Smart Energy International. <https://www.smart-energy.com/renewable-energy/interesting-statistics-on-global-microgrid-projects/>.

[9] P. Asmus, A global tour of contrasting microgrid and VPP markets, Microgrid Knowledge. <https://microgridknowledge.com/microgrid-vpp/>.

[10] Panasonic, Launch of Japan's first microgrid system with a total of 117 homes, Panasonic Newsroom. <https://news.panasonic.com/global/topics/2017/50883.html>.

[11] M. Dupuy, W. Xuan, China takes steps to stimulate distributed renewable energy generation, energypost.eu, 2020. <https://energypost.eu/china-takes-steps-to-stimulate-distributed-renewable-energy-generation/>.

[12] W. Junhong, Policies are very hot and capital is very cold. Can this wave of demonstration projects of the Ministry of Industry and Information Technology drive the development of microgrids? [Chinese], *chuneng*, 2020. <http://chuneng.bjx.com.cn/news/20200221/1045934.shtml>.

[13] National Energy Board, Notice of the National Development and Reform Commission and the National Energy Administration on the Pilot, Development and Reform Energy [2017] No. 1901; Index number: 000019705/2017-00316. <http://zfxxgk.nea.gov.cn/auto87/201711/t20171113_3055.htm>.

[14] Q. Liu, K.M. Kamoto, X. Liu, Microgrids-as-a-service for rural electrification in Sub-Saharan Africa, Comput. Mater. Contin. (2020). Available from: https://doi.org/10.32604/CMC.2020.05598.

[15] Navigant Research, Microgrid Deployment Tracker 4Q2018, 2018.

[16] S. Willette, New microgrid tracker identifies 575 more projects; Rise of Asia Pacific market, Microgrid Knowledge. <https://microgridknowledge.com/microgrid-tracker-rise-asia-pacific/>.

[17] WBCSD, Microgrids for commercial and industrial companies, World Business Council for Sustainable Development, 2017.

[18] E. Meza, PPA signed for solar-wind-storage microgrid on Saint Helena, PV Megazine. <https://www.pv-magazine.com/2020/06/02/a-25-year-ppa-deal-for-hybrid-tech-microgrid-on-british-isle-of-saint-helena/>.

[19] P. Asmus, A. Forni, L. Vogel, Microgrid analysis and case study reportCalifornia Energy Commission, Publication Number: CEC-500-2018-022, 2017.

[20] A. Salgado, A. Anisie, F. Boshell, Pay-As-You-Go Models: Innovation Landscape Brief Irena, International Renewable Energy Agency, 2020.

[21] A. Burger, "Competition heats up in Kenya's off-grid, mobile pay-go solar market, Microgrid Media. <http://microgridmedia.com/competition-heats-kenyas-off-grid-mobile-pay-go-solar-market/>.

[22] P. Maloney, Duke, Schneider pair up to build two microgrids in Maryland, UtilityDIVE. <http://www.utilitydive.com/news/duke-schneider-pair-up-to-build-two-microgrids-in-maryland/435398/>.

[23] J.A. Kliem, DOD update-navy, in: Federal Utility Partnership Working Group Seminar, [Online]. <https://www.energy.gov/sites/prod/files/2018/11/f57/6-fupwg_fall_18_kliem.pdf>.

[24] Constellation, US Marine Corps, Constellation, P&G collaborate to achieve Navy's first 'Net Zero' energy military base. <https://www.constellation.com/about-us/news/archive/2016/US-Marine-Corps-Constellation-PG-collaborate-to-achieve-Net-Zero.html>.

[25] U.S. Navy Office of Information, Navy Signs Lease to Build Renewable Energy Solution in Norco. <https://www.navy.mil/Press-Office/News-Stories/display-news/Article/2284052/navy-signs-lease-to-build-renewable-energy-solution-in-norco/>.

[26] C. Kurnik, P. Voss, Financing microgrids in the federal sector, U.S. Department of Energy's Federal Energy Management Program (FEMP).

[27] Schneider Electric, The financial decision-makers guide to Energy-as-a-Service microgrids copyright, Microgrid Knowledge, 2018.

[28] N.W.A. Lidula, A.D. Rajapakse, Microgrids research: a review of experimental microgrids and test systems, Renew. Sustain. Energy Rev. (2011). Available from: https://doi.org/10.1016/j.rser.2010.09.041.

[29] Guidehouse Insights, DER deployments for microgrids 3Q 2020: regional market forecasts on capacity and revenue for nine DER technologies, 2020. <https://guidehouseinsights.com/reports/der-deployments-for-microgrids>.

[30] Bloomberg NEF, State of the Global Mini-grids Market Report 2020: Trends of renewable energy hybrid mini-grids in Sub-Saharan Africa, Asia and island nations, 2020.

[31] D. Jones, 4 Trends driving the future of microgrids, ICF Energy Insights. <https://www.icf.com/insights/energy/microgrid-database>.

[32] D. Jones, The multi-technology future of microgrids, ICF Energy Insights. <https://www.icf.com/insights/energy/microgrid-technology-trends#>.

[33] E. Konečná, S.Y. Teng, V. Máša, New insights into the potential of the gas microturbine in microgrids and industrial applications, Renew. Sustain. Energy Rev. (2020). Available from: https://doi.org/10.1016/j.rser.2020.110078.

[34] L. Goldstein, B. Hedman, D. Knowles, S.I. Freedman, R. Woods, Gas-fired distributed energy resource technology characterizations, Gas. Res. Inst. Natl. Renew. Energy Lab. (2003).

[35] O. Dessornes, et al., Advances in the development of a microturbine engine, J. Eng. Gas. Turbines Power (2014). Available from: https://doi.org/10.1115/1.4026541.

[36] J. Rathke, Microturbine CHP for microgrids, VERGENT Power Solution, 2017.

[37] D. Akinyele, E. Olabode, A. Amole, Review of fuel cell technologies and applications for sustainable microgrid systems, Inventions (2020). Available from: https://doi.org/10.3390/inventions5030042.

[38] E. Wood, The rise of fuel cell microgrids: special report, Microgrid Knowledge. <https://microgridknowledge.com/fuel-cell-microgrids-fuelcell-energy/>.

[39] IRENA, Renewable Power to Hydrogen Innovation Landscape Brief, Abu Dhabi, 2019.

[40] J.H. Alice Orrell, D. Preziuso, N. Foster, S. Morris, Distributed wind market report, 2018.

[41] Guidehouse Insights, Market data: energy storage for microgrids and remote power systems, 2020. [Online]. <https://guidehouseinsights.com/reports/market-data-energy-storage-for-microgrids-and-remote-power-systems>.

[42] X. Tan, Q. Li, H. Wang, Advances and trends of energy storage technology in microgrid, Int. J. Electr. Power Energy Syst. (2013). Available from: https://doi.org/10.1016/j.ijepes.2012.07.015.

[43] W. Jing, C.H. Lai, S.H.W. Wong, M.L.D. Wong, Battery-supercapacitor hybrid energy storage system in standalone DC microgrids: a review, IET Renew. Power Gener. (2017). Available from: https://doi.org/10.1049/iet-rpg.2016.0500.

[44] E. Wood, ABB microgrids elevated to corporate 'Next Level,' Microgrid Knowledge. <https://microgridknowledge.com/abb-microgrids/>.

[45] Navigant Research, Market Data: Microgrid Controls 4Q 2020: Global Spending Forecasts by Technology Segment and Region, 2020.

[46] Research and Markets, Microgrid Control Systems - Global Market Trajectory & Analytics.

[47] W. Feng et al., A review of microgrid development in the United States — a decade of progress on policies, demonstrations, controls, and software tools, Appl. Energy (2018). Available from: https://doi.org/10.1016/j.apenergy.2018.06.096.

[48] M. Ahmed, L. Meegahapola, A. Vahidnia, M. Datta, Stability and control aspects of microgrid architectures - a comprehensive review, IEEE Access. (2020). Available from: https://doi.org/10.1109/ACCESS.2020.3014977.

[49] J.Y. Kim et al., Cooperative control strategy of energy storage system and microsources for stabilizing the microgrid during islanded operation, IEEE Trans. Power Electron. (2010). Available from: https://doi.org/10.1109/TPEL.2010.2073488.

[50] DoE, DOE microgrid workshop report 2011, Office of Electricity Delivery and Energy ReliabilitySmart Grid Research and Development Program, 2011.

[51] P. Singh, P. Paliwal, A. Arya, A review on challenges and techniques for secondary control of microgrid, in: IOP Conference Series: Materials Science and Engineering, 2019. Available from: https://doi.org/10.1088/1757-899X/561/1/012075.

[52] M. Stadler, C. Marnay, N. DeForest, A green prison: Santa Rita Jail Creeps toward zero net energy (ZNE), in: ECEEE Summer Study, June 6–11, 2011.

[53] L. Meng, M. Savaghebi, F. Andrade, J.C. Vasquez, J.M. Guerrero, M. Graells, Microgrid central controller development and hierarchical control implementation in the intelligent microgrid lab of Aalborg University, in: Conference Proceedings - IEEE Applied Power Electronics Conference and Exposition - APEC, 2015, Available from: https://doi.org/10.1109/APEC.2015.7104716.

[54] W.-J. Lee, D. Wetz, A. Davoudi, Development of a smart MicroGrid testbed: final DOE technical report, DE-OE0000361, Energy System Research Center, Univesity of Texas at Arlington, 2013.

[55] E.O. Schweitzer, D. Whitehead, G. Zweigle, K.G. Ravikumar, G. Rzepka, Synchrophasor-based power system protection and control applications, in: Proceedings - International Symposium: Modern Electric Power Systems, MEPS'10, 2010.

[56] A. Rezai, P. Keshavarzi, Z. Moravej, Key management issue in SCADA networks: a review, Eng. Sci. Technol. Int. J. (2017). Available from: https://doi.org/10.1016/j.jestch.2016.08.011.

[57] R.J. Best, D.J. Morrow, D.M. Laverty, P.A. Crossley, Synchrophasor broadcast over internet protocol for distributed generator synchronization, IEEE Trans. Power Deliv. (2010). Available from: https://doi.org/10.1109/TPWRD.2010.2044666.

[58] DERlab, Activity Report 2018-2020: The Pan-European Smart Grid: Innovative Tools and Demonstration Activities for Future Grid Planning and Operation, 2020.

[59] M. Kermani, D.L. Carnì, S. Rotondo, A. Paolillo, F. Manzo, L. Martirano, A nearly zero-energy microgrid testbed laboratory: centralized control strategy based on SCADA system, Energies (2020). Available from: https://doi.org/10.3390/en13082106.

[60] S.S. Venkata, M.J. Reno, W. Bower, S. Manson, Microgrid protection: advancing the state of the art, Sandia National, 2019.

[61] C.A. McCarthy, Utilities seeking intelligence on electric distribution circuits, Renew Grid, 2011.

[62] S&C Electric Company, PulseClosing technology saves transformers from the stress of damaging fault current, 2017. <https://www.sandc.com/>.

[63] H. Louie, G. Goldsmith, P. Dauenhauer, R.H. Almeida, Issues and applications of real-time data from off-grid electrical systems, in: IEEE PES PowerAfrica Conference, PowerAfrica 2016, 2016. Available from: https://doi.org/10.1109/PowerAfrica.2016.7556577.

[64] S. Kumar, S. Islam, A. Jolfaei, Microgrid communications - protocols and standards, in: Variability, Scalability and Stability of Microgrids, 2019.

[65] J. Cao, M. Ma, H. Li, Y. Zhang, Z. Luo, A survey on security aspects for LTE and LTE-A networks, IEEE Commun. Surv. Tutor. (2014). Available from: https://doi.org/10.1109/SURV.2013.041513.00174.

[66] S. Safdar, B. Hamdaoui, E. Cotilla-Sanchez, M. Guizani, A survey on communication infrastructure for micro-grids, in: 2013 9th International Wireless Communications and Mobile Computing Conference, IWCMC 2013, 2013. Available from: https://doi.org/10.1109/IWCMC.2013.6583616.

[67] Homer-Pro Brochure. <https://www.homerenergy.com/pdf/Homer-Pro-Brochure.pdf>.

[68] About DER-CAM. <https://gridintegration.lbl.gov/der-cam>.

[69] Microgrid Design Toolkit. <https://www.sandia.gov/CSR/tools/mdt.html>.

[70] GridLAB-D. <https://www.gridlabd.org/>.

[71] Renewable Energy Optimization (REopt). <https://www.nrel.gov/docs/fy14osti/62320.pdf>.

[72] RETScreen. <https://www.nrcan.gc.ca/maps-tools-publications/tools/data-analysis-software-modelling/retscreen/7465>.

[73] PV*SOL. <https://pvsol.software/en/>.

[74] M. Pochacker, T. Khatib, W. Elmenreich, The microgrid simulation tool RAPSim: description and case study, in: 2014 IEEE Innovative Smart Grid Technologies - Asia, ISGT ASIA 2014, 2014. Available from: https://doi.org/10.1109/ISGT-Asia.2014.6873803.

[75] ETAP-Microgrid. <https://etap.com/solutions/microgrid>.

[76] Get Started With Simulink. <https://www.mathworks.com/help/simulink/getting-started-with-simulink.html>.

[77] PSCAD Overview. <https://www.pscad.com/software/pscad/overview>.

[78] OPAL-RT Microgrid Overview. <https://www.opal-rt.com/microgrid-overview/>.

[79] RTDS MICROGRIDS & RENEWABLE ENERGY. <https://www.rtds.com/applications/microgrids-renewable-energy/>.

[80] S. Yoneoka, S. Goli, S. Gubbi, Handbook on Microgrids for Power Quality and Connectivity, Asian Development Bank, 2020. Available from: http://doi.org/10.22617/TIM200182-2.

[81] D.T. Ton, M.A. Smith, The U.S. Department of energy's microgrid initiative, Electr. J. (2012). Available from: https://doi.org/10.1016/j.tej.2012.09.013.

[82] W. Bower, I. Llc, J. Reilly, R. Associates, The advanced microgrid integration and interoperability, Sandia Rep., 2014.

[83] N. Hatziargyriou, H. Asano, R. Iravani, C. Marnay, Microgrids, in: IEEE Power and Energy Magazine, 2007. Available from: https://doi.org/10.1109/MPAE.2007.376583.

[84] R. Lasseter et al., The CERTS microgrid concept, white paper on integration of distributed energy resources, California Energy Commission, Office of Power Technologies, Department of Energy, LBNL-50829, 2002. <http//certs.lbl.gov>.

[85] S. Baktiono, J.H. Eto, R.H. Lasseter, D. Klapp, A.S. Khalsa, B. Schenkman, et al., The CERTS microgrid concept, as demonstrated at the CERTS/AEP microgrid test bed, Lawrence Berkeley National Laboratory, 2018.

[86] R. Lasseter, CERTS microgrid concept, CERTS Berkeley Lab, 2019. <https://certs.lbl.gov/initiatives/certs-microgrid-concept>.

[87] B. Lasseter, Microgrids, U.S. Department of Energy Electricity Advisory Committee, 2011. <https://www.energy.gov/sites/prod/files/EAC Presentation - Microgrids 2011-Lasseter.pdf>.

[88] DOE, Microgrids, infrastructure resilience, and advanced controls launchpad, Office of Energy Efficiency and Renewable Energy, Department of Energy. Available from: https://doi.org/DOE/GO-102019-5221.

[89] J. Reilly, MIRACL: microgrids, infrastructure resilience, and advanced controls launch-pad, The National Renewable Energy Laborator. <https://www.nrel.gov/wind/miracl.html#controls>.

[90] Engerati, How artificial intelligence is being used for microgrid management. <https://www.engerati.com/energy-retail/how-artificial-intelligence-is-being-used-for-microgrid-management/>.

[91] NTU Energy Research Institute, About REIDS, Nanyang Technological University. <http://erian.ntu.edu.sg/REIDS/Pages/AboutREIDS.aspx>.

[92] S. Roch Drozdowski, Systems & technologies for a sustainable & affordable energy access-for-all in Southeast Asia, REIDS Renewable Energy Integration Demonstrator – Singapore.

[93] X. Wu, X.G. Yin, X. Song, J. Wang, Research on microgrid and its application in China, Gaoya Dianqi/High Voltage Apparatus, 2013. Available from: https://doi.org/10.4236/epe.2013.54b033.

[94] Chinapower, How is the development of microgrid in China (Chinese), State Grade Info & Telecom Group, 2020. <http://chinapower.com.cn/dlxxh/scyj/20200923/30723.html>.

Chapter 3

Key technical challenges in protection and control of microgrid

3.1 Introduction

The development and implementation of the microgrid has increased rapidly. The concept of the smart microgrid was initially put forward in the early 2000s [1,2]. The capability of high penetration and integration of distributed generators (DGs) and renewable energy sources (RESs) makes it a very attractive option. Such configuration in the power grid provides numerous benefits, such as decreasing substantially the needs for transmission and high-voltage distribution systems, meeting the increase of electricity demand and improving power supply reliability, increasing resiliency and security of power supply to end users, improving the efficiency of energy utilization, and other advantages.

Nevertheless, the interaction between conventional power sources and RESs and DGs of the microgrid will considerably increase the complexity of the power system configuration and raise the generic challenging issues of control, protection, reliability, and power quality. The issues commonly faced in the protection and control of microgrids, which are less prevalent in large grids, include bidirectional flow of power, common occurrences of voltage excursions outside acceptable limits, fault current being supplied from multiple sources in multiple directions, loss of synchronism between multiple sources when a fault happens, potentially limited fault current magnitude, smaller inertia, regular changes in operational configuration due to economic optimization, and intermittency of source-dependent renewable DGs. These issues worsen when the microgrid contains several converter-based generators (CBGs) and operates in the island mode. Therefore, to conform to the requirements of reliable, secure, efficient, and cost-effective operation of the microgrid, and to harness the potential benefits of DGs, these challenges need to be addressed and overcome.

Moreover, the initial characteristics during the occurrence and clearance of faults are very similar to the initial characteristics of transient and dynamic disturbances. Identifying those incidents from each other is very

Microgrid Protection and Control. DOI: https://doi.org/10.1016/B978-0-12-821189-2.00007-3

critical for the normal operation of microgrids. Thus the protection of microgrids is closely related and shall be coordinated with each other.

Scholars and experts have put on many pioneering ideas and technologies to tackle the challenges faced in the microgrids such as devising a wide range of control strategies, power quality improvement measures, energy management schemes, energy optimization techniques, stability analysis tools, and most importantly different innovative protection and fault detection schemes. In this chapter, the technical challenges in control and protection systems are discussed in brief.

3.2 Challenges in control of microgrids

The interaction among DG control systems in the microgrid may commence local fluctuations. Such an occurrence needs a thorough stability analysis. Furthermore, switching of operation between the grid-connected and island modes can create transient instability [3,4]. To this regard, according to the recent studies [5,6] of the direct-current (DC) microgrid interface, significantly simpler control structure, more energy efficient dispatch, and higher current-carrying capacity systems have been developed. The stability constraints in an isolated microgrid have substantial influence on the frequency-droop gains, which in turn lead to instability and lower dynamic performance. Such consideration of gains can be identified using qualitative analysis and small-signal techniques [7]. An adaptive feedforward compensation control strategy was presented by Wu et al. [8] to modify the dynamic coupling between the host microgrid and a distributed resource unit. As a result, the improvement of robustness and system stability to droop coefficients and network dynamic uncertainties is ensured through such a strategy that makes the stability of the microgrid less susceptible and more vigorous to the droop coefficients and network dynamics. An appropriate load sharing in an autonomous microgrid can be ensured by high gain angle droop control strategy especially in weak system conditions, however, as verified by frequency-domain modeling, eigenvalue analysis, and time-domain simulations, it has a negative impact on overall stability. To address this problem, a supplementary loop unit is included in a typical droop control in each DG converter [9]. High gains of the droop controller were also investigated by Choi et al. [10] to meet the stability issues and performance improvements of the microgrid network systems have been demonstrated.

The stability of the microgrid and its control are still critically challenging and different from the conventional grid system due to factors such as:

- Low system inertia
- Short feeder lengths and small X/R ratio
- Higher proportion of intermittent renewable energy resources
- Existence and (in some cases) dominance of CBGs

These issues are more prevalent in the island mode of operation or isolated microgrids resulting in congenital problems to maintain the frequency and voltage levels within the tolerable limits. In case of grid-connected operation, basically, the task of stabilizing the system voltage and frequency lies in the grid and hence less is required from the control functions which contribute to the dynamic stability. Thus control design and implementation of the microgrid manifested with those challenges are discussed hereunder.

3.2.1 Low system inertia

The conventional grid is known to be made up of large generating units with synchronous generators. This fact helps the grid to have a big inertia that enables the network to have higher tolerance to disturbances. Since synchronous generators run at the same frequency as the grid, it provides a natural damping effect to sudden frequency and voltage variations and regulates the frequency and voltage.

However, the case is different in microgrids with them being constituted of smaller units and most of the time CBGs. The power outputs of CBGs are governed by the controllers of their converters which have no rotating mechanical parts. This in turn means the lack of a much needed mechanical inertia that would have contributed a lot for system stability. The issue is more pronounced in isolated microgrids and grid-connected microgrids operating in island mode.

Synchronous generators, though small in size, may still exist in some microgrids. However, those generators have a power output with high frequency or variable frequencies. Thus they are not directly connected to the network, rather interfaced through the power electronic converters, which significantly limits their capability to provide inertia to the system. Microturbines and variable frequency wind turbines (such as direct driven ones) are examples of such units.

The other types of power source in microgrids are naturally direct current (DC) though their outputs are converted to alternating current (AC) though converters. Sources such as photovoltaic (PV) systems, batteries, fuel cells, and other storage components fail into this group. Their outputs are instantly controlled by the controllers of their converters, keeping in mind the minimal delay in control actions due to communication and computation time. DC loads, which are notably becoming a dominant type of load nowadays, are also interfaced with the grid via converters replacing the induction motor dominated load of the old days.

Conventional sources such as diesel generators are also available in microgrids. They do have inertia though not to the level that it can guarantee system stability under disturbances due to the small size of the units and their lower proportion in CBG dominated microgrids.

Due to those factors mentioned earlier, maintaining nominal frequency and voltage of the network has become complicated in microgrids. The low inertia in the system can lead to severe frequency deviations in islanded microgrids and the island mode of operation unless a proper control mechanism is implemented [3].

Inertia of microgrids is expressed as a function of all operating generators. In this regard, since synchronous generators have strong correlation between their rotational speed and electrical frequency, they can play a significant role to increase the overall system inertia. However, islanded microgrids contain directed connected synchronous generators in small size and number and hence have less inertia. Moreover, there has been a trend of replacing conventional generation by PV generation which is basically made with power electronics interfacing to provide an inertialess system with the microgrid, thus leading to lower microgrid inertia. This concept is validated in [4,5] in that since no inertial response is provided with those sources at times of frequency events, the integration of RESs in the network will affect frequency backing to decline. Because of the generally lower system inertia in the microgrid, the operating synchronous generators will be less capable of keeping the frequency value within the stability margins [6]. Keeping this deviation small is very vital in order to avoid destructive vibrations in synchronous machines and untimely load shedding. There are also some loads which are frequency sensitive, thus increasing the requirement for frequency stability. Power quality should also be guaranteed in networks having on-site resources.

Maintaining large enough system inertia of microgrid operation is a critical challenge which confronts reliable microgrid performance [6]. Thus microgrids with less inertia may need to depend on the control of energy storage units, load management, or generation curtailment to keep their stable operation status. In converter-interfaced systems, the frequency is possibly regulated by different mechanisms. For example, synchronverters which are inverters that mimic the synchronous generator enable the provision of frequency control. Such converters can be applied with a battery energy storage, a supercapacitor, or a flywheel, and can provide a restoring capability to the nominal frequency [7].

3.2.2 Low reactance to resistance (X/R) ratio

As there is a necessary requirement in keeping the grid voltage of any power system within the tolerable ranges, voltage stability considered as an essential microgrid performance. However, an inherent challenge in low-voltage (LV) microgrids is attributed to the predominantly resistive line impedances, which render the commonly considered conditions used in the traditional grid droop control method inapplicable [2]. Along with the low of inertia, the resistive feeder characteristics of microgrids invoke complications with regard to voltage regulation.

Microgrids usually involve LV and, on a few occasions, medium voltage distribution lines that are usually short in length leading to low reactance to resistance (X/R) ratios [8]. This therefore results in strong coupling between active and reactive power and consequently voltage and frequency. Thus all system parameters are influenced by any disturbance in microgrids.

Voltage instability and frequency instability are common phenomena and are handled independently in conventional large grids. Whereas in microgrids, voltage and frequency are strongly coupled due to the low X/R ratio of feeders.

Line impedance parameters may strongly affect the control of grid connected inverters [9]. The impact is related to the X/R ratio of the microgrid. In the conventional power system, active power and frequency are correlatively linked with the P/f droop characteristic. Similarly, the relation between reactive power and voltage is defined by the Q/V droop. Thus decoupled controlling is guaranteed based on these relationships. However, for some microgrids which are operating in both grid-connected and island modes, the assumption of inductive line parameters may not be valid. Cables in small-scale microgrids with LV distribution networks (DNs) will be mainly resistive; that is, they usually feature low X/R ratio in LV cables [9]. Moreover, the parameters of LV cables vary with the size of the cable which enables significant alteration of the X/R ratio. The following equations can explain the effect of line parameters on power flow (Fig. 3.1).

Starting with the power flow equations expressed in (3.1) as

$$P_r = \frac{V_s V_r \cos(\theta + \phi - \delta) - V_r^2 \cos(\theta)}{Z}$$

$$Q_r = \frac{V_s V_r \sin(\theta + \phi - \delta) - V_r^2 \sin(\theta)}{Z}$$

(3.1)

where V_s and V_r are respectively sending and receiving voltages for two nodes of the power system; ϕ and δ are power angles, and θ is line impedance angle; and P_r and Q_r are active power and reactive power of receiving end.

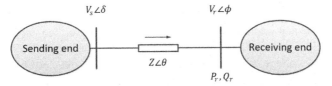

FIGURE 3.1   Simple line diagram to show power flow in the microgrid.

And assuming that the line impedance is inductive $(Z = jX)$, angle $\theta = 90$ degrees. Thus the power equation of (3.1) becomes

$$P_r = -\frac{V_s V_r \sin(\phi - \delta)}{X}$$

$$Q_r = \frac{V_s V_r \cos(\phi - \delta) - V_r^2}{X}$$

(3.2)

For a small angle difference between the two ending voltages, Eq. (3.3) holds.

$$(\phi - \delta) \cong -\frac{X P_r}{V_s V_r}$$

$$V_s - V_r \cong \frac{X Q_r}{V_r}$$

(3.3)

Eq. (3.3) shows that there is a strong correlation between active power and the phase angle difference, and, in the same way, the reactive power and voltage amplitude difference. Therefore traditional P/f and Q/V droop schemes are usually effective in inductive networks, such that frequency control can be achieved through managing the active power flow and voltage control can be achieved through managing the reactive power flow.

Whereas if the impedance is resistive $(Z = R)$, $\theta = 0$ degree, the power equation of (3.1) will be modified to

$$P_r = \frac{V_s V_r \cos(\phi - \delta) - V_r^2}{R}$$

$$Q_r = \frac{V_s V_r \sin(\phi - \delta)}{R}$$

(3.4)

From Eq. (3.4), assuming small angle difference between the two voltages gives us Eq. (3.5).

$$V_s - V_r \cong \frac{R P_r}{V_r}$$

$$(\phi - \delta) \cong \frac{R Q_r}{V_s V_r}$$

(3.5)

Eq. (3.5) verifies that the close linkages between active power and frequency (P/f), and reactive power and voltage (Q/V) are no longer valid. Rather active power flow and reactive power are, respectively, more impacted by voltage amplitude difference and phase angle difference. So, the droop controller of such a system is designed accordingly for resistive impedance, that is, P/V and Q/f droop controllers. Therefore the microgrids behaving with their network impedances varying from dominantly inductive to dominantly resistive are conditional to changing from P/f and Q/V to P/V and Q/f, respectively.

From these results, it is possible to say that the two extreme cases can lead to applying the two different control mechanisms—the conventional droop control mechanism and reversal of droop control mechanism. However, in-between these two cases, the difficulty arises on the coupling of active and reactive powers since clearly unknown line parameters exist. It is also to be noted that the line parameters can be affected by the operating conditions.

3.2.3 Uncertainty and intermittency of renewable sources

Most DGs, such as wind turbines and PVs, have intermittent and fluctuating power generation. This is one of the major distinctive challenges faced in microgrids involving such renewable sources. The consequences of intermittent and fluctuating power generation increases with the greater the renewable penetration rate in the microgrid, specially impacting the frequency and voltage control and normalization [10,11].

The economical operation of microgrids demands an assured level of coordination among various distributed energy resources. This coordination comes to be more challenging in isolated microgrids, where the uncertainty of parameters such as load profile and weather forecast exist. Since there exist reduced number of loads and fluctuating available energy resources in the microgrid, this uncertainty is of a more significant impact than in the large power systems [3]. Thus the demand—supply balancing is an important aspect of the operation and control of microgrid that still needs further studies. Short term and very-short term forecasts of renewable generation and load can be considered as part of the solution in this regard. A higher level of component failure rates in renewable generating units and their pronounced impact on the overall microgrid stability are also a source of concern.

3.2.4 Different modes of operation (grid-connected and island modes)

Microgrids, specially the nonisolated ones, are known to have the capability of operating both in parallel with the grid and independently by themselves. Stable operation should be ensured regardless of the mode of operation. In a grid-connected operation mode, the microgrid can be set to hold to the rules of the DN. Though there are possibilities that the microgrid can provide auxiliary services by offering frequency and/or voltage support to the grid, it is typically expected to be less involved in the operation of the main power system. However, there will still be power exchanges between the microgrid and the grid where the grid could be supplying power to the utility grid or consuming power from the utility grid, based on the whether there is a surplus or deficiency of generation locally in the microgrid. The voltage and frequency of the microgrid are mostly decided by the grid.

Whereas, in island mode of its operation, balancing the operating nominal values of voltage and frequency, ensuring acceptable power quality, balancing of supply and demand, and providing reliable communication among the components are critically essential to keep providing adequate energy to the loads. The frequency and voltage stability of the microgrid is this time controlled locally by the microgrid itself. In island mode, the microgrid will have one or more main DGs operating in U/f mode providing the nominal system frequency and voltage and the reference to the other DGs, so that it serves the role of the utility grid in the case of grid-connected mode. However, the main DG in an island microgrid is not as big as the utility grid and does not have big inertia. This leads to the microgrid being easily susceptible to stability issues in the event of even smaller disturbances.

Aside from the issues in the control of island mode, handling the transition between the two modes of operation is very important. During the event that the microgrid disconnects itself from the grid due to a grid-side fault or some forced reasons, the main DG (U/f source) shall immediately take over from the grid and provide the voltage and frequency reference. Synchronizing the microgrid with the grid is also required when the microgrid goes back to grid-connected mode. Blackstarting is one of the required features of an islandable microgrid. All those issues need to be handled properly to ensure the stable operation of the microgrid in both modes.

3.2.5 Existence and (in some cases) dominance of CBGs

Unlike the conventional grid, due to the existence of CBGs in the microgrid, the disturbance problem is a significant challenge. For instance, forced or unintended series of voltage and current changes will be exhibited while the natural feature of DG intermittency and operational load variation happen in the microgrid. Such changes can be characterized by the variations of voltage magnitude and its phase beyond the normal operating limits and will continue for milliseconds to seconds. Such a phenomenon can lead to maloperation or component damage.

3.3 Challenges in protection of microgrids

Protection of microgrids is a challenging task and a theme that is attracting many researchers and scientists of the power system world. Reclosers, fuses, and overcurrent (OC) relays have been the commonly used protective devices for distribution systems and somehow continued to be functional in the protection of microgrids. To identify fault directions, directional OC relays are extensively implemented. OC relays' coordination with their optimal setting selection has also been designed and developed so far.

However, the protection of microgrids is significantly different from that of the distribution system in the traditional grid. As a result, the protection based on OC relays, fuses, and reclosers is less effective in microgrids, although not totally obsolete.

Many protection schemes have been studied and developed so far. An OC-based protection scheme is commonly applied for distribution systems with radial structure. But, fault occurrence, fault location, and direction detections are not effective in this method. A voltage total harmonic distortion scheme was developed by Al-Nasseri and Redfern [12] to discriminate between the faulted zones in the network. But finding the fault location is complex since the voltages in all buses substantially decline. The voltage-drop method was also presented Al-Nasseri et al. [13] so that the relays effectively identify the faults. However, the fault location and the coordination of the relays remain a challenge. In order to effectively detect the fault occurrence, the negative sequence current protection strategy has been applied, but still the detection of fault location and direction is not effectively confirmed. Differential current protection scheme [14], differential negative sequence scheme [15,16], and phase-angle comparison of currents scheme [17] have been studied to effectively detect fault occurrence, fault location, and direction. Nevertheless, these schemes rely on the communication system for fault detection and coordination, which is complicated and incurs significant cost. The mismatches of the current transformers might also affect the accuracy of these schemes.

With the island mode of microgrid operation, the short circuit level falls considerably. New challenges have arisen from the system parameters' variations, such as the fault current's magnitude and its direction reversal, when the DGs are integrated into DNs. Coordination problems between different protective equipment, that is, between relay, autorecloser, and sectionalizer can also take place. The interfacing scheme type and the DG penetration level have a basic impact on the protection scheme, since the value of the short circuit current is importantly determined by these factors. Thus, as we have seen from the challenges in microgrid control, its protection development faces its own inherent and even tougher challenges too. Fast operation of the protection system is, at the same time, vital to ensure the stability of microgrids. Thus the critical protection challenges which should be taken into consideration in a novel scenario process include different short circuit currents in microgrid operating modes, reduction in reach of impedance relays, bidirectional power flow, and voltage profile. These challenges are discussed in this section.

3.3.1   Different short-circuit current in island and grid-connected modes

In fact, the protection of the grid-connected operation mode of microgrid is potentially simplified by the large fault currents supplied from the main grid.

However, the fault current is usually smaller in island operation mode. This is because the sources are commonly interfaced through power electronics converters whose fault current levels are limited by the current carrying capacity of their power electronic switches. The different short circuit current values in the two operation modes means it will be impossible to use the same protection settings values used in the grid-connected mode in the island mode. The low value of fault current in island mode may also be not enough to make protection systems, such as OC relays, trip.

Thus it requires a special consideration that setting values or the protection system can be adjusted during changes of operation modes. One of the suggested solutions is an adaptive protection system which can adjust the settings of the relays by adapting to the current operating condition of the system to ensure higher reliability of microgrid system protection [18,19]. Communication network-based digital relays are also capable of ensuring reliable protection [20]. Two current-limiting algorithms, which are used to prevent the large line current flow during a voltage sag, and are implemented in a voltage-source inverter (VSI) are presented in [21]. For limiting the large line current during such a phenomenon, a bulky virtual L or RL impedance is imitated by the VSI. The VSI is introduced in series between the microgrid and the utility grid.

3.3.2 Reduction in reach of impedance relays

Reach of impedance relay refers to the maximum fault distance, which matches with a detectable minimum fault current or maximum fault impedance, that makes a relay activate in a definite impedance zone, or operating time [22]. When the fault takes place downstream of the point of common coupling where DGs are connected, the measurement of upper relay impedance will be greater than the real fault impedance. This impendence seemingly reveals a risen fault distance which is because of the voltage increase caused by extra infeed at the common bus. In other words, the DG with the main source contributes to the fault current. Consequently, this can have an impact on the grading of relays, which results in delayed or no triggering action. Such a phenomenon is referred to as underreaching of a relay.

3.3.3 Bidirectional power flows and voltage profile change

Even though distribution feeders were purposely designed for supplying power with unidirectional flow, the reverse power flows are common phenomena in distribution systems consisting of microgrids due to the existence of DGs. As local generation surpasses the local consumption, the power flow reverses its direction in the case of distributed networks embedded with DGs [23]. This situation leads to more difficulties in power flow patterns, fault current distribution, and protection coordination processes [3]. The fact that faults may be fed from multiple directions make the directional protection units less useful in microgrids because their operating principle of using a reverse flow of current as a signal for fault occurrence is no more valid in microgrids.

To address such issues, a voltage-based fault detection strategy is presented in Ref. [13]. In this work, the strategy is capable for various types of faults within the microgrid being reliably and speedily detected. To distinguish and isolate faults happening within the protection zone, a state observer is also proposed in [24]. In Ref. [25] various fault-detection methods along with the concerns related to the grounding protection system are discussed.

3.3.4   Dominant existence of CBGs

CBGs are sources whose output is AC power while they naturally are DC sources, or AC sources whose frequency is different from the system frequency, and are connected to the power system through a power converter.

With CBGs, the fault current is generally only slightly larger than the rated current of the converter, whereas a synchronous machine can provide three to five times its rated current. The short circuit current level is dictated by the current ratings of the switching devices such as insulated gate bipolar transistors (IGBTs) in the converter. Depending on the operating mode, the minimum short-circuit current could be close to or even below the maximum load current, especially for feeders close to main busbars (where sources are connected) or feeders between busbars with DGs connected. This is not generally an issue for feeders supplying smaller loads, as the ratio between the minimum short-circuit current and the rated load current is high enough to allow discrimination between fault and load currents. The low level of fault current supplied by CBGs in a microgrid results in a small overall fault current in microgrids operating in island mode. This will lead to the current magnitude based protection schemes being less effective. Some advanced methodologies are being suggested to address this. However, it will continue being a central point of research and development work on the power system protection.

References

[1]   C. Marnay, F.J. Rubio, A.S. Siddiqui, Shape of the microgrid, in: Proceedings of the IEEE Power Engineering Society Transmission and Distribution Conference, 2001.

[2]   R.H. Lasseter, MicroGrids, 2003.

[3]   D.E. Olivares, A. Mehrizi-Sani, A.H. Etemadi, C.A. Cañizares, R. Iravani, M. Kazerani, et al., Trends in microgrid control, IEEE Trans. Smart Grid (2014).

[4]   P. Tielens, D. Van Hertem, Grid inertia and frequency control in power systems with high penetration of renewables, in: Young Researchers Symposium in Electrical Power Engineering, 2012.

[5] S. Sharma, S.H. Huang, N.D.R. Sarma, System inertial frequency response estimation and impact of renewable resources in ERCOT interconnection, in: IEEE Power and Energy Society General Meeting, 2011.

[6] A. Ulbig, T.S. Borsche, G. Andersson, Impact of low rotational inertia on power system stability and operation, in: IFAC Proceedings Volumes (IFAC-PapersOnline), 2014.

[7] Y.S. Kim, E.S. Kim, S. Il Moon, Frequency and voltage control strategy of standalone microgrids with high penetration of intermittent renewable generation systems, IEEE Trans. Power Syst. (2016).

[8] X. Wu, X.G. Yin, X. Song, J. Wang, Research on microgrid and its application in China, Gaoya Dianqi/High Volt. Appar. (2013).

[9] A. Toliyat, Energy Storage Sizing for Low-Inertia Microgrids, and Lessons Learned From a Potential Microgrid, The University of Texas at Austin, 2016.

[10] J. Choi, S. Oh, R. Horowitz, Distributed learning and cooperative control for multi-agent systems, Automatica (2009).

[11] T. Morstyn, B. Hredzak, V.G. Agelidis, Distributed Cooperative Control of Microgrid Storage, IEEE Trans. Power Syst. (2015).

[12] H. Al-Nasseri, M.A. Redfern, Harmonics content based protection scheme for micro-grids dominated by solid state converters, in: 2008 12th International Middle East Power System Conference, MEPCON 2008, 2008.

[13] H. Al-Nasseri, M.A. Redfern, F. Li, A voltage based protection for micro-grids containing power electronic converters, in: 2006 IEEE Power Engineering Society General Meeting, PES, 2006.

[14] X. Liu, M. Shahidehpour, Z. Li, X. Liu, Y. Cao, W. Tian, Protection scheme for loop-based microgrids, IEEE Trans. Smart Grid (2017).

[15] E. Casagrande, W.L. Woon, H.H. Zeineldin, D. Svetinovic, A differential sequence component protection scheme for microgrids with inverter-based distributed generators, IEEE Trans. Smart Grid (2014).

[16] S. Kar, S.R. Samantaray, M.D. Zadeh, Data-mining model based intelligent differential microgrid protection scheme, IEEE Syst. J. (2017).

[17] H.F. Habib, T. Youssef, M.H. Cintuglu, O.A. Mohammed, Multi-agent-based technique for fault location, isolation, and service restoration, IEEE Trans. Ind. Appl. (2017).

[18] M.R. Islam, H.A. Gabbar, Study of micro grid safety & protection strategies with control system infrastructures, Smart Grid Renew. Energy (2012).

[19] H.J. Laaksonen, Protection principles for future microgrids, IEEE Trans. Power Electron. (2010).

[20] E. Sortomme, S.S. Venkata, J. Mitra, Microgrid protection using communication-assisted digital relays, IEEE Trans. Power Deliv. (2010).

[21] D.M. Vilathgamuwa, P.C. Loh, Y. Li, Protection of microgrids during utility voltage sags, IEEE Trans. Ind. Electron. (2006).

[22] M. Geidl, Protection of power systems with distributed generation: state of the art, Technology (2005).

[23] N. Jenkins, et al., Embedded Generation, 2000.

[24] M. Esreraig, J. Mitra, An observer-based protection system for microgrids, in: IEEE Power and Energy Society General Meeting, 2011.

[25] D. Salomonsson, L. Söder, A. Sannino, Protection of low-voltage DC microgrids, IEEE Trans. Power Deliv. (2009).

Chapter 4

Short-term renewable generation and load forecasting in microgrids

4.1 Introduction

Forecasting is a very important tool used in a variety of fields, such as economics, finance, logistics, weather, insurance, and power system, so as to make decisions based on informed estimates of future trends. Forecasting can be generally defined as "a practice of gathering and analyzing data of the past and the present for predicting what will happen in the future." In the fields of economics and business, forecasts of market trends, exchange rates, demand for the goods and services, and many other variables are utilized to determine the future plans of budgets allocation, human resources, business starting, and expansion. Forecasting is also quite an important tool in the power system. The increased integration of renewable resources, adaption of load management methods, deregulation of the energy market, requirements for efficient utilization of energy, and implementation of different smart grid technologies are some of the driving factors for the increasing demand for forecasts of different quantities in the power system sector.

The main parameters to be forecasted in power system applications are the power demand, supply, and prices. Such forecasts are also needed in planning and operational management of microgrids. Forecasts of different time horizons and time intervals are used for different applications.

Out of the different types of generation in the power system, a special case with the need for forecasting is the generation from renewable energy sources (RES). RES includes sources such as solar photovoltaic (PV), solar thermal, wind, geothermal, ocean, and hydropower. These energy sources are based on natural sources or processes that are constantly replenished. They are known for being clean energy sources. However, most of them have an intermittent and fluctuating nature and, as a result, are usually referred to as "nondispatchable" sources. Such characteristics have led to a greater need for forecasting their power outputs. Solar and wind energy are the most commonly used RES in microgrids, typically exhibiting intermittency and fluctuation in their outputs.

Microgrid Protection and Control. DOI: https://doi.org/10.1016/B978-0-12-821189-2.00005-X

57

The other aspect where forecasting is important in the power system is the power demand. The load pattern of a single end user or a cluster of users may differ from time to time based on the type, number, and power consumption of electric and heating loads. Forecasting the load in a power system or a microgrid is critical for optimal planning and operation of the system.

There are different ways of classifying forecasts in the power system. One way of classification is as "deterministic" and "probabilistic." Probabilistic (or sometimes called categorical) forecasting is intended to predict occurrence/ magnitude probability of an event while deterministic forecasting can provide explicitly defined information on the occurrence or magnitude of an event. There also various ways of presenting forecasts. Some of the formats are point, quantile, intervals, densities, trajectories, etc. A point forecast represents the expected value of the forecasted variable by the mean or the median of the variable. Hence, point forecasts are analogous to deterministic forecast. The other presentation formats are more suited to probabilistic forecasts.

Practitioners in power systems usually need to consider a single forecast value and not a probability distribution. Thus deterministic point forecasts have been more widely adopted in operational planning, decision-making, or power trading. Similarly, a larger part of the forecasting literature is based on such forecasts. However, there is a growing debate recently on the preference of probabilistic or deterministic forecasts.

Different actors in the power system may be interested in different types of forecasts of renewable generation and load. With respect to microgrids, the forecasts may be used by microgrid owners, energy traders, end-users or customers, utilities or grid operators, and research and development institutions. The specifics of the different types and techniques in the forecasting of renewable generation and load are discussed in the upcoming sections of this chapter.

The energy and load forecasting system of a microgrid may be used to forecast the power generation from distributed generation (DG) resources, the different loads (electric, heat, and cold) and the gas and water supply demand of the microgrid according to the meteorological information, historical data, and real-time data. Fig. 4.1 shows the basic steps and components in a typical energy and load forecasting system of a microgrid.

4.2 Basics and classification of renewable generation forecasting

Higher penetration of RES in the energy market is being encouraged in recent years through the implementation of energy polices in many countries due to increasing environmental concerns. Among a wide range of RES utilized in the power system, PV and wind energy systems are getting higher attention by researchers, energy policy makers, and utilities due to their economic and environmental benefits.

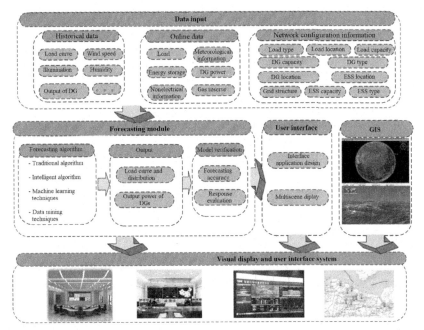

FIGURE 4.1 Microgrid renewable energy and load forecasting system. ESS, Energy Storage System; GIS, Geographic Information System.

PV and wind are also very common components of microgrids. Forecasting the power output of PV and wind systems is, as a result, one of the most discussed topics. The power outputs of PV and wind systems are highly variable as they are dependent on meteorological conditions. The inherent variability of output powers from PV and wind systems creates different issues in the power grid and microgrids, such as difficulty to achieve optimal dispatching, power systems security issues, excessive generation, curtailment, and increase of the levelized cost of energy. Short-term forecasting of renewable generation is, thus, a very useful and indispensable tool to address those issues. Generation forecasting is generally considered critical to facilitate the integration of higher penetrations of renewable resources with variable generation into the power system [1].

Renewable generation forecasts are primarily classified based on the involved timescale. The basic and most common ones are short-term, medium-term, and long-term forecasts. There is also another group used in some literature for a much shorter timescale than short-term forecasts, labeled as very-short-term forecasting.

1. *Very-short-term forecasting* is forecasting for a very short period, usually for a period less than few hours. A period of 4 hours is specified as the upper boundary for very-short-term forecasts in most of the literature [2,3],

while it is expected to be updated every 15 minutes [4] and have 5-minute [5,6] (or generally 1−15-minute [7]) time resolution. Other terms such as ultrashort-term forecasting, intrahour forecasting, or sometimes nowcasting are used to refer to this class of forecasting. The forecasts are used for power smoothing, market clearing, and online optimal operation of power systems. Immediate forecasting of energy production on a minute by minute or second by second scale is also becoming useful for managing the operation of wind and solar plants and intelligent energy networks such as microgrids with the objectives of ensuring reliable power supply and improving system efficiency. Forecasting models based on climate models, satellite data, and historical generation data are inadequate for a very-short-term forecasting due to their low spatial and temporal resolution. Nowcasting models are usually based on current observations and extrapolation techniques. Hence, most of the very-short-term forecasting models are based on sky images for solar power forecasting [8] or wind profiles based on cloud motion [9] and variational data assimilation systems such as Variational Doppler Radar Analysis System [10] for wind power forecasting. Different postprocessing methods to blend disparate models are also applied in very-short-term forecasts.

2. *Short-term forecasting* refers to a forecasting time horizon of less than a week, ranging from a period of hours to a few days (2 or 3 days in most cases). The forecasts involve temporal resolution of 15 minutes−1 hour and are usually updated on an intraday timescale. The forecasts are used for hours-ahead and day-ahead operational decisions in microgrids or generally in power system with specific functions, such as unit commitment, scheduling, load following, programming backup, reserves allocation, and congestion management. [11,12]. Different techniques, such as statistical or time series methods, physical models, and hybrid or ensemble methods are applied for short-term forecasts of renewable generation.

3. *Medium-term forecasting* refers to forecasts with time horizons in the range of a few days to a week. Such forecasts may be applicable for trading on medium-term markets, congestions forecast, week-ahead planning, medium-term operations and maintenance plans, and scheduling of fuel supply [13].

4. *Long-term forecasting* is forecasting usually performed on time horizons of weeks, months, quarters, a year, or even multiple years and is applicable to make decisions on long-term investments and reinforcements in a power system. Functions such as grid expansion planning, generation units construction planning, resource planning, contingency analysis, and maintenance planning can be served by long-term forecasts.

Most of the renewable generation forecasting models in the literature and applied in the industry use recent power generation measurements, recent measurements of metrological parameters, calendar variables, weather forecasts for the coming period, and possibly some extra information on the site and the RES units. Tables 4.1 and 4.2 provide compilations of different input variables used in short-term PV and wind power forecasting literature.

4.3 Basics and classification of load forecasting

Microgrids consist of a variety of loads mainly divided into electricity, cold, heat, water, and gas loads. Different load types require different forecasting methods. As cold and heat are both manifestations of heat energy in physics, their basic characteristics are the same. Hence, cold and heat load forecasts can implement similar approaches. On the other hand, water consumption

---

TABLE 4.1 Compilation of input variables used for short-term photovoltaic (PV) power forecasting considered in literature [14–17].

- Solar irradiance (W/m²)
- PV power (W)
- Ambient temperature (°C)
- Wind speed (m/s)
- Humidity (%)
- Wind direction
- Cloud shapes
- Heights of clouds (m)
- Visibility (m)
- Ground temperatures (°C)
- Surface ground temperature (°C)
- Lowest temperature (°C)
- Highest temperature (°C)
- Cloudiness (okta)
- Zenith angle (degrees)
- Date number
- Precipitation duration (h)
- Minimum humidity (%)
- Precipitation (mm)
- Sunshine duration (h)
- Continued sunshine duration (h)
- High, middle, and low cloud amount (okta)
- Time
- Day of year
- Hour of day
- Short-wavelength radiation (W/m²)
- Surface pressure (hPa)
- Relative humidity (%)
- Azimuth angle (degrees)
- Extraterrestrial irradiance (W/m²)

TABLE 4.2 Compilation of input variables used for short-term wind power forecasting considered in literature [18–21].

- Wind speed (m/s)
- Wind direction
- Wind power output (W)
- Humidity (%)
- Outdoor temperature (°C)
- Nacelle temperature (°C)
- Atmospheric pressure (Pa)
- Manufacturer's power curve
- Blade pitch angle (degrees)
- Rotor speed (m/s)
- Air density (kg/m³)

load basically presents periodic changes due to the lifestyle and production mode; is strongly regulated; and does not basically involve transformation of energy forms. As a result, prediction of water load is not to be discussed in this chapter. The energy uses and conversions in the load side of microgrids are shown in Fig. 4.2.

As electrical loads are the most dominant type of load in microgrids, the focus will be on this load type. Electrical loads can be classified based on different factors. Based on the functions they provide, loads may be classified as commercial, industrial, civil, traffic, medical, education and scientific research, agricultural, etc. On the other hand, according to controllability and influencing factors, loads can be divided into baseline load, planned load and controllable load.

- *Baseline load* refers to the load that occurs naturally according to the needs of life and work style. It is mandatory, random and difficult to control.
- *Planned load* is a type of load with relatively fixed start−stop and operation duration, which is not easy to control, but whose operation time can be flexibly adjusted.
- *Adjustable load* usually has long-term operation and is affected by temperature and other climate factors while its power can be adjusted or it can even be interrupted sporadically.

Further classification can be done based on the different characteristics of the load pattern as shown in Table 4.3.

Classifying the type of load forecasts is not something precisely standardized. However, the same approach as used in classifying renewable generation

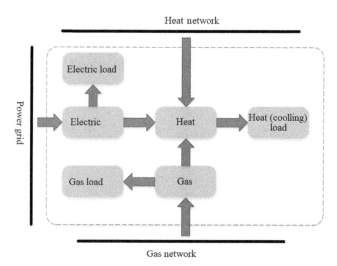

FIGURE 4.2 Energy utilization and conversion in microgrid loads.

TABLE 4.3 Electrical load classification.

Classification	Load characteristic index			Main features	Major industries involved
	cycle	Average load	Peak–valley difference		
Class I	Small	Small	Small	The power consumption is small and the fluctuation is frequent, but the fluctuation capacity is not strong and the average load level is relatively small. It can be considered as a part of base load.	Light industries, transportation, postal service, storage and logistics, education and scientific research.
Class II	Small	Small	Large	Small power consumption, frequent fluctuation, strong fluctuation, and difficult to predict and control; important reasons for the multi peak characteristics of early load.	Light industries
Class III	Large	Small	Small	The power consumption is small, the fluctuation is not frequent, and the fluctuation capacity is not large; It can be considered as part of base load.	Information Communication Technology computer services, software industry, part of manufacturing industry; and education and scientific research industry
Class IV	Large	Small	Large	The power consumption is small and the fluctuation is not frequent, but the fluctuation capacity is large.	Various industries
Class V	Small	Large	Small	The power consumption is large, the fluctuation is not frequent, and the change is small; It can be considered as a part of base load.	Residential electricity and information transmission computer service and software industry

(Continued)

TABLE 4.3 (Continued)

Classification	Load characteristic index			Main features	Major industries involved
	cycle	Average load	Peak–valley difference		
Class VI	Small	Large	Large	It is difficult to predict and control because of large power consumption and strong fluctuation, which is the main reason for peak load characteristics.	Wholesale and retail, metal products manufacturing and metallurgical industry
Class VII	Large	Large	Small	Power consumption is large, fluctuation is not frequent, and the fluctuating capacity is not strong; it belongs to important base load.	Education and scientific research institutions, finance and various light industries
Class VIII	Large	Large	Large	The power consumption is large and the fluctuation is not frequent, but the fluctuation capacity is strong and easy to predict and control. It is the main reason for the peak load characteristics, and a key optimization element.	Wholesale and retail, real estate

forecasts may still be applicable. This classification includes four categories in very-short-term, short-term, medium-term, and long-term forecasts [22].

1. *Very-short-term load forecasting* comprises forecasts of a time horizon ranging from few minutes to an hour. Again similar with renewable generation forecasting, this group of forecasts are sometimes called as ultrashort-term forecasting, interhour forecasting, or nowcasting. These forecasts are important for real-time control of microgrid or power system.

2. *Short-term load forecasting* represents forecasts of the next hour to next week and are applicable for day-to-day routine operation of microgrid or power system specially for scheduling, energy trading and optimization purposes. More discussion is provided on short-term forecasts in this chapter.

3. *Medium-term load forecasting* usually covers a period of 1 week up to a year, and typically useful for maintenance scheduling and planning fuel purchases.

4. *Long-term load forecasting* covers a long time horizon which may go as far as 10 or more years, and fundamentally used for strategic planning, such as construction of generation plants and planning of transmission and distribution system construction and upgrading.

As already stated earlier, the main focus will be on short-term forecasts. The influencing factors for the load pattern are discussed in Table 4.4.

TABLE 4.4 Influencing factors for uncertainty in different forecasts of electric loads.

Type of forecast	Major factors	Collection way
Long-term forecast	National economy, population, unit consumption, industrial structure and electricity price	Preliminary research and statistics
Medium-term forecast	User production plan, meteorological conditions, adjustment of industrial structure and electricity price	Preliminary research and statistics
Short-term forecast	Type of day (working days, weekends, holidays), meteorological factors, such as temperature, humidity, rainfall, and electricity price	Meteorological information collection and prediction
Ultrashort-term forecast	Real-time load information, meteorological factors	Supervisory Control and Data Acquisition (SCADA) system

With the power system facing changes in form of integration of dynamic demand, supply, price, and regulation and many other smart grid features, short-term forecasting is becoming more essential and the need to develop forecasting methods that fit the ongoing changes is critical at the same time. One such change in the load forecasting sector may be the need to move to probabilistic forecasts in addition to the point forecasts famously applied up to recent times. The fact that the loads are becoming more flexible and dependent on the generation variation in the DG scheme means it may be useful to have load forecasts showing ranges of expected values conforming to the uncertainties in generation from RES-type DGs and interaction between demand and electricity prices. It is also expected that load forecasting may need to be more integrated with renewable generation forecasting and electricity price forecasting.

4.4 Short-term renewable generation and load forecasting techniques

4.4.1 Introduction

There is a wide range of techniques applied in the short-term forecasting of renewable generation and loads. They can generally be grouped into two main categories as physical methods and time series methods. There could also be a hybrid of the two.

a) Physical methods

These methods are usually applicable for forecasting of renewable generation. Physical forecasting models are based on a set of mathematical equations that describe the physical state and dynamic motion of the atmosphere [23]. These methods are mainly based on numerical weather prediction (NWP), sky imagery and satellite imaging, and cloud movement tracking models. They usually rely on weather forecasts and specification of the solar panels and the wind turbines to generate the power forecasts. This type of model has the drawbacks that it is usually available for a resolution of wide geographical area (in the range of 16–50 km [24]) and the performance greatly depends on the stability of the weather variables. They may be less effective in events of sharp changes in weather conditions [25].

This group consists of several models, which employ mathematical descriptions of the physical processes so as to translate the NWP forecasts at a certain grid point to a power forecast at the considered site of the wind or PV system. Qualitative or subjective forecasting methods, in which the future load is predicted based on experts opinion and structured approaches without using historical data, and multifactor or cross-sectional forecasting methods, which focus on the search for the causal relationships between different influencing factors and forecasted load, may be considered to fall into this category to some extent. Subjective curve fitting, the Delphi method, and technological comparisons are examples of qualitative/subjective forecasting methods applied in load forecasting [22].

b) Time series methods

This group refers to forecasting methods which are based on a sequence of observations or measurements of one or more parameters at successive instants in time. This type of model may also be referred to as data driven methods. Time series methods are applied for renewable generation as well as load forecasts. The models make use of historical records of the power generation or power demand and some exogenous variables, usually meteorological, social and calendar variables.

There is also a trend of farther classifying time series methods into statistical methods, intelligent (artificial intelligence based or machine learning) methods and hybrid methods [22]. The statistical models, which are also sometimes named as traditional time series methods, include methods such as similar day, persistence methods, regression analysis, decomposition methods, exponential smoothing, moving average methods, Box−Jenkins models (including autoregressive, AR; moving average, MA; autoregressive moving average, ARMA; autoregressive integrated moving average, ARIMA; ARMA with exogenous variables, ARMAX; and ARIMA with exogenous variables, ARIMAX), and Kalman Filtering Algorithms.

The second group, intelligent methods, consists of models that emulate the relation between meteorological predictions and historical measurements with the generation output or load demand through models whose parameters are adjusted through a learning procedure based on the historical data. Different machine learning techniques, such as different types of neural networks (NNs), fuzzy systems, support vector machines (SVMs), and k-nearest neighbor, and their combinations are some of the methods in this category. Some of these methods will be explained later in this chapter.

4.4.2 Physical models

4.4.2.1 Numerical weather prediction models

Most of the established wind and solar power prediction systems usually use numerical weather models as input, in particular, the forecasted wind speeds and solar radiation. The leading NWP models are developed and operated by national weather centers of countries and regions. There could be slight variations in the different NWP models of different developers. The way that NWP models work is in such a way that the world is divided into a grid of cells, which means the atmosphere is split into little boxes or grid cells with finite spatial extension. The averages of the meteorological parameters are calculated and used to represent the cell with all the points within the cell sharing the same values as the center of each cell. That is where the drawback of the NWP models lies: no details of variations within the grid cell are reflected. Interpolation of the center point values may be used to estimate forecasts at a specific location other than the center of the grid cell. The size of the cells may vary between developers of the respective NWP models. However, it may be generally stated that the horizontal size is between a few hundred meters in case of the highest resolution ones to tens of kilometers in those with lower resolution [13].

In the case of wind power forecasting, there is also an additional step where the forecasted wind speeds at a standard height (10 m in most cases) need to be converted to the equivalent at the specific hub height of the wind turbine. There are verities in the way this vertical interpolation is performed in different forecasting models. The physical wind power forecasting models also involve the use of the power curve (specification of the wind turbine) to convert the wind speed forecasts to forecasts of generated wind power.

In the case of PV power forecasting, the global horizontal irradiance and ambient temperature forecasts (and sometimes other weather parameters) of the NWP models are first treated to be converted to tilted plane values and then applied to the PV system specification models to give the PV generation forecasts.

4.4.2.2 Sky imagery–based forecasts

Sky imagery is usually used for forecasting of PV power generation. PV generation forecasting that makes use of sky imagery (or sometime called as whole sky imagery) analysis involves four major steps [26]. The first stage is the acquisition of a sky image in close proximity to the PV array using some instrumentation such a Whole Sky Imager [27]. Analyzing the recorded sky image data to identify the availability and thickness of clouds follows. What comes next is systematic mapping of the successive images to estimate the cloud motion vectors. The last stage involves the use of cloud location and motion vector data for estimation of the required weather parameters such as cloud cover, irradiance, and generated power.

Sky imagery–based forecasting models have the advantage of very detailed information regarding the amount, structure, and motion of clouds. Such methods are mostly applicable for very-short-term and sometimes short-term forecasts with low spatial resolution. However, it is also possible to enlarge the spatial scale through the utilization of multiple imagers at different locations. One issue with such a method is that upper level clouds could potentially be covered by lower level clouds in the case of multiple cloud layers, which may cause forecast errors.

4.4.3 Time series methods

We have earlier stated that time series methods can be divided into statistical models, artificial intelligence–based models, and hybrid models. Statistical methods are used to analyze historical data, find out the internal rules, and apply them to forecast the target parameter. This approach includes regression techniques, variations of AR and MA models, time trend extrapolation methods, etc. Artificial intelligence−based methods include artificial neural networks (ANNs), hybrid NNs, artificial immune networks, SVM, and Bayesian approach, and their integration with fuzzy theory, wavelet analysis, and so on. The hybrid prediction method applies a combination of the statistical method and artificial intelligence−based methods. Fig. 4.3 outlines some of the different time series techniques that may be used in renewable generation and load forecasting.

4.4.3.1 Artificial neural network

ANN is the most popular of the artificial intelligence−based techniques used in forecasting studies. ANN emulates the way that the human biological neural

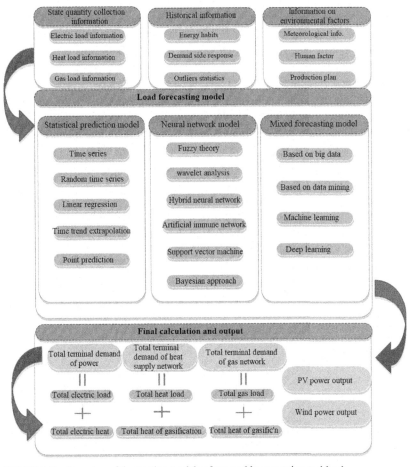

FIGURE 4.3 Overview of forecasting models of renewable generation and load.

system functions with the analogy of processing complex and nonlinear information in the form of working in parallel, distributed, and local processing and adaptation. The building blocks of an ANN model are the simple elements called neurons which operate in a parallel to the way the biological nervous system works. It is this small unit, the neuron, which plays a role in determining the function and operation of the network. The mathematical model of a neuron is defined in Eq. (4.1) and is graphically shown in Fig. 4.4.

$$u_k = \sum_{j=1}^{n} w_{kj}x_j \quad \text{and} \quad y_k = \varphi(u_k) + b_k \tag{4.1}$$

There are a variety of activation functions such as linear function, fixed limiter function, sigmoid function, and bipolar sigmoid function that may be used in neurons. An ANN architecture consists of three layers: the **input layer** accepting the input data such as the weather forecasts and the other parameters; the **hidden layer** which processes the input layer outputs using arithmetic or mathematical functions; and a final layer called **output layer**. The pattern of connections between the neurons defines the type of the ANN as in the two basic types: feedforward and feedback NNs. There are also variations based on the specific activation and training functions used within those two broad categories. Levenberg—Marquardt, Bayesian Regularization, and Scaled Conjugate Gradient [16]. Conventional Feedforward Network, Backpropagation ANN (BPANN), Radial Basis Network, and Nonlinear Autoregressive NNs are some of the commonly used ANN models in forecasting applications.

4.4.3.2 Adaptive neuro-fuzzy inference system

Adaptive neuro-fuzzy inference system (ANFIS) is another tool in the artificial intelligence—based forecasting models family that effectively integrates features of ANN with fuzzy systems adopting the reasoning capabilities of fuzzy logic and the learning capabilities of ANN [28]. An ANFIS model optimizes

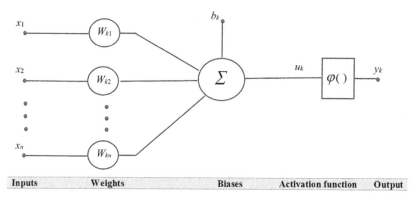

| Inputs | Weights | Biases | Activation function | Output |

FIGURE 4.4 Mathematical model of a neuron.

the distribution of membership functions though analysis of input−output mapping relationship and using fuzzy if−then rules of a Sugeno-type fuzzy inference system [29]. The architecture of ANFIS model is similar to that of a multilayer feedforward ANN consisting of nodes and directional links. Each node performs a particular function on the input signals and its input−output relation is defined by the learning rules [30]. The basic structure of a typical ANFIS model involves five layer of the following functions.

- *Layer 1* maps the input parameters using membership functions.
- *Layer 2* performs "AND" operation to determine the firing strength.
- *Layer 3* performs normalization.
- *Layer 4* performs adaptively with the output.
- *Layer 5* gives weighted average of all rule outputs.

4.4.3.3   Support vector machine

SVM is the other artificial intelligence−based mode widely used in short-term renewable generation and load forecasting applications. The method's core operational principle is attempting to find a hyperplane that divides the two classes with the largest margin by dividing the data in a higher dimensional space through a kernel function. Translating this back to the original feature space results in a nonlinear system. A set of support vectors that define each variation in the final model is involved and that defines how much influence a specific support vector would have on the final decision. SVM is known for being a nonparametric technique that estimates a regression problem using a set of linear functions and minimizing the risk function which is defined as a combination of the empirical error and a regularization term.

4.4.3.4   Deep neural network

Deep neural networks (DNNs) are improved versions of the conventional ANN with multiple layers. The DNN models are recently becoming very popular due to their excellent performance to learn not only the nonlinear input−output mapping but also the underlying structure of the input data vectors [31].

When implementing DNN training in short-term generation or load forecasting, a set of training samples including an input vector consisting of the weather, calendar, or other variables and a target vector, which is the generated power (for generation forecasting) or the load power (for load forecasting), need to be collected and a regression problem for optimization will be formulated. The DNN parameters are accordingly estimated by minimizing the sum-of-squares error function calculated from DNN outputs. Starting from an initialization stage where the model parameters are set to an initial set of values, a stochastic gradient descent algorithm is continuously run to reduce the error function until it converges to a specified lowest value [32]. The DNN training involves two passes based on the error backpropagation algorithm, which are the forward pass and the backward pass. In the former one, the affine transformation and nonlinear activation are calculated layer by layer from the input to the output layer. In the later one, the derivatives of the error function with respect to individual weights are calculated in a reverse order, that is, from the output layer to the input layer.

Considerations of weather factors and the historical data are used in renewable generation forecasts. Those data combined with the power consumption habits or pattern of users, influence of the price policy and the demand side response schemes can be used to establish load forecasting models (Fig. 4.5).

Based on the theory of deep belief network of the limited Boltzmann machine, the features are relearned in the upper layer of the network, and the

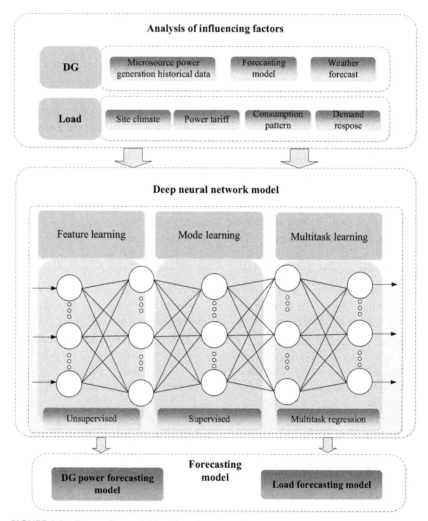

FIGURE 4.5 Forecasting model based on deep neural network.

multitask fitting learning in the lower layer of the network is applied in the regression layer to build the distributed power and load of the DNN.

In the day-ahead and ultrashort-term prediction models, the unsupervised greedy layer by layer training algorithm is used to train the network, and the comprehensive test and error analysis are carried out through the test samples.

4.4.3.5   Kernel function extreme learning machine

Kernel function extreme learning machine is a single-layer feedforward NN algorithm, which has a good performance in solving regression prediction problems. Compared with the SVM algorithm, it has similar or better prediction accuracy and faster calculation speed. The basic extreme learning machine algorithm is also known as a single hidden layer feedforward NN, which belongs to a kind of feedforward NN (Fig. 4.6).

4.5   Accuracy enhancement techniques in generation and load forecasting

4.5.1   Forecast accuracy metrics

Accuracy is the most important index of forecasts. Before any forecasting model can be considered applicable with confidence, there shall be adequate

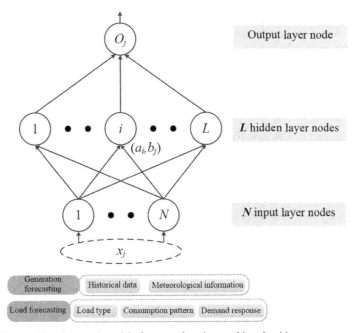

FIGURE 4.6   Neural network model of extreme learning machine algorithm.

validation or assessment of accuracy and magnitude of errors that may result from the use of the models. Model validation involves measuring the accuracy of a forecasting model through the comparison between forecasted and measured values.

The accuracy of a forecasting model can be measured or expressed in different ways while it basically revolves around quantifying the forecast error, which is the difference between forecasted value and actual values. The error values are used to produce forecast accuracy metrics, which allows to anticipate the level of uncertainty in the forecasts and compare the different forecasting methods.

The accuracy measuring metrics can be grouped into two types as summary measures and difference measures. Summary measures are intended to describe the quality of the forecaster while difference measures try to locate and quantify the forecasting errors. Most of the metrics used in the renewable energy and load forecasting studies are summary measures. Another way of classifying accuracy measuring metrics is described by Shcherbakov et al. [33]. It classifies the metrics as absolute error−based, percentage error−based, symmetric error−based, relative error−based, and scaled error−based measures.

There are five requirements stated by the National Research Council [34] which accuracy metrics are expected to meet. They are reliability, ease of interpretation, measurement validity, clarity of presentation, and support of statistical analysis.

Some of the forecast accuracy metrics used in the literature are correlation coefficient (r) and its square, coefficient of determination (r^2); mean bias error (MBE); mean absolute error (MAE); mean square error (MSE); root mean square error (RMSE); mean absolute percent error (MAPE); index of agreement (d); modified versions of those and others. Some of the statistical measures of forecast accuracy used in the literature are briefly presented hereunder. In the forthcoming equations, M is used to refer to "Measure values," F to "Forecasted values," and N to "the number of forecasted or measured points."

1. Correlation coefficient (r) and coefficient of determination (r^2)

Pearson's correlation coefficient, commonly called the correlation coefficient, is one of the most popular measures of dependence between two variables. It is a measure of the strength and direction of the linear relationship between the two variables which, in the case of forecast accuracy, are the forecasted and measured values. Correlation coefficient is originally defined as the ratio of covariance between two variables to the product of their standard deviations which can be simplified as Eq. (4.2).

$$r = \frac{N\left[\sum_{i=1}^{N} (F_i M_i)\right] - \left(\sum_{i=1}^{N} (F_i)\right)\left(\sum_{i=1}^{N} (M_i)\right)}{\sqrt{N\sum_{i=1}^{N} F_i^2 - \left(\sum_{i=1}^{N} F_i\right)^2 \left[N\sum_{i=1}^{N} M_i^2 - \left(\sum_{i=1}^{N} M_i\right)^2\right]}} \tag{4.2}$$

Correlation coefficients assume values between -1 and 1 and values outside this range are attributed to errors in the measurement. A correlation of "-1" refers to a perfect negative correlation, while a correlation of "1" represents a perfect positive correlation. Values in-between the two indicate a lesser level of relationship, while a correlation of "0" refers to no linear relationship between their movements.

The Pearson's correlation coefficient ultimately measures the linear relationship between the measured and forecasted dataset and can fail to reflect possible nonlinear relationships even in cases of a significantly strong one-to-one relationship between individual values of the two sets. The other issue with this metric is being highly influenced by outlying pairs of values. Hence, different modified versions of it, such as Spearman's rank correlation coefficient and Kendall Tau rank correlation coefficient, have been developed and applied.

As the naming would suggest r^2 is the square of the correlation coefficient, r, and can be represented by simplified form of Eq. (4.3). It illustrates how differences in one variable (the measured data) can be explained by differences in the second variable (the forecasted data). In other words, the metric can show how many forecasted data points fall within lines formed by the regression equation based on the forecasting input variables. The squaring component discards the negative values in correlation coefficient and hence the values of r^2 lie in the range $0-1$. However, it may assume negative values sometimes and that would usually refer to the forecasting model being a poor fit for the dataset. An r^2 value of "0" would mean none of the observed variation is explained by the forecasting model's inputs while an r^2 value of "1" tells the opposite, indicating a better goodness of fit.

$$r^2 = 1 - \sqrt{\frac{\sum_{i=1}^{n}(F_i - M_i)^2}{\sum_{i=1}^{n}(F_i - \overline{F})^2}} \tag{4.3}$$

Some studies have demonstrated that magnitudes of r and r^2 cannot consistently reflect the accuracy of a forecast [35] and their values may rise when a predictor variable is added to the regression model. This may result in the possibility of misleadingly high values for overfitted forecasting models with too many input variables [36]. The r and r^2 metrics are more suited and famous for input variables selection procedure which will be discussed later in this chapter.

A variation of r^2 which is adjusted for the number of predictors in the regression model, called adjusted r^2 (r^2_{adj}) is devised to address the issue with dependence on the number of predictors. It is defined as:

$$r^2_{adj} = 1 - \frac{N-1}{N-P-1}(1 - r^2) \tag{4.4}$$

2. Mean bias error

MBE or mean error (ME) captures the average bias in the forecast and is calculated as [37]:

$$ME = \frac{1}{N} \sum_{i=1}^{N} (F_i - M_i) \tag{4.5}$$

This metric indicates whether the model is systematically under- or overforecasting the predicted parameter. It is primarily used to estimate the average bias in the model and hence is majorly important to make a decision on the use of extra steps to correct the bias in the forecasting model. MBE is not a popular choice in forecasting due to the fact that high individual errors may also produce a low MBE [36].

3. Mean absolute error and mean absolute percent error

MAE measures the average accuracy of forecasts without considering error direction. It is said to be a natural measure of average error [38]. MAE represents the ratio of the 1 norm of the error vector $(F-M)$ to the number of points in the forecasted dataset and is defined as:

$$MAE = \frac{1}{N} \sum_{i=1}^{N} |F_i - M_i| \tag{4.6}$$

There are suggestions by Willmott [35] and Willmott and Matsuura [38] that average-error measures based on absolute values of differences, like MAE, are generally preferable to those based on squared differences, like MSE and RMSE, which will be discussed later. MAE is less sensitive to extreme values than RMSE. The general recommendation is that dimensionless indices of model performance are better to be based on absolute values of differences.

One issue with the MAE is that the relative size of the error is not always obvious. MAPE is devised to resolve this issue as it allows comparison of forecasts of different series in different scales. It is defined by Eq. (4.7). MAPE is one of the most often used summary measures of forecast error and works best if there are no extremes and no zeros. It meets most of those requirements stated earlier that an accuracy metric is expected to satisfy with one important exception, the lack of measurement validity [39].

$$MAPE = \frac{1}{N} \sum_{i=1}^{N} \left| \frac{F_i - M_i}{M_i} \right| \tag{4.7}$$

Since both MAE and MAPE are based on ME, the impact of big but infrequent errors may be understated. The other shortcomings of MAPE are inapplicability for data with zero values due to division by zero; luck of upper limit to the percentage error and the heavier penalty put on negative errors.

4. Mean square error and root mean square error

MSE represents the ratio of the square of the two norms of the error vector $(F-M)$ to the number of points in the dataset and is mathematically expressed as:

$$\text{MSE} = \frac{1}{N} \sum_{i=1}^{N} (F_i - M_i)^2 \qquad (4.8)$$

Though the preference to absolute valued metrics has been discussed earlier, there are also instances when useful information can be garnered from a set of squared differences, as in MSE and RMSE. MSE can encompass systematic and unsystematic components. The systematic components can be either additive or proportional, detailed analysis of which could illuminate the sources or types of error [35].

RMSE is nothing but the square root of the MSE, as shown in Eq. (4.9). It quantifies the dispersion between forecasted and measured data and has the feature that higher weights are given to large errors. This feature is actually similar to MSE and is basically related to the squaring component. It is a function of the variability in error distribution, the dataset size, and the average-error magnitude (MAE) [38].

$$\text{RMSE} = \sqrt{\frac{1}{N} \sum_{i=1}^{N} (F_i - M_i)^2} \qquad (4.9)$$

The MBE, MAE, and RMSE are related as $\text{MBE} \leq \text{MAE} \leq \text{RMSE} \leq \sqrt{N} \text{ MAE}$ [36]. MSE and RMSE are often preferred for convenience to conduct theoretical analysis on the error measure due to the ease of applying derivatives and other analytical measures [36]. Furthermore, RMSE is easy to interpret since it has the same unit as the forecasted or measured parameter. It is also advantageous in that it reflects the actual size of the error, unlike r or r^2 metrics in which a large error could be covered by high values of standard deviations or a small error could appear larger due to low magnitudes of standard deviation [35]. However, RMSE may sometimes fail to truly represent the average performance of the forecasting model due to being highly influenced by larger errors and its upper limit being variable depending on the number of data points ($\sqrt{N}$) which could lead to different interpretations for different sizes of dataset [38].

MSE and the other metrics discussed above, such as MBE and MAE, have the same unit as the forecasted variable and thus cannot be used to make comparisons for different variables with different scales. Normalized root mean square error (NRMSE) provides a solution for that though normalization of the error as shown in Eq. (4.10). NRMSE can be defined as the ratio of the RMSE to the mean of the measured dataset or the range of the data.

$$\text{NRMSE} = \sqrt{\frac{\sum_{i=1}^{N}(F_i - M_i)^2}{\sum_{i=1}^{N}(M_i - \overline{M})^2}} \quad (4.10)$$

where $\overline{M}$ is the mean of measured values. There is also another variation of RMSE called scaled root mean square error (SRMSE), defined as [40]:

$$\text{SRMSE} = \frac{1}{\overline{M}} \sqrt{\frac{1}{N} \sum_{i=1}^{N}(F_i - M_i)^2} \quad (4.11)$$

5. Index of agreement (d)

One of the summary measures for accuracy of a forecast model is the d, sometimes referred to as "the Willmott index," which was first defined by Willmott [35] as:

$$d = 1 - \frac{S(F - M)}{S(|F'| + |M'|)} \quad (4.12)$$

where $F' = F - \overline{M}$ and $M' = M - \overline{M}$; S stands for standard deviation. Eq. (4.12) can be written in expanded form as:

$$d = 1 - \frac{\sum_{i=1}^{n}(F_i - M_i)^2}{\sum_{i=1}^{n}(|F_i - \overline{M}| + |M_i - \overline{M}|)^2} \quad (4.13)$$

d is a dimensionless measurement of forecasting model accuracy bounded by 0 (no agreement) and 1 (a perfect fit). Though it is primarily meant to be used to assess relative superiority of alternative models, it can also be used as a descriptive measure of a forecasting model's performance.

4.5.2 Factors affecting forecasting accuracy

The accuracy of short-term forecasting of renewable generation and load can be affected by different factors. Some of those factors are shortly explained below.

1. *Volume and quality of historical data:* The size of available historical data to be used in a forecasting task determines the level of accuracy in the forecast to a great extent. The general rule is that the larger the historical data available for training, the better the forecasting model accuracy would be. However, a larger size of historical data cannot always guarantee an absolute forecast or even a better accuracy. There are some reports that an increase in data size after some level would not cause any more improvement in accuracy or at times lead to a negative impact based on the type of model and the data storage and computation speed requirements. Just as influential as the data volume is the quality of the available data. If the input data is bad in quality, the forecasts tend to be bad too.

2. *Socioeconomic factors:* This factor is more important in load forecasts. As electricity is a substantial part of people's daily life, changes in economic status and social lifestyle of an area and a specific customer would in turn influence the usage of electricity. Economic factors are more critical in long-term forecasting. However they also can impact the load curve for short-term load forecasting [41]. The price of electricity and the purchasing capacity of users are also other important economic factors that may influence the usage of electricity and hence the forecast accuracy.

3. *Meteorological factors and meteorological forecast accuracy:* Some weather parameters affect the forecasting of both generation and load. In most forecasting studies, the main factor for the accuracy level of forecasts is considered to be the prediction model, and the other secondary factors are ignored. However, the accuracy of weather prediction has a very significant role on the accuracy of the forecasted generated power or the load demand. Sudden and unpredicted changes in weather parameters (those listed in Tables 4.1 and 4.2) will have an immediate impact on the forecast accuracy of renewable generation and load.

4. *Time factor:* The periodic and seasonal changes of time, holidays, and other time factors affect the performance of a forecasting model. The impact of this factor is more critical and visible in load forecasts. People's daily life activities vary in the different times of a day and different types of days. The demand during working time is different from that of leisure time or sleeping time. Hence, any adjustments to those timings may alter the accuracy of the forecasting model.

5. *Sampling factors:* The sampling factors influencing short-term power forecasting include sample quantity, sample quality, and sample range. In the short-term and ultrashort-term power forecasting, we should not only consider the time and historical data factors, but also consider all kinds of factors comprehensively. At the same time, we should make quantitative and qualitative analysis and then select the best sample to make the forecast more accurate.

6. *Forecasting model:* The forecasting model is the core of a forecasting task and the most influential one in terms of deciding the accuracy of the forecasts. As it has already been discussed in the earlier subtopics in this chapter, a large number of models and their ensembles have been suggested in the literature. As extensive as the discussions on the relative advantages of one model over the other could be, it still is not possible to put forward a definite recommendation on a specific model or a combination of some that can guarantee an accurate forecast in all applications. So, we should make the choice of the best forecasting model according to the convenience of the model for the specific application and system characteristics.

7. *Internal characteristics of the generation units:* Different internal characteristics of the generating units, the PV array and the wind turbines, influence the accuracy of the generation forecast. Some characteristics related to aging, weather dependence, performance ratio tests, AC yield capacity, and so on, could cause alterations between forecasted generation and actual generation.

8. *Environmental issues:* Obstructing obstructing structures, such as buildings, trees, and moving bodies, could affect the generation from PV and wind systems. Soiling (accumulation of snow, dirt, dust, leaves, pollen, and bird droppings on the surfaces of PV modules), PV module cleaning schedule and methods, physical damages to turbine components, and so on, may contribute to possible accuracy issues in forecasts of generation from the respective systems.

9. *Other factors:* The implementation of the time of use electricity price in peak−valley periods and other demand response schemes has an impact on the load curve and hence the load forecasts. With the generation and demand being highly interlinked in the current power system and more importantly in microgrids, such an impact can be extended to the power generation forecasts too. It is difficult to determine the number of samples reflecting the periodicity, trend, and the relationship with the influencing factors. Some emergencies, such as power rationing, load impact, maintenance outage, and major activities, may have a great impact on the forecast accuracies as well (Table 4.5).

TABLE 4.5 Causes of uncertainty in short-term forecasting of renewable generation and load [42−46].

Causes of uncertainty in short-term photovoltaic power forecasting	
• Uncertainties in measurements of irradiance • Module properties (indoor and outdoor, performance ratio tests, Alternating Current (AC) yield) • Weather forecasts • Degradation • Shading	• Angle of solar incidence • Efficiency • Temperature • Balance of system (BOS) components • Long-term assumptions • Cloud-induced ramps • Ramps due to simultaneous inverter trips

Causes of uncertainty in short-term wind power forecasting	
• Uncertainties in forecasts of meteorological variables • Correction to the hub height • Power curve of wind turbine • Influence of the terrain	• Local extreme wind gusts that cause turbines to switch off • Nonpredicable extreme events • Direction changes within short time • Long-term wind resource quality

Causes of uncertainty in short-term load forecasting	
• Forced outages of system generators • Weather forecast quality • Special days and holidays	• Price elasticity • Extreme weather events, • New economic activities • Ad hoc dummy variables

While some of the factors affecting forecasting accuracy are not controllable at all, additional effort and more information could help to reduce the impact of some of the factors. In general, it is recommended to have as much extensive information as possible about the geographic, weather condition, socioeconomics, electricity-related regulations, and other related information for the microgrid whose generation and load are to be forecasted. In addition, having the most recent and updated information on system parameters and environmental conditions would be helpful. The use of the most recent data of weather parameters and power data is important.

As the forecasting model is the most critical player in deciding the forecasting accuracy, there are different suggestions on how to come up with accurate forecasts of generation and load. Some of the techniques used in short-term forecasting of load and renewable generation can be categorized into the following groups:

1. Input variable selection
2. Data preprocessing
3. Choosing best method and architecture
4. Output processing or ensembling

The choice of the best method for a forecasting model is not a simple straightforward suggestion and there are a wide range of options, including some already discussed, that have been applied in the literature. One technique usually applied in the studies is the incorporation of optimization techniques such as genetic algorithm (GA), particle swarm optimization, ant colony optimization, and the like to externally optimize the parameters (usually weights and biases) of the models in the forecasting procedure. The other techniques that may be applied to enhance the accuracy of forecasts of load and renewable generation are discussed below.

4.5.3 Input variable selection methods

It may be a customary recommendation that the use of more input parameters can improve the forecast accuracy. However, that should be based on a strategic decision-making about the selection of the important variables. Hence, devising an appropriate method for the selection of the input parameters plays a decisive role in achieving better accuracy. This procedure is particularly common and important in artificial intelligence—based forecasting models. Some of the commonly used techniques are:

1. Correlation analysis
    Correlation analysis has been discussed earlier as a metric for forecast accuracy measurement. It is also one of the oldest and popular methods applied in input variables selection tasks for forecasting. Slightly different from the definition used earlier in the case of the correlation between

forecasted and measured data, the correlation coefficient between an input set of variables (matrix X) and output variable (vector Y) can be mathematically expressed as in Eq. (4.14).

$$r = \frac{\sum_i \sum_j (X_{ij} - \overline{X})(Y_{ij} - \overline{Y})}{\sum_i \sum_j (X_{ij} - \overline{X})^2 (Y_{ij} - \overline{Y})^2} \qquad (4.14)$$

where $\overline{X} = mean(A)$ and $\overline{Y} = mean(B)$

It was already discussed that the correlation coefficient measures the strength and direction of the linear relationship between two variables. It is simple and has well-tested performance as an input selection method in many forecasting studies. However, the fact that it solely reflects the linear relationship between the two parameters means it may lead to the unnecessary exclusion of some essential candidate input parameters which are related to the target variable in a nonlinear way.

The r^2 value, which as discussed earlier in this chapter, is the square of the correlation coefficient, can also be used as an input variable selection tool.

2. Garson's algorithm

This method of input variable selection is implemented through the use of an ANN. The method measures the relative importance of an input variable through the sum of the connection weights of the NN model linked to the specific input variable. The relative importance (RI_{ij}) of an input variable X_i with respect to a target variable Y_j can be evaluated using an ANN of H hidden layer neurons as shown in Eq. (4.15) [47]:

$$RI_{ij} = \sum_{k=1}^{H} |W_{ik}.W'_{kj}| \qquad (4.15)$$

where W_{ik} is the connection weight between the input neuron linked to input X_i and the hidden neuron h_k; and W'_{kj} is the connection weight between the hidden neuron h_k and the output neuron connected to output Y_j.

3. Mutual information

Mutual information (MI) is one the tools to evaluate the statistical dependence between two time series which is based on the information theory introduced by Shannon [48]. MI between two time series X and Y can be defined as [49]:

$$MI(X, Y) = \sum_{i,j} P_{x,y}(x_i, y_j) \log_2 \left[\frac{P_{x,y}(x_i, y_j)}{P_x(x_i) P_y(y_j)} \right] \qquad (4.16)$$

where $P_{x,y}(x_i, y_j)$ is the joint probability density of x and y evaluated at (x_i, y_j), and $P_x(x_i)\, P_y(y_j)$ are the marginal probability densities of x and y evaluated at x_i and y_j, respectively.

4. Sensitivity analysis

This method is based on applying some changes to the input variable and measuring the resulting change in forecast accuracy using the same model as the one to be used for the actual forecasting task. The following types of changes are applied for the same purpose in [16]. The technique evaluates an input variable as more important when the change in MSE due to a change in the respective parameter is larger.

- *Constant row replacement* replacing each raw by mean value of the original row.
- *Permutation* replacing each raw by a random permutation of values in the raw.
- *Perturbation* adding a small noise to each input value.
- *Profile method* letting each input to evolve in equal intervals between its minimum and maximum value.

4.5.4 Data preprocessing

Preprocessing the data to be used for developing a forecasting model (including weather information, recorded power data, and others) is an important step. The most commonly applied method is normalizing the data, which usually is automatically performed in most of the artificial intelligence-based models. There are also other signal processing tools used in renewable generation and load forecasting studies. Those tools are primarily used to process the target data and extract components that can be forecasted easily, and process the input data to more informative inputs for forecasting. Some of the techniques are reviewed below.

4.5.4.1 Fourier transform

Fourier transform is the most popular and the oldest of the signal processing tools in different science fields that is typically convenient for analysis of periodic signals. Fourier transform decomposes a signal into a combination of sine waves of different amplitudes and frequencies and is mathematically defined by Eq. (4.17) or as in Eq. (4.18) for a discrete form called discrete Fourier transform (DFT).

$$\text{Fourier transform } (f(t)) = \int_{-\infty}^{+\infty} f(t)e^{-jwt}dt = \frac{1}{2\pi}\int_{-\infty}^{+\infty} F(jw)e^{jwt}dw \quad (4.17)$$

$$\text{DFT } (f[n]) \sum_{n=-\infty}^{+\infty} f[n]e^{-jwn} = \frac{1}{2\pi}\int X(e^{jw})e^{jwn}dw \quad (4.18)$$

Fast Fourier transform (FFT) is a form of DFT usually used in the forecasting studies with a lower level of computation required while essentially having the same principle. FFT is used to detect the seasonal components in hourly electrical load data for short-term load forecasting in Musbah and El-Hawary [50].

4.5.4.2 Wavelet transform

Wavelet transform (WT) is another signal analysis tool that enables to keep information of both time and frequency in the decomposed signal unlike Fourier transform which is basically targeted at the frequency component. WT is employed in the data preprocessing stage of artificial intelligence−based forecasting models, where the trend and high-frequency information in the original data series will be filtered into a group of constitutive series, called "approximation" and "detail" components, as shown in Fig. 4.7. Wavelet analysis is the process of decomposing an original signal into shifted and scaled versions of a mother wavelet. Wavelet analysis has the capability of unveiling the trends, discontinuities, or sharp changes and easily denoising a signal without degradation, which makes it a convenient tool for use in short-term renewable generation and load forecasting [51]. WT can be of different types, mainly continuous and discrete. A more detail description of WT is provided in Chapter 5, Fault and Disturbance Analysis in Microgrid.

The basic procedure involved in WT-based (usually artificial intelligence) models for short-term forecasting of load and renewable generation is summarized in Fig. 4.8. The first stage is where the recorded load or generation data (signal) is decomposed into approximation and detail components. The main issues in this procedure are the choice of the mother wavelet and the decomposition level. Though mother wavelets such as Haar and some families of Daubechies are utilized in most studies, there still is not a definite recommendation on an objective way to choose the best mother wavelet for an application. The level of decomposition is also application specific and

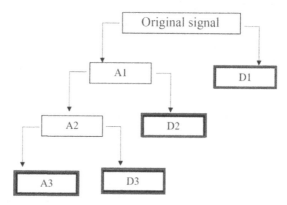

FIGURE 4.7 Wavelet decomposition of signals.

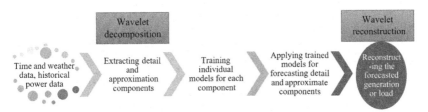

FIGURE 4.8   Procedure involved in wavelet transform—based forecasting models.

usually decided on a trial and error basis. The usual trend in most WT-based forecasting models is to forecast the approximate and detail components independently, sometimes using different techniques. There is also a practice of omitting the most detailed component in the forecasting procedure. The final forecast of the load or renewable generation are usually achieved through wavelet reconstruction or sometimes using an additional forecasting model which forecasts the actual target from its forecasted components.

4.5.4.3   Empirical mode decomposition

Empirical mode decomposition (EMD) is a decomposition algorithm, which is a direct, self-adaptive, and intuitive method first proposed by Huang et al. [52]. The core idea behind the method is that the possibility any signal can be composed of and decomposed into a finite number of stationary frequency components called intrinsic mode functions (IMFs) and a residual. EMD allows decomposition without leaving out the time domain information and is suitable for processing data series that are nonstationary and nonlinear such as wind power, PV power, and load demand. The procedure provides insight into various signals contained within the data and hence allows for easier and more accurate forecasts of the constituting components. Each of the IMFs are expected to satisfy two basic requirements of having only one extreme value between zero crossings and a mean value of zero. EMD-based decomposition of an original signal $x(t)$ into N IMFs $x_i(t)$ and a residual $r(t)$ can be reconstructed as:

$$x(t) = \sum_{i=1}^{N} x_i(t) + r(t) \qquad (4.19)$$

The process used in EMD to decompose a signal into IMFs is called the shifting process. It involves steps of identifying the local extrema in the data and a repeated process of connecting them by a cubic spline line to produce the upper and lower envelopes.

EMD has been applied in a number of studies in short-term forecasting of renewable generation [53] and load [54,55]. The most oscillating or high-frequency components (usually the first IMF) in the decomposition may be excluded during the forecasting procedure as they may neither be easy to forecast nor significantly affect the final forecast since they are mostly random in nature and low in magnitude.

When EMD is combined with Hilbert spectral analysis it is called Hilbert—Huang transform, which is also another method applied in the short-term renewable generation and load forecasting as in [56,57].

4.5.4.4 Variational mode decomposition

Variational mode decomposition (VMD) is one of the latest signal decomposition techniques, similar to EMD, first proposed by Dragomiretskiy and Zosso [58]. The method is robust to sampling and noise which enables perfect reconstruction of the original signal from the decomposed IMFs which are smoothed after demodulation into the baseband. VMD decomposes a signal $X(t)$ into N narrowband IMFs $x_i(t)$ and a residual $r(t)$, as in Eq. (4.19).

The basic characteristic of IMFs in VMD is that each of them will have a cosine function wave shape, slowly varying and positive envelopes, and an instantaneous frequency that varies slowly in a nondecreasing pattern. The procedure of decomposing a signal using VMD involves Wiener filtering, Hilbert transform, frequency mixing, and heterodyne demodulation. The core process is finding a discrete set of IMFs $x_i(t)$ and respective central frequencies $w_i(t)$ that minimize the constrained variational problem defined by Eq. (4.20) [58].

$$\min_{\{x_i\},\{w_i\}} \left\{ \sum_i \left\| \partial_t \left[\left(\delta(t) + \frac{j}{\pi t} \right)^* x_i(t) \right] e^{-jw_i t} \right\|_2^2 \right\} \tag{4.20}$$

where $\{x_i\} = \{x_1, x_2, \ldots, x_N\}$ is the set of all modes, $\{\omega_i\} = \{w_1, w_2, \ldots, w_N\}$ is the set of central frequencies of all modes, δ is Dirac function, $\|.\|_2$ represents the L2 distance, and * represents the convolution operation.

One of the challenges in the utilization of VMD in short-term forecasting or other studies is the critical choice of two important parameters: the number of decomposition modes is the penalty factor. A wrong choice of those parameters might lead to modal aliasing, noisy IMFs, or loss of important information (or IMF), which may influence the predictability of the IMFs and accuracy of the final forecast. VMD is applied in some studies in short-term load forecasting, such as [59,60], and short-term renewable generation forecasting, such as [61,62].

4.5.5 Output processing or ensembling methods

Since it is always difficult to suggest a single method which can perform outstandingly under different circumstances, some forecasting studies prefer to employ an ensemble of more than one technique. There are different techniques of ensembling forecasts by different models.

4.5.5.1 Simple averaging

This is the simplest method of ensembling where the final forecast is an arithmetic mean of the alternative forecasts assigned with equal weights. Its simplicity is a factor behind the approach being utilized in a considerable number of studies. However, the approach suffers from high sensitivity to extreme values.

4.5.5.1.1 Regression

This is also another statistical averaging technique where the linear combination of individual candidate forecasts is performed with weights given based on performance of the individual models. This approach has an upper hand to simple averaging in the fact that best performing models have higher weights, while those with lesser performance are given lower weights. This feature allows to achieve better overall performance of the ensemble forecast.

4.5.5.1.2 Using an additional model

This method involves an additional model where the candidate forecasts from the individual models are provided as inputs to come up with the final forecast. This method is adopted in many studies in both renewable energy and load forecasting. Such a technique is usually referred to as a hierarchical method as there are forecasting procedures of more than one level. It allows for the use of relatively less complicated models at the lower stage forecast while still delivering on the main objective of accurate forecasts.

4.6 Application examples

4.6.1 Short-term wind forecasting using EMD and hybrid artificial intelligence technique

This application example for short-term forecasting of wind power generation is performed based on a 2.5 MW wind turbine which is part of a demonstrational microgrid in China. The data source for the historical power output record is the SCADA of the microgrid while the weather information (including historical record and future forecasts) is from an NWP model. Data for a period of 2 years is used for training and testing the model and the objective is to forecast the power output of the wind turbine for the next day (24 hours) on an hourly basis.

The method used in this example is summarized in Fig. 4.9. As can be seen from the figure, some of the techniques discussed in this chapter are used in the model.

The input variable selection procedure is based on correlation analysis, where the correlation coefficients of a candidate list of parameters with the

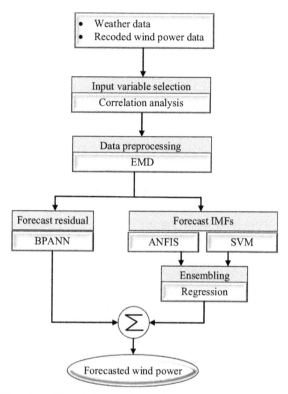

FIGURE 4.9 Flowchart of the wind power forecasting model.

generated power are calculated and evaluated to select the important ones. The selected parameters, as a result, are wind speed and air density. EMD is used to decompose the target data (wind power) to three IMFs and a residue. The decomposed components are forecasted using BPANN, ANFIS, and SVM models. A combination of ANFIS and SVM are used to forecast each IMF. The individual forecasts of the IMFs are ensembled using regression analysis. The forecasted IMFs and residue are added to come up with a forecast of the generated wind power. As the graphical illustration in Fig. 4.10 shows, the model is able to forecast the wind power with a good fit to the measured values, which is also verified by an MAE of 24.34 kW for the selected 10 testing days.

4.6.2 Day-ahead PV forecasting using VMD-GA-ANN

This application example for short-term PV power forecasting is also undertaken based on the same microgrid as in Section 4.6.1, which also constitutes

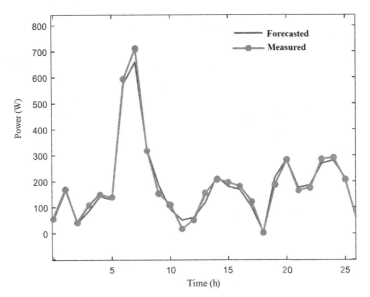

FIGURE 4.10   Forecasted and measured PV power for a test day.

a 100 kWp PV array. The data sources are also similar, with SCADA of the microgrid providing the PV power record while an NWP model is the source for the weather data. The method applied in this example is based on a VMD decomposition technique and an ANN whose parameters are optimized by GA (as a result named VMD-GA-ANN). The overall procedure is shown in the flowchart below (Fig. 4.11).

The input dataset includes weather information (solar radiation, air temperature, and humidity) and calendar variables (day of the year, hour of the day, and time). The input variable selection strategy utilized in this case is Garson's algorithm. The IMFs as a result of VMD decomposition of the PV power data are shown in Fig. 4.12. The Garson's algorithm evaluates comparative importance of the input parameters for forecasting of each of the IMFs. The GA-ANN model is an improved ANN model whose weights and biases are externally optimized by GA. Forecasted PV power for 3 test days looks as shown in Fig. 4.13 with RMSE of 0.032.

4.6.3   Short-term load forecasting using wavelet transform and LSTM

In this example, a hybrid approach composed of WT and long short-term memory (LSTM) model is used to forecast hourly load pattern of the next day. The input variables used are air temperature, hour of the day, hourly electricity price, and type of day. The forecasting method (shown in Fig. 4.14) starts with breaking down the load data through wavelet

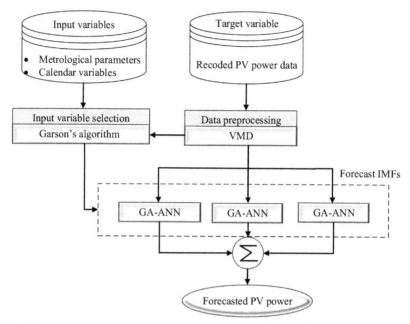

FIGURE 4.11 Flowchart of day-ahead PV power forecasting using VMD-GA-ANN.

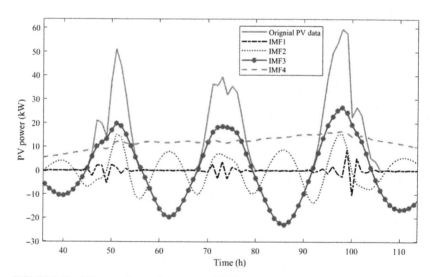

FIGURE 4.12 PV power data and decomposed IMFs.

decomposition into components which can be grouped as temporal and weather-dependent components that will be forecasted independently using respective LSTM models. Those forecasts are merged to give the load forecast using wavelet reconstruction.

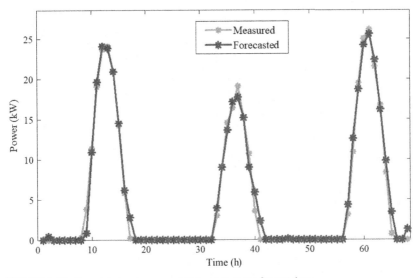

FIGURE 4.13 Forecasted and measured PV output power for test days.

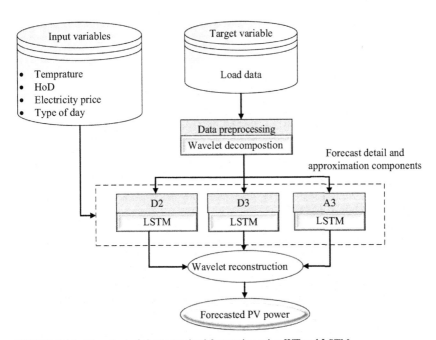

FIGURE 4.14 Flowchart of short-term load forecasting using WT and LSTM.

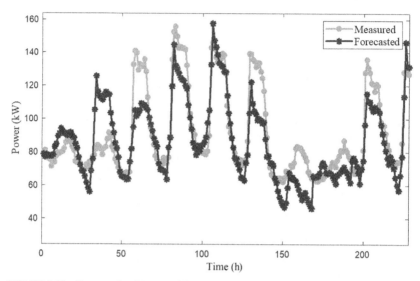

FIGURE 4.15   Forecasted and measured load data for test days.

LSTM is a type of recurrent neural network (RNN) which is also one of the deep machine learning techniques. RNNs have the advantage over the conventional feedforward NNs in that they have internal loops that allow for information to persist or learn long-term dependencies. LSTMs are a special kind of RNN which have the capability to remove or add information to the cell state (i.e., selectively remember or forget things) based on regulation by structures called gates.

The devised forecasting technique is tested using a yearlong actual load data and weather data of a commercial type load in the microgrid mentioned in the earlier two examples. The result showed effectiveness of the method, as seen in Fig. 4.15, and an r^2 value of 84.2%.

References

[1] M.L. Ahlstrom, L. Jones, R. Zavadil, W. Grant, The future of wind forecasting and utility operations, IEEE Power Energy Mag. (2005) 57−64. Available from: https://doi.org/10.1109/MPAE.2005.1524621.

[2] R. Liu, M. Peng, X. Xiao, Ultra-short-term wind power prediction based on multivariate phase space reconstruction and multivariate linear regression, Energies (2018). Available from: https://doi.org/10.3390/en11102763.

[3] A.M. Foley, P.G. Leahy, A. Marvuglia, E.J. McKeogh, Current methods and advances in forecasting of wind power generation, Renew. Energy (2012) 1−8. Available from: https://doi.org/10.1016/j.renene.2011.05.033.

[4] Z. Wang, M.M. Hayat, N. Ghani, K.B. Shaban, Optimizing cloud-service performance: efficient resource provisioning via optimal workload allocation, IEEE Trans. Parallel Distrib. Syst. (2017) 1689−1702. Available from: https://doi.org/10.1109/TPDS.2016.2628370.

[5]   B. Chen, P. Lin, Y. Lai, S. Cheng, Z. Chen, L. Wu, Very-short-term power prediction for PV power plants using a simple and effective RCC-LSTM model based on short term multivariate historical datasets, Electron (2020). Available from: https://doi.org/10.3390/electronics9020289.

[6]   M.K. Behera, N. Nayak, A comparative study on short-term PV power forecasting using decomposition based optimized extreme learning machine algorithm, Eng. Sci. Technol. Int. J. (2019) 156−167. Available from: https://doi.org/10.1016/j.jestch.2019.03.006.

[7]   R. J. Bessa, J. Dowell, and P. Pinson, Renewable Energy Forecasting, in Smart Grid Handbook, S.-J. L. Chen-Ching Liu, Stephan McArthur, Ed. John Wiley & Sons, Ltd, 2016, 639−659.

[8]   J. Zhang, R. Verschae, S. Nobuhara, J.F. Lalonde, Deep photovoltaic nowcasting, Sol. Energy (2018). Available from: https://doi.org/10.1016/j.solener.2018.10.024.

[9]   S. E. Haupt, Short-range forecasting for energy, in Weather and Climate Services for the Energy Industry, T. A., Ed. Palgrave Macmillan, Cham, 2018.

[10]   W.P. Mahoney, et al., A wind power forecasting system to optimize grid integration, IEEE Trans. Sustain. Energy (2012). Available from: https://doi.org/10.1109/TSTE.2012.2201758.

[11]   T.H.M. El-Fouly, E.F. El-Saadany, M.M.A. Salama, One day ahead prediction of wind speed and direction, IEEE Trans. Energy Convers. (2008) 191−201. Available from: https://doi.org/10.1109/TEC.2007.905069.

[12]   A. Chaouachi, R.M. Kamel, K. Nagasaka, Neural network ensemble-based solar power generation short-term forecasting, J. Adv. Comput. Intell. Intell. Inform. (2010) 69−75. Available from: https://doi.org/10.20965/jaciii.2010.p0069.

[13]   M. Zieher, M. Lange, U. Focken, Variable renewable energy forecasting—integration into electricity grids and markets—a best practice guide, vRE Discussion Series, Paper 6, 2015.

[14]   M. Malvoni, M.G. De Giorgi, P.M. Congedo, Forecasting of PV power generation using weather input data-preprocessing techniques, Energy Procedia (2017) 651−658. Available from: https://doi.org/10.1016/j.egypro.2017.08.293.

[15]   M.H. Chung, Estimating solar insolation and power generation of photovoltaic systems using previous day weather data, Adv. Civ. Eng. (2020). Hindawi papers only have Article IDs. Available from: https://doi.org/10.1155/2020/8701368.

[16]   S. Netsanet, D. Zheng, L. Zhang, M. Hui, Input parameters selection and accuracy enhancement techniques in PV forecasting using artificial neural network, in: 2016 IEEE International Conference on Power and Renewable Energy, ICPRE 2016, 2017. Available from: https://doi.org/10.1109/ICPRE.2016.7871139.

[17]   S.P. Durrani, S. Balluff, L. Wurzer, S. Krauter, Photovoltaic yield prediction using an irradiance forecast model based on multiple neural networks, J. Mod. Power Syst. Clean. Energy (2018) 255−267. Available from: https://doi.org/10.1007/s40565-018-0393-5.

[18]   D.B. De Alencar, C. De Mattos Affonso, R.C.L. De Oliveira, J.L.M. Rodríguez, J.C. Leite, J.C.R. Filho, Different models for forecasting wind power generation: case study, Energies (2017). Available from: https://doi.org/10.3390/en10121976.

[19]   K. Pan, Z. Qian, N. Chen, Probabilistic short-term wind power forecasting using sparse Bayesian learning and NWP, Math. Probl. Eng. (2015). Hindawi papers only have Article IDs. Available from: https://doi.org/10.1155/2015/785215.

[20]   R.P. Shetty, A. Sathyabhama, P.S. Pai, Comparison of modeling methods for wind power prediction: a critical study, Front. Energy (2018) 347−358. Available from: https://doi.org/10.1007/s11708-018-0553-3.

[21]   S. Netsanet, J. Zhang, D. Zheng, R.K. Agrawal, F. Muchahary, An aggregative machine learning approach for output power prediction of wind turbines, in: 2018 IEEE Texas

Power and Energy Conference, TPEC 2018, 2018. Available from: https://doi.org/10.1109/TPEC.2018.8312085.

[22] M.A. Hammad, B. Jereb, B. Rosi, D. Dragan, Methods and models for electric load forecasting: a comprehensive review, Logist. Sustain. Transp. (2020) 51−76. Available from: https://doi.org/10.2478/jlst-2020-0004.

[23] C. Monteiro, L.A. Fernandez-Jimenez, I.J. Ramirez-Rosado, A. Muñoz-Jimenez, P.M. Lara-Santillan, Short-term forecasting models for photovoltaic plants: analytical versus soft-computing techniques, Math. Probl. Eng. (2013). Hindawi papers only have Article IDs. Available from: https://doi.org/10.1155/2013/767284.

[24] M. Diagne, M. David, P. Lauret, J. Boland, N. Schmutz, Review of solar irradiance forecasting methods and a proposition for small-scale insular grids, Renew. Sustain. Energy Rev. (2013) 65−76. Available from: https://doi.org/10.1016/j.rser.2013.06.042.

[25] M.Q. Raza, M. Nadarajah, C. Ekanayake, On recent advances in PV output power forecast, Sol. Energy (2016) 125−144. Available from: https://doi.org/10.1016/j.solener.2016.06.073.

[26] K.D.B. Sophie Pelland, J. Remund, J. Kleissl, T. Oozeki, Photovoltaic and solar forecasting: state of the art, IEA-PVPS 14 (2013) 1−36.

[27] J.E. Shields, M.E. Karr, T.P. Tooman, D.H. Sowle, S.T. Moore, The whole sky imager—a year of progress, in: Presented at the Eighth Atmospheric Radiation Measurement Science Team Meeting, March 23−27, 1998, Tucson, AZ, 1998.

[28] H.Y. Wang, et al., Leuconostoc mesenteroides growth in food products: prediction and sensitivity analysis by adaptive-network-based fuzzy inference systems, PLoS One (2013). Available from: https://doi.org/10.1371/journal.pone.0064995.

[29] J.S.R. Jang, ANFIS: adaptive-network-based fuzzy inference system, IEEE Trans. Syst. Man Cybern. (1993) 665−685. Available from: https://doi.org/10.1109/21.256541.

[30] S. Zaferanlouei, D. Rostamifard, S. Setayeshi, Prediction of critical heat flux using ANFIS, Ann. Nucl. Energy (2010) 813−821. Available from: https://doi.org/10.1016/j.anucene.2010.02.019.

[31] J. Karhunen, T. Raiko, and K. Cho, Unsupervised deep learning, in Advances in Independent Component Analysis and Learning Machines, S. K. E. Bingham, Ed. Academic Press, 2015, 125−142.

[32] J.-T. Chien, Model-Based Source Separation, in Source Separation and Machine Learning, J.-T. Chien, Ed. Academic Press, 2019, 21−52.

[33] M.V. Shcherbakov, A. Brebels, N.L. Shcherbakova, A.P. Tyukov, T.A. Janovsky, V.A. evich Kamaev, A survey of forecast error measures, World Appl. Sci. J. (2013) 171−176. Available from: https://doi.org/10.5829/idosi.wasj.2013.24.itmies.80032.

[34] National Research Council, Estimating Population and Income for Small Places, National Research Council, Washington, D.C., 1980.

[35] C.J. Willmott, Some comments on the evaluation of model performance, Bull. Am. Meteorol. Soc. (1982) 1309−1313. Available from: https://doi.org/10.1175/1520-0477(1982)063 < 1309: SCOTEO > 2.0.CO;2.

[36] R. Pal, Validation methodologies, in Predictive Modeling of Drug Sensitivity, 1st ed., R. Pal, Ed. Academic Press, 2017, pp. 109−120.

[37] D.G. Fox, Judging air quality model performance, Bull. Am. Meteorol. Soc. (1981) 599−609. Available from: https://doi.org/10.1175/1520-0477(1981)062 > 0599:jaqmp < 2.0.co;2.

[38] C.J. Willmott, K. Matsuura, Advantages of the mean absolute error (MAE) over the root mean square error (RMSE) in assessing average model performance, Clim. Res. (2005) 79−82. Available from: https://doi.org/10.3354/cr030079.

[39] D.A. Swanson, J. Tayman, T.M. Bryan, MAPE-R: a rescaled measure of accuracy for cross-sectional subnational population forecasts, J. Popul. Res. (2011) 225−243. Available from: https://doi.org/10.1007/s12546-011-9054-5.

[40] M. Dust, et al., Simulation of water and solute transport in field soils with the LEACHP model, Agric. Water Manag. (2000) 225−245. Available from: https://doi.org/10.1016/S0378-3774(99)00093-1.

[41] M.U. Fahad, N. Arbab, Factor affecting short term load forecasting, J. Clean Energy Technol. (2014) 305−309. Available from: https://doi.org/10.7763/jocet.2014.v2.145.

[42] C. Reise, B. Müller, IEA, Uncertainties in PV system yield predictions and assessments, Report IEA-PVPS T13-12:2018, 2018.

[43] A. Mills, et al., Understanding variability and uncertainty of photovoltaics for integration with the electric power system, Electr. J. (2009) 18.

[44] C. Möhrlen, Uncertainty in wind energy forecasting, PhD Thesis, University College Cork, Ireland, 2004.

[45] B.-M. Hodge, et al., Wind power forecasting error distributions: an international comparison, in: 11th Annual International Workshop on Large-Scale Integration of Wind Power into Power Systems as well as on Transmission Networks for Offshore Wind Power Plants Conference, Lisbon, Portugal, 2012.

[46] I. González-Aparicio, A. Zucker, Impact of wind power uncertainty forecasting on the market integration of wind energy in Spain, Appl. Energy (2015) 334−349. Available from: https://doi.org/10.1016/j.apenergy.2015.08.104.

[47] J.D. Olden, D.A. Jackson, Illuminating the 'black box': a randomization approach for understanding variable contributions in artificial neural networks, Ecol. Model. (2002) 135−150. Available from: https://doi.org/10.1016/S0304-3800(02)00064-9.

[48] C.E. Shannon, The mathematical theory of communication, MD Comput. (1997) 125. Available from: https://doi.org/10.2307/410457.

[49] T.M. Cover, J.A. Thomas, Elements of Information Theory, John Wiley & Sons, 2005.

[50] H. Musbah and M. El-Hawary, SARIMA model forecasting of short-term electrical load data augmented by fast Fourier transform seasonality detection, in: 2019 IEEE Canadian Conference of Electrical and Computer Engineering, CCECE 2019, 2019. Available from: https://doi.org/10.1109/CCECE.2019.8861542.

[51] J.P.S. Catalão, H.M.I. Pousinho, V.M.F. Mendes, Hybrid intelligent approach for short-term wind power forecasting in Portugal, IET Renew. Power Gener. (2011) 251−257. Available from: https://doi.org/10.1049/iet-rpg.2009.0155.

[52] N.E. Huang, et al., The empirical mode decomposition and the Hubert spectrum for non-linear and non-stationary time series analysis, Proc. R. Soc. A Math. Phys. Eng. Sci. (1998) 903−995. Available from: https://doi.org/10.1098/rspa.1998.0193.

[53] X. An, D. Jiang, M. Zhao, C. Liu, Short-term prediction of wind power using EMD and chaotic theory, Commun. Nonlinear Sci. Numer. Simul. (2012) 1036−1042. Available from: https://doi.org/10.1016/j.cnsns.2011.06.003.

[54] J. Bedi, D. Toshniwal, Empirical mode decomposition based deep learning for electricity demand forecasting, IEEE Access (2018) 49144−49156. Available from: https://doi.org/10.1109/ACCESS.2018.2867681.

[55] M.R. Haq, Z. Ni, A new hybrid model for short-term electricity load forecasting, IEEE Access (2019) 125413−125423. Available from: https://doi.org/10.1109/ACCESS.2019.2937222.

[56] J. Shi, W.J. Lee, Y. Liu, Y. Yang, and P. Wang, Short term wind power forecasting using Hilbert-Huang transform and artificial neural network, in: DRPT: 2011 Fourth

International Conference on Electric Utility Deregulation and Restructuring and Power Technologies, 2011. Available from: https://doi.org/10.1109/DRPT.2011.5993881.

[57] Z. Liu, W. Bai, and G. Chen, A new short-term load forecasting model of power system based on HHT and ANN, Lecture Notes in Computer Science book series (including Subseries in Lecture Notes in Artificial Intelligence and Lecture Notes in Bioinformatics), 2010. Available from: https://doi.org/10.1007/978-3-642-13318-3_56.

[58] K. Dragomiretskiy, D. Zosso, Variational mode decomposition, IEEE Trans. Signal. Process. (2014) 531−544. Available from: https://doi.org/10.1109/TSP.2013.2288675.

[59] F. He, J. Zhou, Z.-K. Feng, G. Liu, Y. Yang, A hybrid short-term load forecasting model based on variational mode decomposition and long short-term memory networks considering relevant factors with Bayesian optimization algorithm, Appl. Energy (2019) 103−116. Available from: https://doi.org/10.1016/j.apenergy.2019.01.055.

[60] J. Cui, R. Yu, D. Zhao, J. Yang, W. Ge, X. Zhou, Intelligent load pattern modeling and denoising using improved variational mode decomposition for various calendar periods, Appl. Energy (2019) 480−491. Available from: https://doi.org/10.1016/j.apenergy.2019.03.163.

[61] H. Zang, et al., Hybrid method for short-term photovoltaic power forecasting based on deep convolutional neural network, IET Gener. Transm. Distrib. (2018) 4557−4567. Available from: https://doi.org/10.1049/iet-gtd.2018.5847.

[62] A.A. Abdoos, A new intelligent method based on combination of VMD and ELM for short term wind power forecasting, Neurocomputing (2016) 111−120. Available from: https://doi.org/10.1016/j.neucom.2016.03.054.

Chapter 5

Fault and disturbance analysis in microgrid

5.1 Introduction

Microgrids accommodate different configurations and integrates various types of distributed energy resources, such as energy storage systems (ESS), renewable energy sources, and other distributed generators (DGs) as well as loads. It is not avoidable that microgrids, the same as any power system, face faults and disturbance. Microgrids may be more vulnerable to such events due to their constituent components and their operational behaviors. Handling of faults and disturbances is critical to enable both the microgrid and the individual elements meet their operational constraints. Thus, effective fault and disturbance analysis techniques should be pursued to ensure reliability and power supply security in microgrids.

Fault analysis in conventional power systems has been discussed in different works. However, some of the special nature of microgrids, which leads to the special characteristics of faults and disturbances, means it is worth reviewing and evaluating the fault analysis techniques for application in microgrids.

Additionally, the initial characteristics of voltage and current waveforms during the occurrence and clearance of faults are similar to those of other disturbances. Distinguishing the fault incidents from disturbances is very crucial for ensuring the normal operation of microgrids. In the next sections, the concept of fault analysis is briefly presented; different types of fault analysis algorithms applied in microgrids are discussed; and the means of distinguishing the faults from the disturbances are explained.

5.2 Distinguishing faults from dynamic and transient disturbances

Even though the occurrence and clearance of faults could be similar to other types of faults, which presents a challenge in identifying one from the other, faults should be accurately distinguished from the other disturbances. Such identification helps to allow the appropriate protection schemes to operate with their higher degree of selectivity and sensitivity in the system.

Microgrid Protection and Control. DOI: https://doi.org/10.1016/B978-0-12-821189-2.00003-6

Transient disturbances are sudden and severe voltage and current changes caused by large DG or heavy load switching actions or faults in the micro-grids and characterized by a large magnitude and a phase change in a time duration of 0–50 ms. Dynamic disturbance, on the other hand, refers to a series of voltage and current changes caused by DG intermittency, load variation, and faults and characterized by magnitude and phase changes smaller than that of the transients while still beyond the normal operating limits, and continuing for a period of 50 ms to 2 s. Based on these concepts, therefore, faults and disturbances should be handled differently. Transient disturbances are expected to be handled by the transient disturbance control system (TDCS), usually through the use of power intensive type ESS; and the task of stabilizing dynamic disturbances shall be the duty of the dynamic disturbance control system, again mostly through the use of energy intensive type ESS. The details of transient and dynamic disturbances and their respective control systems are addressed in Chapter 7, Dynamic Control of Microgrids, and Chapter 8, Transient Control of Microgrids.

One strategic way of identifying and appropriately handling faults and other disturbances is achieved through the coordination of the control systems and the protection system, as expressed in the flowchart in Fig. 5.1. The first time any kind of disturbance is sensed from the local measurements and extracted features of the electrical parameters, the dynamic and transient control systems shall act to try to stabilize the system. There should not be any protection (circuit breaker tripping) action during this period. The expectation is that if the type of event causing the disturbance was anything other than a fault, the disturbance indicating parameters will return to the normal operation range immediately. However, in the case that fault was what caused the disturbance, the exhibited changes in the electrical parameters will still stay out of the normal range. Thus the appropriate fault detection stage of the protection system shall judge the event as a fault, and locate and isolate the faulted section based on the satisfaction of the tripping criteria.

A practical example of the above-discussed scenario is given in Figs. 5.2 and 5.3. For a microgrid of the topology shown in Fig. 5.2 two events are exhibited in the system and how the control and protection systems respond accordingly are shown in Fig. 5.2. The first event is the sudden switching of a big load in the form of motor load 1 and the second event is a three-phase to ground fault on the line connecting the second motor load (F1). The busbar voltage profiles during the two events are shown in Fig. 5.3A and B. As shown in Fig. 5.3A the TDCS was able to act immediately after motor load 1 is switched off and bring the system back to stability through the use of the power intensive ESS (a supercapacitor in this case). Whereas during the fault event (F1), the normal voltage waveform (as shown in Fig. 5.3B) stays changed for longer period though the TDCS tries to stabilize it. Hence, a fault is anticipated and an action from the protection system is expected (the protectionaction is not shown in the figure). Thus, we can see how the transient control

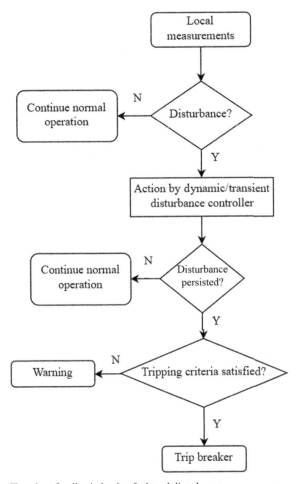

FIGURE 5.1 Flowchart for discriminating fault and disturbance.

system and the microgrid protection system (MGPS) can coordinate effectively to address the issues of disturbance and fault in the microgrid properly.

5.3 Fault analysis

A power system, along with its components, is purposely designed to operate a necessary function. The operation should be continuously performed, excluding some actions such as preventive/scheduled maintenance period or other intended actions. However, faults may cause a failure to perform this function.

Faults, due to their randomness in nature, may appear at any location of a power system network at any time. These phenomena may thus lead to cutting of supply in some areas, power blackouts and/or brownouts, destruction

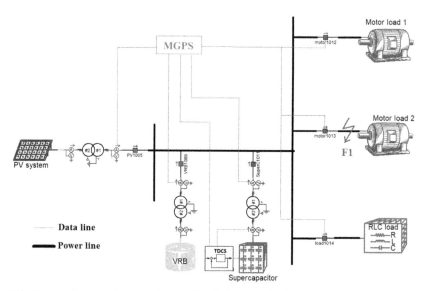

FIGURE 5.2 An example microgrid with disturbance control and protection systems.

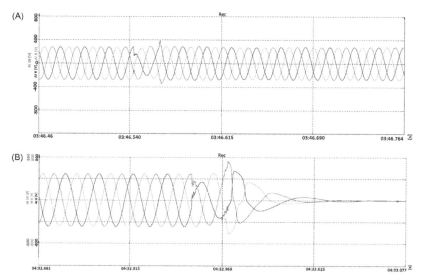

FIGURE 5.3 Busbar voltage profiles during (A) big motor switching and (B) fault.

of the power system network, loss of property, destruction of industrial, commercial, and domestic activities, and even the loss of lives.

One of the reasons that necessitated the demand for microgrids is their capability to be isolated from the grid in the event of a fault happening on the grid side and still ensuring reliable power supply to local loads. However, this does not mean that microgrids are just excused from or naturally immune to faults within themselves. There could be faults occurring inside the microgrids themselves, usually labeled as internal faults, while faults from the other side of the point of common coupling can be called external faults.

Microgrids should be furnished with protection systems which help them to identify external faults and disconnect the microgrid and locate the internal faults and isolate the specific faulted branch. The protection systems employ techniques to detect, identify the type, and locate the faults. The main requirements of a protection system are dependability, security, speed, and selectivity. Fault analysis is the method of analyzing the voltage and current waveforms so as to detect the faults and locate them. The method greatly decides whether the protection system satisfies those basic requirements.

Reliable and accurate protection of the microgrids against faults is challenging, particularly in island operation mode. All the faults inside the microgrid must be cleared in all operating modes. Therefore, during fault conditions, determining the values of system voltages and currents is very vital. This helps the protective systems to detect the faults instantly and to minimize the devastating consequences.

A power system is usually said to be dynamic in nature because of possible alterations during its operation which could be related to the switching of the generators and loads [1]. The dynamicity is even much higher in the case of microgrids. A microgrid may operate in grid-connected or island mode and the individual components may be not only turned on and off but also could be of varying power output, especially in renewable DGs. Thus the fault's current magnitude and characteristics could also be variable under different configurations and operation modes.

In order to achieve proper fault detection, fault location, and fault type identification during fault incidents in microgrids, fault analysis is required to provide appropriate information for the selection of switches/breakers, setting of relays, and stability of system operation.

Faults may be generally classified as open circuit and short circuit faults. As the names may clearly imply, open circuit faults represent faults due to an opening of a circuit (broken conductor) resulting in the interruption of flow of current. Such faults may affect power supply availability; however, they do not cause significant damage to the components or harm to lives. They are also easier to detect and fix. Short circuit faults are the main concern in power systems and this chapter is focused on such faults.

Short circuit faults are caused by conductors of a power system being shorted or connected to each other or to the ground due to factors such as insulation failure, flashover, physical damage, human error, or other causes. In three-phase systems, short circuit faults are classified as either symmetrical faults or unsymmetrical faults based on the number of conductors involved in the short circuit. Symmetrical or balanced faults refer to faults where all the three phases are involved. Such faults are less frequent in power systems. The other type of fault, asymmetrical or unbalanced faults, involve one or two-phase lines and possibly the neutral line. Overall, short circuit faults may be one of the following:

- single line to ground fault (LG);
- line to line fault (LL);
- double line to ground fault (LLG);
- three-phase line to line fault (LLL); or
- three-phase line to ground fault (LLLG).

The most commonly applied fault analysis or short circuit analysis technique is based on Thevenin equivalent circuit and symmetrical components analysis. Thevenin equivalent circuit is based on Thevenin's theorem which states that any linear network containing any number of voltage sources and impedances can be represented by an equivalent single voltage source and an impedance. Such an ideal circuit with a voltage source and an impedance representing a more complicated power system is called a Thevenin equivalent circuit. The Thevenin equivalent circuit simplifies the calculations in fault analysis, which is made after obtaining the open circuit voltage and network impedance as seen from the fault point.

The other important concept in the conventional fault analysis is symmetrical components. Any set of unbalanced three-phase voltages (or currents) can be transformed into three symmetrical components called positive sequence, negative sequence, and zero sequence components (Fig. 5.4). The symmetrical components do not have a separate existence; they are just mathematical components of unbalanced currents (or voltages) which exist in a system.

The positive sequence components are displaced from each other by 120 degrees and have the same phase order as the original signal. A subscript "1" is used to represent the positive sequence components (as U_{a1}, U_{b1}, U_{c1} for voltage signals U_a, U_b, U_c). On the other hand, negative sequence components have an opposite phase sequence to the original signal while still displaced from each other by 120 degrees. The negative sequence components are represented with a subscript of "2" (as U_{a2}, U_{b2}, U_{c2}). Zero sequence components are all equal in magnitude and in phase with each other, usually denoted by a subscript of "0" (as U_{a0}, U_{b0}, U_{c0}).

In fault analysis based on symmetrical components, the positive, negative, and zero sequence components of current and voltage signals are computed and analyzed to judge the occurrence and type of fault.

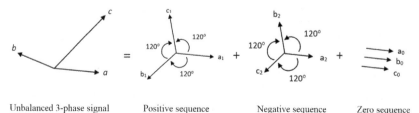

| Unbalanced 3-phase signal | Positive sequence | Negative sequence | Zero sequence |

FIGURE 5.4 Symmetrical components of an unbalanced three-phase system.

The phase components are the vector sum of the symmetrical components. The relationship between phase voltage and its symmetrical components, for example, can be written as follows.

$$U_a = U_{a1} + U_{a2} + U_{a0}$$
$$U_b = U_{b1} + U_{b2} + U_{b0}$$ (5.1a)
$$U_c = U_{c1} + U_{c2} + U_{c0}$$

where,

$$U_{abc1} = U_{m1}\cos(\omega t) + U_{m1}\cos\left(\omega t - \frac{2\pi}{3}\right) + U_{m1}\cos\left(\omega t + \frac{2\pi}{3}\right)$$

$$U_{abc2} = U_{m2}\cos(\omega t) + U_{m2}\cos\left(\omega t + \frac{2\pi}{3}\right) + U_{m2}\cos\left(\omega t - \frac{2\pi}{3}\right)$$ (5.1b)

$$U_{a0} = U_{a0} = U_{a0} = \left(\frac{1}{3}\right)\{U_a + U_b + U_c\}$$

The residual current will be three times the zero sequence current of one phase. In the case of a fault involving the ground, the positive and negative sequence currents are in equilibrium while the zero sequence currents flow through the ground and the neutral wire. Thus symmetrical components are importantly useful to analyze such faults. In a symmetrical/balanced fault where the three lines are short-circuited with each other or earthed, only a positive sequence component is applied. The last case is where asymmetrical faults happen, in which case negative sequence components are involved.

To perform calculations in the power system conveniently, per-unit value or percentile expression are often used to state power system quantities, such as power, voltage, current, and impedance by taking into consideration common base values. All quantities are expressed as ratios of some base value or values. These expressions are quite uniform throughout, although they have different voltage and power ratings in the power system network. This approach is independent of voltage changes and phase shifts through transformers where the base voltages in the winding are proportional to the number of turns in the windings. Additionally, manufacturers commonly stipulate the impedance of device in per-unit or percent on the bases of the power and voltage nameplate ratings. Thus fault analysis can be carried out by applying the per-unit expressions. In other words, such expressions are very helpful to avoid different voltage levels existing in the system network while carrying out fault calculations. As some parameters denoted in per-unit values come from the same range, the per-unit quantities can be more easily comparable, which also allows obtaining computational errors.

Therefore the per-unit quantity is defined as in (5.2).

$$\text{Per-unit quantity} = \frac{\text{Actual value}}{\text{Base value}}$$ (5.2)

The base value and the actual value quantities have the same units, making the per-unit quantity become dimensionless.

5.4 Advanced algorithms

As explained earlier, some special characteristics of microgrids result in conventional current magnitude-based protection systems and the conventional fault analysis techniques being less effective. Thus a variety of algorithms are suggested in different literature and some are tested experimentally. Though most of them are on the research stage and yet to be used in market available products, they shall be proved useful in the near future, possibly through utilization in parallel with the conventional and more proven techniques. We will discuss some of the available options in the literature shortly.

5.4.1 Voltage and current THD-based algorithm

Total harmonics distortion (THD) is the ratio of the RMS value of the harmonic content to the RMS value of the fundamental component or the reference fundamental component of an alternating quantity. The algorithm developed based on THD is one of the algorithms deployed in the application of MGPS, which enables the protection of the microgrid against faults that are both internal and external to protection zones.

The THD of a given signal waveform is computed by

$$\text{THD} = \frac{\sqrt{\sum_{n=2}^{\infty} U_n^2}}{U_1} \tag{5.3}$$

where U_n is the amplitude of the signal with harmonic component at nth order and U_1 is the fundamental component.

THD can also be computed using the RMS value of the voltage as in (5.4).

$$\text{THD} = \sqrt{\frac{2}{3}\left(\frac{U_{rms}^2}{U_1^2}\right) - 1} \tag{5.4}$$

For doing so, during fault incidence, the time domain of the output voltage RMSs of the DGs are required to be transferred into the frequency domain. Such a transformation is carried out through discrete Fourier transform (DFT). The DFT is computed for each phase signal separately. As the signals are transients, applying a windowing function in the process of DFT is vital to extract required features.

In this algorithm, the THD of the terminal voltage and/or branch currents are supposedly monitored by the protection relay. If the THD exceeds a threshold value, the relay starts to trip the breakers to isolate or shutdown the inverter. The reason is that, in normal operation, the distribution network behaves as a stiff voltage source, which has low impedance, enabling the maintenance of the network with a low distortion voltage level on the inverter terminals, which records the THD being close to zero [2]. Thus when any type of fault occurs, an increase in voltage THD is anticipated. Following the fault occurrence, the low impedance distribution network is disconnected and only the local load remains. As a result, the impedance at the inverter terminals increases, and so the current harmonics in the inverter output will cause the voltage harmonics in the terminal voltage to increase their levels.

This algorithm can perform two main functions. One is identifying the type of the fault, whether it is phase-to-ground, phase-to-phase, or three-phase. This is performed by checking the amplitude of the fundamental frequency deviation from its nominal value, which will drop for the faulted phase compared to the unfaulted phase. So, the calculated THD for the faulted phase will be greater than the unfaulted phase.

The other function is detection of fault location and direction. The discrimination amongst the relays to identify the fault location is carried out by calculating the THD, having greater THD for the faulted branches than for the unfaulted ones. The communication link is needed to decide which of the configured relays should trip.

An example of a flowchart for a protection scheme based on THD calculations is depicted in Fig. 5.5. It explains the two functions discussed above regarding how the THD algorithm performs. As per the flowchart, the first part is intended to detect the occurrence of a fault. The later one on the right-hand side is dedicated to discriminating amongst the faulted zones and later the fault type, based on which the relay will trip and isolate the faulted section and line, leaving the rest of the network intact.

To show the effectiveness of this algorithm, simulation and experimental examples are presented. In the simulation results shown below a protection system based on THD method is designed and tested for a microgrid shown in Fig. 5.6. The total harmonic disturbances in the busbar voltages and branch currents (THD_U and THD_I) are examined to detect and locate faults. In the case of faults happening in a branch, both THD_U and THD_I will show and increase, while the rise in THD_I in the specific branch will be significantly changed with respect to the healthy branches which will enable discrimination of the faults.

The following figures show the sequence of events and subsequent circuit breaker actions for detecting and clearing temporary (staying for a period of 0.2 seconds) single phase to ground and phase to phase faults using the THD-based method. The sequence of actions in Figs. 5.7−5.10 is as follows:

- Fault occurs at T1 (3 seconds)
- Circuit breaker opens at T2 (3.1 seconds)
- Fault disappears at T3 (3.2 seconds)
- Circuit breaker recloses at T4 (3.21 seconds)

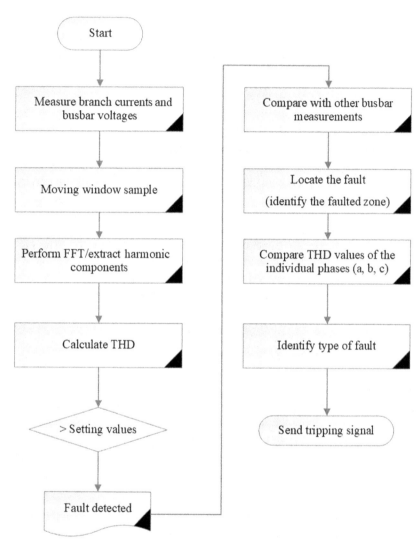

FIGURE 5.5 Flowchart for THD algorithm-based protection scheme.

The effectiveness of the THD-based method to detect and locate faults is experimentally tested and verified on an operational microgrid and the results are illustrated in Figs. 5.7 and 5.8. As it can be seen in the figures, this technique has had its effectiveness verified in detecting and locating the different types of faults that occur in the microgrid system. In all the figures, the sequence of events is provided in such a way that the fault occurs at 1.4 seconds and the circuit breaker is triggered to trip after a delay at 2.58 seconds.

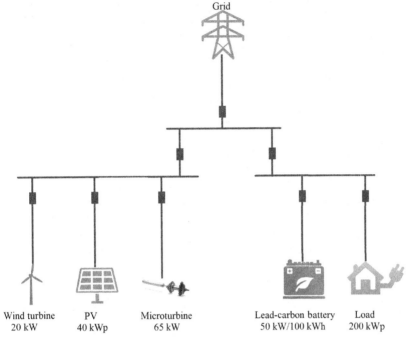

FIGURE 5.6 Test microgrid topological.

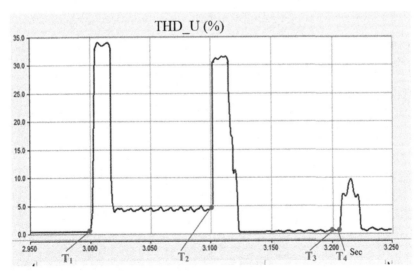

FIGURE 5.7 Simulation result of THD-based protection for single phase to ground fault (voltage THD).

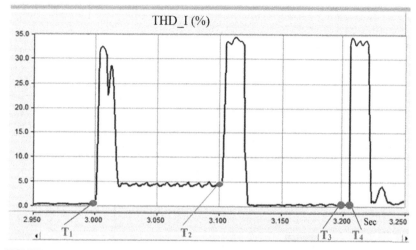

FIGURE 5.8 Simulation result of THD-based protection for single phase to ground fault (current THD).

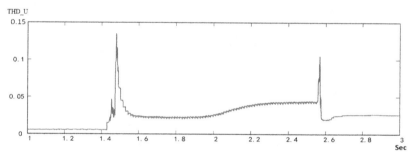

FIGURE 5.9 Experimental testing result of THD-based protection for single phase to ground fault (voltage THD).

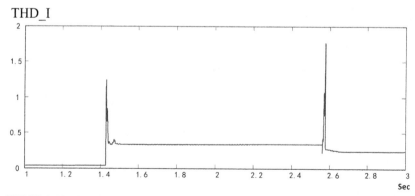

FIGURE 5.10 Experimental testing result of THD-based protection for single phase to ground fault (current THD).

5.4.2 Park transformation-based algorithm

Protection schemes of a microgrid system based on Park transformation are presented in Refs. [3,4]. Park transformation, sometimes called direct-quadrature-zero (qd0) transformation, is a mathematical transformation used to convert the stationary *abc* frame three-phase parameters into a rotating reference frame (direct, quadrature, and zero) components. Though commonly called Park transformation for short, the process of *dq0* transformation is composed of the Clarke transformation and Park transformation. The Clarke transformation is defined by (5.5):

$$[U_{\alpha\beta\gamma}] = T_C \begin{bmatrix} U_a(t) \\ U_b(t) \\ U_c(t) \end{bmatrix} \tag{5.5}$$

where the Clarke transformation matrix $T_C = \frac{2}{3} \begin{bmatrix} 1 & -1/2 & -1/2 \\ 0 & \sqrt{3}/2 & -\sqrt{3}/2 \\ 1/2 & 1/2 & 1/2 \end{bmatrix}$

Suppose the general form of the three-phase system as in (5.6),

$$U_a = V_m \cos(\omega t + \varphi)$$
$$U_b = V_m \cos\left(\omega t + \varphi - \frac{2\pi}{3}\right) \tag{5.6}$$
$$U_c = V_m \cos\left(\omega t + \varphi + \frac{2\pi}{3}\right)$$

Thus the Clarke transformation is expressed as the following equations:

$$[U_{\alpha\beta\gamma}] = \frac{2}{3} \begin{bmatrix} 1 & \dfrac{-1}{2} & \dfrac{-1}{2} \\ 0 & \dfrac{\sqrt{3}}{2} & -\dfrac{\sqrt{3}}{2} \\ \dfrac{1}{2} & \dfrac{1}{2} & \dfrac{1}{2} \end{bmatrix} \begin{bmatrix} U_a(t) \\ U_b(t) \\ U_c(t) \end{bmatrix} \tag{5.7}$$

$$U_\alpha = V_m \cos(\omega t + \varphi)$$
$$U_\beta = V_m \sin(\omega t + \varphi) \tag{5.8}$$
$$U_\gamma = 0$$

The later stage, Park transformation, converts vectors in the $\alpha\beta\gamma$ reference frame to the $dq0$ reference frame. The reference frame of the Park transformed vectors rotates at an arbitrary frequency. Thus, with the application of the $dq0$ transformation, the frequency spectrum of a signal with an arbitrary frequency now appears as "DC" and the old "DC" appears as the negative of the arbitrary frequency.

The Park transformation is defined by the Park transformation matrix of Eq. (5.9).

$$[T_P] = \begin{bmatrix} \cos(\theta) & \sin(\theta) & 0 \\ -\sin(\theta) & \cos(\theta) & 0 \\ 0 & 0 & 1 \end{bmatrix} \tag{5.9}$$

where θ is the instantaneous angle of an arbitrary ω frequency.

$$\begin{aligned} [U_{dq0}] &= [T_P][U_{\alpha\beta\gamma}] \\ &= [T_P][T_C][U_{abc}] \\ &= [T_{dq0}][U_{abc}] \end{aligned} \tag{5.10}$$

where T_P and T_C are the Park and Clarke transformation matrices respectively and T_{dq0} is defined as:

$$[T_{dq0}] = \sqrt{\frac{2}{3}} \begin{bmatrix} \cos(\theta) & \cos\left(\theta - \dfrac{2\pi}{3}\right) & \cos\left(\theta + \dfrac{2\pi}{3}\right) \\ \sin(\theta) & \sin\left(\theta - \dfrac{2\pi}{3}\right) & \sin\left(\theta + \dfrac{2\pi}{3}\right) \\ \dfrac{\sqrt{2}}{2} & \dfrac{\sqrt{2}}{2} & \dfrac{\sqrt{2}}{2} \end{bmatrix} \tag{5.11}$$

Substituting (5.11) and (5.6) into (5.10), the (5.12) holds

$$\begin{aligned} U_d &= \sqrt{\frac{3}{2}} V_m \cos(\theta - \omega t - \varphi) \\ U_q &= \sqrt{\frac{3}{2}} V_m \sin(\theta - \omega t - \varphi) \\ U_0 &= 0 \end{aligned} \tag{5.12}$$

For a synchronous reference frame, the $dq0$ frame rotates with the same angular speed as the system frequency, that is, $\omega t = \theta$. Thus Eq. (5.12) becomes (5.13).

$$\begin{aligned} U_d &= \sqrt{\frac{3}{2}} U_m \cos(\varphi) \\ U_q &= \sqrt{\frac{3}{2}} U_m \sin(-\varphi) \end{aligned} \tag{5.13}$$

As clearly seen in the expressions of Eq. (5.13), any disturbance in the busbar voltage will be reflected as disturbance in the dq values. Thus these perturbed dq values are used to draw the disturbance signal (U_{dist}) out, which helps to characterize the deviation of the bus voltage from a nominal reference value (U_{ref}) as expressed in (5.14). Setting the reference values as U_{ref}, the error signal $U_{dqerror}$ is computed as

$$
\begin{aligned}
U_{derror} &= U_{dref} - U_d \\
U_{qerror} &= U_{qref} - U_q
\end{aligned}
\tag{5.14}
$$

The error signal $U_{dqerror}$ represents the disturbance signal $U_{dq,dist}$; that is, $U_{d,dist} = U_{derror}$ and $U_{q,dist} = U_{qerror}$.

The dq voltage values are DC components. As per (5.14), if the measured values of U_{dq} are different from the reference values, $U_{dq,dist}$ exists, which confirms the incidence of the faults and the characteristics of the disturbance signal depending on the nature of the fault. This algorithm is useful to detect not only the disturbance value $U_{dq,dist}$ but also the location, direction, and different types of the fault, such as three-phase, two-phase, and single-phase faults.

There are a variety of ways that the Park transformation is used in fault detection studies. Both $U_{d,dist}$ and $U_{q,dist}$ are digitally filtered and processed by a dual hysteresis comparator to be the fault detection signals in Ref. [3]. Only the q-axis component ($U_{q,dist}$) is assumed relevant in Ref. [4] and used as a solid fault detection signal. The Park transformation is applied to the symmetrical components in Ref. [5]. The other method using the Park transformed disturbance signals is the use of a composite voltage vector $U_{dq} = \sqrt{U_d{}^2 + U_q{}^2}$ as in Ref. [6].

5.4.2.1 Symmetrical fault detection

A symmetrical fault is said to happen when either the three phases are short-circuited, or the three phases are jointly earthed with or without impedance. When such a fault occurs, the three-phase voltages are balanced, and can be expressed as

$$
\begin{aligned}
U_a &= U_m \cos(n\omega t + \varphi) \\
U_b &= U_m \cos\left(n\omega t + \varphi - \frac{2\pi}{3}\right) \\
U_c &= U_m \cos\left(n\omega t + \varphi + \frac{2\pi}{3}\right)
\end{aligned}
\tag{5.15}
$$

where n indicates the nth harmonic order during the fault incidence.

Transforming (5.15) to $dq0$ gives to Eq. (5.16).

$$
\begin{aligned}
U_d &= \sqrt{\frac{3}{2}} U_m \cos(\theta - \omega t - \varphi) \\
U_q &= \sqrt{\frac{3}{2}} U_m \sin(\theta - \omega t - \varphi)
\end{aligned}
\tag{5.16}
$$

According to Ref. [4], for a balanced operation of a machine, the rotor angle $\theta = \omega_r t = \omega t$; thus disturbance in Eq. (5.17) is computed as:

$$U_d = \sqrt{\frac{3}{2}}\, U_m\cos(\varphi)$$
$$U_q = \sqrt{\frac{3}{2}}\, U_m\sin(-\varphi)$$
(5.17)

The dq values in (5.17) are constants and the subtracted results from their respective reference values also remain constant. Hence, for a symmetrical fault, $U_{dq, dist}$ is pure DC. An example fault analysis is conducted using PSCAD simulation environment for a microgrid shown in Fig. 5.6. In the simulation tests whose results are displayed below, a centralized protection based on Park transformation is designed and tested. The local voltage and current measurements are sent to the central protection and control unit where Park transformation is applied and the transformed values are compared against an updatable setting matrix and decisions are made on the occurrence and location of the fault.

Fig. 5.11 shows the simulation results for a symmetrical fault (three phase to ground) occurrence in microgrids. As shown in the figure, $U_{q, dist}$ is constant or DC. In other words, alternating three-phase voltage in the ABC reference frame due to a symmetrical fault appears as a change in the constant or DC value in the dq reference frame.

5.4.2.2 Asymmetrical fault detection

According to Ref. [7], the $dq0$ transformation method can again be used to analyze the unsymmetrical faults. Starting with considering Eq. (5.17), the module of the vector U_{dq} can be computed by (5.18).

$$\|U\| = \sqrt{\frac{3}{2}}U_m$$
(5.18)

Normalizing the three-phase voltages simplifies the fault analysis. That is achieved by dividing the phase voltages by the module of the dq vector.

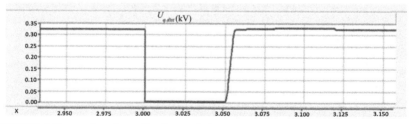

FIGURE 5.11 q-axis voltage disturbance during a symmetrical fault occurring at the 3rd second.

Suppose u_{iN} represents the normalized value for the three-phase voltages, as defined by (5.19).

$$u_{iN} = \frac{U_i}{\|U\|} = \sqrt{\frac{2}{3}} \frac{U_i}{U_m} \tag{5.19}$$

where $i = a, b, c$.

The values of the normalized parameter u_{iN} always lie within the range of ± 1, and it is stated as in (5.20) with its average absolute value expressed as in (5.21).

$$u_{iN} = \begin{cases} u_{aN} = \cos(n\omega t + \varphi) \\ u_{bN} = \cos\left(n\omega t + \varphi - \frac{2\pi}{3}\right) \\ u_{cN} = \cos\left(n\omega t + \varphi + \frac{2\pi}{3}\right) \end{cases} \tag{5.20}$$

$$\langle |u_{iN}| \rangle = \frac{\omega_s}{2\pi} \int_0^{\omega_s/2\pi} u_{iN} dt \tag{5.21}$$

When the power generating units are operating in the normal condition, the three-phase voltages become sinusoidal waves. Then, a constant value $\zeta = 2/\pi$ can be used for the sake of computing and analyzing the signal flexibly, which is the normalized average absolute value $\langle |u_{iN}| \rangle$ during normal condition.

A typical fault detection algorithm based on $dq0$ transformation involves three variables, which are the errors of the normalized values, expressed as

$$e_i = \zeta - \langle |u_{iN}| \rangle \tag{5.22}$$

During the normal operating conditions of the microgrid, the variables e_i are all close to zero. In the times of fault incidence, however, one of the values from the variables will assume a unique value that can be used to detect the faults. Further explicit analysis for this method is also presented as follows.

As explained above, asymmetrical faults mean the existence of an unbalanced three-phase system. Thus the Park transformation computational analysis can be supported by utilizing the symmetrical components.

Considering Eq. (5.1a and b) and the zero sequence component subtracted from the three-phase system:

$$U_a' = U_a - U_0 = U_{1m}\cos\left(\omega t + \varphi_1\right) + U_{2m}\cos\left(\omega t + \varphi_2\right)$$
$$U_b' = U_b - U_0 = U_{1m}\cos\left(\omega t + \varphi_1 - \frac{2\pi}{3}\right) + U_{2m}\cos\left(\omega t + \varphi_2 + \frac{2\pi}{3}\right)$$
$$U_c' = U_c - U_0 = U_{1m}\cos\left(\omega t + \varphi_1 + \frac{2\pi}{3}\right) + U_{2m}\cos\left(\omega t + \varphi_2 - \frac{2\pi}{3}\right)$$

$$\tag{5.23}$$

Applying the $dq0$ conversion matrix of (5.11) for (5.23), we can obtain (5.24).

$$U_d = \sqrt{\frac{3}{2}} \left\{ V_{1m}\cos\left(\theta - \omega t - \varphi_1\right) + V_{2m}\cos\left(\theta + \omega t + \varphi_2\right) \right\}$$

$$U_q = \sqrt{\frac{3}{2}} \left\{ V_{1m}\sin\left(\theta - \omega t - \varphi_1\right) + V_{2m}\sin\left(\theta + \omega t + \varphi_2\right) \right\}$$

(5.24)

For a synchronous reference frame, $\omega t = \theta$, thus Eq. (5.24) becomes (5.25).

$$U_d = V_{1m}\cos\left(\varphi_1\right) + V_{2m}\cos\left(2\omega t + \varphi_2\right)$$
$$U_q = V_{1m}\sin\left(-\varphi_1\right) + V_{2m}\sin\left(2\omega t + \varphi_2\right)$$

(5.25)

Hence, for asymmetrical faults, $U_{d,dist}$ and $U_{q,dist}$ will not only have a DC component, but also have a ripple component oscillating with twice the fundamental frequency ($2\omega t$).

The following simulation examples are presented for asymmetrical faults based on the microgrid topology in 5.5 for a fault happening at the 3rd second.

As verified in the simulation results shown in Figs. 5.12−5.14, alternating three-phase voltage in the abc reference frame due to asymmetrical fault appears as a DC component plus an oscillating component (AC ripple) with twice the fundamental frequency in the dq synchronous reference frame. For the single line to ground fault, $U_{q,dist}$ is an oscillating signal between some lower voltage value and the peak voltage value. For the double line to ground fault, $U_{q,dist}$ oscillates between zero and some intermediate voltage value. For the line to line fault, $U_{q,dist}$ is an oscillating signal between zero and the peak value.

5.4.3 Wavelet transform-based algorithm

To study composite signals with an inconsistent pattern, it is crucial to extract and analyze the constituting time−frequency components. Wavelet transform (WT) is one of the methods that can be used for this purpose and

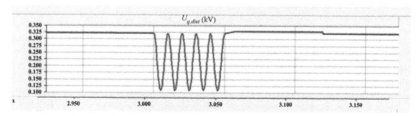

FIGURE 5.12 q-Axis voltage disturbance during single phase to ground fault.

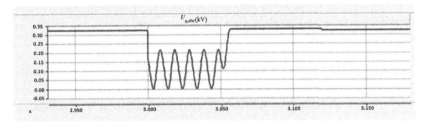

FIGURE 5.13 q-Axis voltage disturbance during double phase to ground fault.

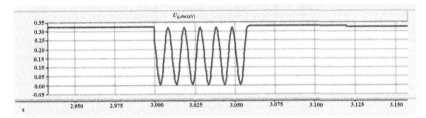

FIGURE 5.14 q-Axis voltage disturbance during phase to phase fault.

is considered as a good choice. The WT decomposes an original signal into detailed and approximation components with the application of low-pass and high-pass filters [8]. The approximation component is a high-scale, low-frequency component, and the detail components, which may be more than one based on the level of decomposition, are low-scale and high-frequency components [9]. The approximation and detail components will appear as a pair in every stage of decomposition. When the signal is analyzed with the implementation of the WT method, the connection of signal and a wavelet is measured by the coefficient vectors [10]. WT characterizes the signal in the time−frequency domain and provides the time localization of the signal [11].

Since a frequency content of a transient signal varies with time, the WT will be a credible mechanism for studying such a signal. It decomposes wideband and transient signals into specific time−frequency resolutions. When comparing WT with the Fourier transform, the former has the advantage in that it provides both time and frequency information, which are important to extract transient features from the signals [12].

A wavelet is a function in a space of square-integrable functions on $\mathbb{R}$, $\varphi(t) \in L^2(\mathbb{R})$. It provides a zero average:

$$\int_{-\infty}^{+\infty} \varphi(t) \, dt = 0 \qquad (5.26)$$

Its normalization is denoted as $\| \varphi(t) \| = 1$, and centered about $t = 0$.

WT involves a set of elemental functions which can be generated by scaling and translation of a distinctive mother wavelet $\{\varphi(t)\}$, as shown in (5.27).

$$\varphi_{p,q}(t) = \frac{1}{\sqrt{q}}\varphi\left(\frac{t-p}{q}\right) \tag{5.27}$$

where p and q represent the time translation and scaling parameters, respectively.

5.4.3.1 Continuous wavelet transform

Continuous wavelet transform (CWT) can be defined as the correlation between the signal $x(t)$ with the mother wavelet for each p and q parameter and expressed as in (5.27) [13]. In other words, the WT of a signal $f(t) \in L^2(\mathbb{R})$ at time translation of p parameter and frequency scaling of q parameter is stated as:

$$W_{p,q}(t) = \left\langle f, \varphi_{p,q} \right\rangle = \int_{-\infty}^{+\infty} f(t)\frac{1}{\sqrt{q}}\varphi^*\left(\frac{t-p}{q}\right)dt \tag{5.28}$$

where $p, q \in \mathbb{R}$, $q \neq 0$, and $\varphi_{p,q}^*(t)$ is the complex conjugate of the family of wavelet. Basically, CWT performs an analysis of the redundant data of the signal.

5.4.3.2 Discrete wavelet transform

Discrete wavelet transform (DWT) enables the development of the practical application of WT [8]. It is used to overcome data redundancy and reduce computational time, which are challenging in CWT [12]. DWT is processed by discretizing the parameters p and q, which are set to powers of 2, offering dyadic sampling which corresponds to the sampling of the frequency axis.

Let $f(t)$ be a continuous time signal over the domain [0,1]. Suppose $f[n]$ is a discrete signal which is obtained by implementing a low-pass filter for continuous signal f with a regular sampling interval $N-1$. The DWT can be computed only over $N-1 < q < 1$. It is computed for $q = s^i$, with $s = 2^{1/p}$, which provides q intermediate scales in each octave set $[2^i, 2^{i+1}]$.

Let $\varphi(t)$ be a wavelet whose support is included in $[-K/2, K/2]$. For $1 \leq s^i \leq NK^{-1}$, as per Ref. [12], a discrete wavelet scaled by s^i is defined by

$$\varphi_i[n] = \frac{1}{\sqrt{s^i}}\varphi\left(\frac{n}{s^i}\right) \tag{5.29}$$

Therefore Ks^i nonzero values on $[-N/2, N/2]$ are found in this discrete wavelet. The scale s^i should be larger than 1, if not the sampling interval might be larger than the wavelet support.

In a multiresolution analysis, DWT is applied to decompose the original signal into approximations and details. The approximation coefficients α_ℓ and the detail coefficients d_ℓ are used to synthesize the low- and high-frequency components of the signal so as to provide each wavelet decomposition level ℓ. The coefficients can be computed as in (5.30) [10]:

$$\alpha_\ell(r) = \sum_{\ell_r} w_0(\ell_r - 2r)\alpha_{\ell-1}(\ell_r) \ \text{ and } \ d_\ell(r) = \sum_{\ell_r} w_h(\ell_r - 2r)\alpha_{\ell-1}(\ell_r)$$

$$(5.30)$$

where w_0 is the low-pass filter, and w_h the high-pass filter.

Energies exist from the wavelet coefficients. The hidden features of the analyzed signal are held by these energies. They are computed by (5.31).

$$Y_{\alpha\ell} = \sum_r |\alpha_\ell(r)|^2 \ \text{ and } \ Y_{d\ell} = \sum_r |d_\ell(r)|^2 \qquad (5.31)$$

where $Y_{\alpha\ell}$ represents energy for approximation wavelet coefficients and $Y_{d\ell}$ for detail wavelet coefficients. These energies are affected directly by the choice of the wavelet function and the length of the filtering process. This influence may affect the effectiveness of the WT application. At times of analysis, different wavelets with various properties may be used. However, a notable improvement in the performance for a specific application may be achieved due to a careful choice of the wavelet function [10].

5.4.3.3 Wavelet transform-based fault detection

WT is one of the signal processing tools used in the power system fault detection studies. An illustration for such a method in microgrid protection is provided by Netsanet et al. [14] who used features extracted from windowed WT of the current and voltage signals (changes in approximate and detail wavelet energies (ΔE_a and ΔE_d) and number of peak values of wavelet coefficients (ΔN_p)) as fault detection signals (Fig. 5.15). An example of how the wavelet energies of current and voltage waveforms change during the

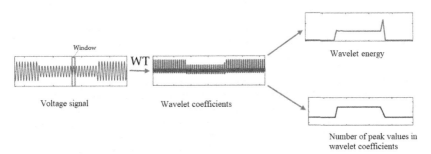

FIGURE 5.15 Use of wavelet transformation for fault detection in microgrids.

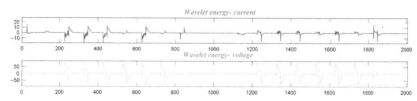

FIGURE 5.16 Changes in wavelet energy of approximation components of current and voltage signals during fault incidents.

incidence of faults at different locations in a microgrid at different times is shown in Fig. 5.16.

Wavelet transformation is applied in coordination with $dq0$ decomposition in Ref. [15] to extract the high-frequency details of current and voltage waveforms so as to be used for high-impedance fault (HIF) detection and location in islanded microgrids. The phase displacements between CWTs of zero sequence voltage and current signals are specifically used as fault detection signal for HIF. The same principle is applied in Ref. [16] for HIF detection in distribution networks.

WT is used to detect and locate a fault in microgrids in Ref. [17]. The fault location is achieved through the use of the first-level power high-frequency detail, which is the product of the respective values branch currents and busbar voltage. A fault is detected when the cumulative sum of this parameter exceeds a threshold value.

The fault detection schemes based on WT are different in the procedures involved and the features they use as the detection signal. However, the core principle is applying WT on the measured voltage and/or current signals to extract features that can depict the occurrence of a fault and compare those parameters to some threshold values. One of the challenging aspects in the application of this and other recent methods, including the THD-based method discussed earlier in this chapter, is the difficulty to set a threshold value for the judgment of fault occurrence. Approaches such as adaptive setting adjustments and use of artificial intelligence or machine learning-based tools are alternatives used to address this issue.

5.4.4 Other applicable algorithms

Some other algorithms that can be used in the application of microgrid system for fault analysis and diagnosis may include the trend-relationship of adjacent fold-lines, fuzzy logic, and neural networks.

Asymmetric interferences, such as overcurrent and bias components in the power system, may cause uncertain fluctuation, false alarm, and erroneous triggering in the protection systems. The trend-relationship of adjacent fold-lines algorithm, therefore, can cope with such interferences. This algorithm is utilized to extract the fault features. The trend feature is encoded, visualized, calculated, identified, and stored. From the encoded feature, two indexes are computed to represent the degree of abnormal and location information [7]. The fuzzy logic approach can also identify multiple open circuit faults and faults as in Ref. [18]. Neural network algorithms have been widely implemented in the application of fault protection schemes in the microgrid. A deep neural network algorithm is, for example, deployed for extracting higher level features from input data to detect and to identify the fault inception in the microgrid [9]. An artificial neural network (ANN) algorithm along with the WT technique can be implemented for fault detection. ANN has the ability to infer the underlying nonlinear and complex relationships between input and output data. It utilizes a soft criterion for feature comparison. The ANN algorithm is provided through an iterative training procedure with the sets of input data and output labels. Hence the ANN weights and biases are adjusted by an error signal in a way that the network output tries to follow the desired output [13].

References

[1] J. Izykowski, Power System Faults. Wrocław: Wrocław University of Technology, 2011.

[2] H. Al-Nasseri, M.A. Redfern, Harmonics content based protection scheme for microgrids dominated by solid state converters, in: 2008 12th International Middle East Power System Conference, 2008, MEPCON (2008). Available from: https://doi.org/10.1109/MEPCON.2008.4562361.

[3] H. Al-Nasseri, M.A. Redfern, F. Li, A voltage based protection for micro-grids containing power electronic converters, in: 2006 IEEE Power Engineering Society General Meeting, PES, (2006). Available from: https://doi.org/10.1109/pes.2006.1709423.

[4] D. Zheng, A.T. Eseye, J. Zhang, A communication-supported comprehensive protection strategy for converter-interfaced islanded microgrids, Sustainability (2018). Available from: https://doi.org/10.3390/su10051335.

[5] X. Hou, C. Hu, A study of voltage detection based fault judgment method in micro-grid with inverter interfaced power source, in: Proceedings of the 2009 6th International Conference on Electrical Engineering, pp. 1−5.

[6] R.M. Tumilty, M. Brucoli, G.M. Burt, T.C. Green, Approaches to network protection for inverter dominated electrical distribution systems, 2006, Available from: https://doi.org/10.1049/cp:20060183.

[7] Z. Huang, Z. Wang, A fault diagnosis algorithm for microgrid three-phase inverter based on trend relationship of adjacent fold lines, IEEE Trans. Ind. Inform. (2020). Available from: https://doi.org/10.1109/TII.2019.2918166.

[8] W. Gao, J. Ning, Wavelet-based disturbance analysis for power system wide-area monitoring, IEEE Trans. Smart Grid (2011). Available from: https://doi.org/10.1109/TSG.2011.2106521.

[9] S. Samal, S.R. Samantaray, M.S. Manikandan, A DNN based intelligent protective relaying scheme for microgrids, in: 2019 8th International Conference on Power Systems: Transition towards Sustainable, Smart and Flexible Grids, 2019, ICPS (2019). Available from: https://doi.org/10.1109/ICPS48983.2019.9067600.

[10] T.S. Abdelgayed, W.G. Morsi, T.S. Sidhu, A new approach for fault classification in microgrids using optimal wavelet functions matching pursuit, IEEE Trans. Smart Grid (2018). Available from: https://doi.org/10.1109/TSG.2017.2672881.

[11] S. Mallat, A Wavelet Tour of Signal Processing, Academic Press, 1999.

[12] D.P. Mishra, S.R. Samantaray, G. Joos, A combined wavelet and data-mining based intelligent protection scheme for microgrid, IEEE Trans. Smart Grid (2016). Available from: https://doi.org/10.1109/TSG.2015.2487501.

[13] D.K.J.S. Jayamaha, N.W.A. Lidula, A.D. Rajapakse, Wavelet-multi resolution analysis based ANN architecture for fault detection and localization in DC microgrids, IEEE Access (2019). Available from: https://doi.org/10.1109/ACCESS.2019.2945397.

[14] S. Netsanet, J. Zhang, D. Zheng, Bagged decision trees based scheme of microgrid protection using windowed fast Fourier and wavelet transforms, Electronics (2018). Available from: https://doi.org/10.3390/electronics7050061.

[15] K. Lai, M.S. Illindala, M.A. Haj-Ahmed, Comprehensive protection strategy for an islanded microgrid using intelligent relays, IEEE Trans. Ind. Appl. (2017). Available from: https://doi.org/10.1109/TIA.2016.2604203.

[16] M. Michalik, W. Rebizant, M.R. Lukowicz, S.J. Lee, S.H. Kang, High-impedance fault detection in distribution networks with use of wavelet-based algorithm, IEEE Trans. Power Deliv. (2006). Available from: https://doi.org/10.1109/TPWRD.2006.874581.

[17] P. Kanakasabapathy, M. Mohan, Digital protection scheme for microgrids using wavelet transform, in: Proceedings of the 2015 IEEE International Conference on Electron Devices and Solid-State Circuits, EDSSC, 2015, (2015). Available from: https://doi.org/10.1109/EDSSC.2015.7285203.

[18] H. Yan, Y. Xu, F. Cai, H. Zhang, W. Zhao, C. Gerada, PWM-VSI fault diagnosis for a PMSM drive based on the fuzzy logic approach, IEEE Trans. Power Electron. (2018). Available from: https://doi.org/10.1109/TPEL.2018.2814615.

Chapter 6

Protection of microgrids

6.1 Introduction

A protection system is very important for the proper utilization of electric networks and a lack of protection system will cause serious danger to the public and the electrical equipment. Thus the history of protection system starts with the history of the power system. As the power networks have been growing in size and complexity, the protection systems and relays have also been growing in variety, tasks, operation principles, and technology.

A *protection system* is defined by International Electrotechnical Commission (IEC) as an arrangement of one or more protection equipment, and other devices intended to perform one or more specified protection functions [1]. The definition also states that a protection system may include protection equipment, instrument transformers [current transformers (CT) and voltage transformers (VT)], wiring, tripping circuits, auxiliary power supply, communication system, and automatic reclosing equipment, as shown in Fig. 6.1.

The protected circuits or components usually involve high voltage and current levels. The level of current can be even higher during fault events. Thus it will not be possible for the protection relays to directly measure and process the actual current and voltage magnitudes. Instrument transformers (CT and VT) are used for this purpose. CT, as shown in Fig. 6.1, are inserted in the circuit in series and measure the current level in Alternating Current (AC) circuits. CTs reduce the high current level to a lower level (mostly in the range of $0-5$ A). VT, sometimes called as potential transformers (PT), measure voltage level in AC circuits by reducing the actual voltage levels to a lower equivalent (mostly in the range of $0-120$ V). A protection system also consists of auxiliary power supply units that provide the required energy for the operation of the operation relay and possibly a communication system.

Relays are the basic component in a protection system that constitutes what is stated as the "protection equipment" in the above definition. They detect faults or other abnormal conditions in a power system and send commands to open or close a circuit under certain conditions. The first protective relays were developed in the 1830s. The relays of the time were mostly based on electromagnets. There are also other types of relays such as solid-state relays and microprocessor-based relays. Electromagnetic relays are made up of input coil that accepts a particular voltage signal and a set of

Microgrid Protection and Control. DOI: https://doi.org/10.1016/B978-0-12-821189-2.00006-1

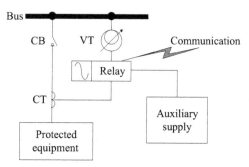

FIGURE 6.1 Components of a protection system.

contacts that rely on an armature activated by the energized coil to open or close an electrical circuit. Mechanical contacts are replaced with semiconductor output in semiconductor relays. Microprocessor-based relays use digital technology in the form of microprocessors for switching mechanism.

Though the earliest electromagnetic relays continue to be applicable until this day, solid-state and microprocessor-based relays are getting the nod recently. Solid-state relays have some advantages over electromagnetic relays such as nonexistence of moving parts, absence of sparking between contacts, energy conservation, etc. Microprocessor (or sometimes called microcomputer)-based relays offer higher precision, flexibility and celerity. Microprocessor-based relays have a more important feature in that different protection functions which would have been built as independent relays in the other types of relays can be grouped into and be encompassed in one relay. Each of those functional blocks can act independently or coordinate effectively for a more reliable protection. Such advantages and the growing complexity especially in protection of microgrids are making microprocessor-based or digital relays the better choices in microgrid applications.

Aside from the construction of the relays, the operating principles behind the relays have also evolved through time. The most popular and conventional relays are known to rely on magnitudes of current and/or voltage. The typical example is an overcurrent (OC) relay where the relay opens a circuit breaker when the current through a branch exceeds a preset current level. The other common protection relays are differential, distance, overvoltage, and undervoltage.

The fault detection methods in protection relays are basically intended to evaluate the changes in some electrical parameters and judge the occurrence of a fault in the power network. The most common phenomena in the occurrence of faults are the increase in magnitude of currents and the decrease in magnitude of voltages. Most relays are thus based on this concept and the detection and comparison of those changes. However, there are other changes in electrical parameters, such as phase angles, harmonic components, active power amount, reactive power amount, system frequency, etc. [2,3]. There are different operating principles that protection relays function with. One approach is detecting the level of a parameter such as voltage (as in undervoltage relay), current (as in OC relay), or phase angle (as in directional relays) and comparing with a threshold. The comparison can also be between two measured parameters like in a current balance relay. Differential relays also use differences of two measured currents to detect a fault in a specific protection zone or device. Other approaches such as measurement or calculation of distance, harmonic contents, frequency, rate of change, etc. can also be used in protection relays [2,3].

Protection relays may also be classified according to their timing characteristics as instantaneous, definite time (DT) lag, inverse time lag, and inverse definite minimum time (IDMT) lag relays. Their basic difference lies on their time of operation as reflected in their names.

6.2 Requirements of microgrid protection

Though the currently available protection relays are well tested and proven to be effective in the conventional power system, they may not be as effective in fulfilling the requirements of a protection system in microgrid applications. That is why there is a lot of research and development going on into the protection of microgrids. The basic requirements of a protection system in microgrids (most of which are applicable to a power system in general) are:

1. *Reliability*: made up of two concepts: *dependability* and *security*. There could be two types of incorrect operations of a protection system. Those are failure to operate while needed and operating when not needed. The two requirements correspond to the two incidents. Thus dependability refers to the ability to perform correctly when needed and security represents the ability to avoid unnecessary operation.

   The main components of a microgrid such as busbars, transformers, DERs, and cables shall be equipped with reliable primary protection such as differential protection, to implement instantaneous protection with full coverage. There shall also be a well-designed backup protection so that when the primary protection fails, the corresponding backup protection can reliably operate.

2. *Speed or celerity*: the ability to clear faults as fast as possible so that damage to the protected equipment or public can be avoided and system stability can be maintained. Instantaneous protection action with no intentional delay shall be provided for faults occurring within the coverage range (protection zone) of a protection system. The action time of a primary protection shall be short enough so that the system stability tolerance or the equipment withstand capability limits are not surpassed. However, instantaneous fault clearance may not be achievable or even necessary for the high impedance grounded or ungrounded microgrids.

3. *Selectivity*: the ability to identify and disconnect the faulty section and/or phase(s) of a power system to assure maximum service continuity with

minimum system disconnection. A selective protection system is able to avoid risk of isolating healthy parts of the power system or a microgrid which would have resulted in unnecessary outage. One way of achieving selectivity is through time delays such that protective relays operate as fast as possible for faults within their primary protection zone, but operate with some intentional delay while they are playing a backup role.

Microgrid protection shall accurately identify external faults from internal faults. When an external fault occurs, the fault should be cleared by external devices or protection at the point of connection (POC), and internal microgrid protection shall not have unwanted operation. When an internal fault occurs, the faulted branch shall be accurately identified and the fault shall be cleared by the corresponding circuit breaker with protections of nonfaulted branches not having unwanted operation.

4. *Sensitivity*: the ability to detect even the smallest faults within the concerned protection zone. Sensitivity is an important feature of protection in microgrids as the fault level may vary based on the mode of operation and configuration. There is also the possibility of confusion with nonfault disturbances whose initial characteristic may resemble that of faults. To ensure the sensitivity of protection in microgrids, the protection system may need to adjust protection settings according to operation mode, power source operation characteristic, and state. All types of faults, such as single phase-to-ground, phase-to-phase and three-phase fault, occurring within the coverage range of a microgrid protection, should be accurately identified, and the protection should trip or send alarms according to related setting principles. Another issue is the case of high impedance faults where the fault current or change in the fault detection parameter may be very small. There are different techniques being investigated or suggested to address this and other issues for the protection of microgrids.

There could also be other requirements such as *economics*, which refers to the provision of maximum protection at minimum cost, and *simplicity* of a protection system so that a minimum amount of equipment, circuitry, and sequence of operations are involved. Though these two requirements are not the most discussed in the traditional power system, they are critical in microgrids due to their relatively small size and total investment.

6.3 Differences between protection of traditional power system and microgrids

Protection of the power system is as old and well matured technology as the power system itself. However, recent developments in the sector in the form of the smart grid and its different features have led to the need for reconsidering the protection system. One of the new concepts introduced to the power system is the microgrid. The definitions and basic introduction of the microgrid are provided in the Chapter 1, The Concepts of the Microgrid and Related Terminologies.

The conventional distribution network protection systems are designed to consider radial architectures and usually are based on voltage arresters to respond to transient surges and fuses or current limiters to respond to faults. The typical protection scheme of traditional distribution networks is such that the main feeder adopts a two-stage protection composed of current quick-break protection and OC protection, and is equipped with a one-time reclosing device. The current quick-break is set according to the sensitivity of the line end fault, and the OC protection is set according to the full length of the protection line plus the time limit.

The most common radial type distribution networks, as stated earlier involve a single direction of power flow, which significantly eases the protection system design. The protection system in such cases is constituted of simple devices such as fuses, OC relays, and reclosers. A time delay-based coordination between these components is implemented to achieve the right level of selectivity and sensitivity. The coordination within fuses or relays allows for the smallest possible section of the network to be disconnected. Coordinating the fuses and reclosers makes sure that reclosers act faster and clear temporary faults so that the fuses may be blown only when the fault persists.

Radial networks have the drawback that power flows only in one way from one source and the whole downstream section will be cut from the supply of power in the event of protection action at a point. Thus ring networks are advised and implemented specially for critical loads. In ring-type networks, there will be two paths for flow of current which will ensure the possibility of power supply after protection action on a branch. This will, at the same time, imply that a fault may be fed from two directions. As such, the protection system design will not be as straightforward as in radial systems. The use of directional relays will be effective in this case, especially when the network is supplied from a single source, although there exist multiple paths. The case may be more complicated when there is more than one source that supplies power to the network. Differential protection could come as an option in this case though it is relatively costly.

The recent distribution networks are becoming more complicated in their architecture than radial or ring ones and a greater number of sources are being involved. This makes the conventional protection schemes less effective. Microgrids, which are becoming a fixture in the modern distribution network, have some peculiarities that distinguish them from the conventional interconnected gird. The special technical challenges related to protection and control of microgrids are discussed in Chapter 3, Key Technical Challenges in Protection and Control of Microgrid. The most relevant feature of microgrids that causes special considerations for microgrid protection is the fact that integration of distributed generation (DG) units changes the distribution network from a radial network with a single power source to a complex network with bidirectional power flow. This results in the need for:

- coordination between DG units and the original distribution network protection;
- assessing and addressing the influence of DGs on line reclosing; and
- monitoring islanding of the distribution system (microgrid) and devising antiislanding measures.

In order to cope with the bidirectional energy flows caused by multiple distributed generators, and to adequately clear internal faults, especially when operating in island mode, new protection schemes are necessary for microgrids. A compromise between reliability and cost imposes a series of restrictions when implementing solutions, as it must be taken into account that very complex and expensive protection systems will not be economically viable in the low-voltage (low size) environment. One of the possible solutions is the use of simple protection functions, enhanced by the use of communication system.

In general, a microgrid, which is not isolated from the grid, can operate both in parallel to the distribution network and autonomously in island mode. Thus it is essential to protect it in both modes of operation, and from both faults that occur in the upper network and those that are internal to the microgrid itself.

Protection of microgrids is more complicated than that of the conventional distribution networks. Some of the determining factors for the uniqueness of the protection of microgrids are:

- changes in the structure of distribution networks due to presence of generation in medium and low voltage levels which lead to the bidirectional energy flows;
- possibility of two modes of operation in grid-connected and island modes;
- continuous topological changes due to connection and disconnection of generators, storage devices, and loads;
- intermittency and fluctuation of some DGs, essentially the renewable ones, which depend on the availably of sunshine, wind, waves, etc.;
- limited short circuit current from DGs that are connected to the low-voltage network by means of power electronics devices (converters). (Note: These devices are not capable of providing more than $1.2-2$ times the nominal current in the event of a short circuit, compared to much higher values that can be provided by conventional synchronous generators.);
- reduced maximum allowable times for clearing faults due to the need to maintain the stability of microgrids; and
- higher level of power supply reliability demanded by critical loads requiring higher level of selectivity from the protection system.

The main objective in designing protection systems of microgrids is to maintain the security and stability of the microgrid in both grid-connected and island modes of operation. The operation of microgrids in island mode is the one that presents the most difficulties in designing the protection system of a microgrid. Advanced protection schemes are essential for adapting to the changing configurations and conditions of the microgrid. It can be said that protections are going to become an integral part of distribution automation. Problems that were previously only relevant for transmission systems are now becoming relevant for distribution systems, especially those constituting microgrids, as microgrids are usually intended to serve very critical loads that do not tolerate interruptions of a minimal time period. The bidirectionality of the power flow also requires more complex protections, which at least requires that they have directional feature. But the main workhorse in the protection of microgrids, for internal faults, is the variability of the circuit currents depending on the configuration of the moment, a particularly significant phenomenon when they are operating on an island. Basically, the ideal protection system of microgrids must respond both to faults in the distribution system and to internal faults.

6.4 Design of protection system for microgrids

6.4.1 Overcurrent protection

The IEC online dictionary defines OC protection as "a protection intended to operate when the current is in excess of a predetermined value" [1]. OC protection is critical to safety and protection from hazardous conditions that can result from extreme overloads or faults [4].

OC protection can be classified into two types, that is, phase OC protection and ground OC protection, based on the current measured from which line is considered in detecting the abnormality [5]. The relays used in such schemes may also be either directional or nondirectional [5]. The directional relays operate for events in one direction (in-front events) while the later ones operate regardless of the direction. The choice of relays depends on the structure of the microgrid. Nondirectional OC relays are mostly used in simple radial networks. As there is a single power source in such systems, the flow of fault current is only in one direction and the change in the current magnitude is a good enough indication of fault. The current magnitude in the conventional distribution networks of interconnected grids is quite large, reaching up to 20 or more times the nominal current.

Adoption of the microgrid concept and involvement of DGs resulted in two major changes to those premises that made OC protection a simple, economical, and yet effective protection in distribution networks. They are the existence of more than one power source and hence multiple directions of flow of fault currents, and the possibility of microgrids to operate in two modes of operation (island and grid-connected). The former issue may prevail in distribution networks with nonradial architecture as well. The issue may be addressed with the addition of directional features to the OC relays. It is suggested to use directional OC relays for the OC protection of microgrids [6]. The latter is the more challenging due to the significant change in the fault current magnitude in the two modes of operation.

As nondirectional OC relays are prone to false tripping due to microgrid operations and fault current feedback from the DGs in the microgrid, there have been attempts to modify them for application in microgrids [7]. As stated earlier, one of the attempts is the addition of directional components to the single relay setting protection [8]. The other latest improvement is the dual setting directional OC relay that operates more effectively for meshed microgrids or distribution networks with DGs [9,10]. These relays are furnished with two inverse time−current characteristics (TCCs) corresponding to the forward and reverse fault directions [11].

OC protection, as the other types of protection, may be implemented through either electromechanical or digital relays. Characteristics of most of the electromechanical OC relays depend on the performance of an induction disk, which makes it difficult to modify the TCC [12]. However, the modern microprocessor-based relays allow for modification of their characteristic curves for better protection coordination [13].

Based on their TCCs, we may have three types of OC relays [14]. They are instantaneous, DT, and inverse-time.

1. Instantaneous OC relays: a trip signal is produced instantly when the current amplitude exceeds the predefined threshold value. In this type of relay the only protection setting to be adjusted is the pickup current.
2. DT OC relays: a trip signal is produced after a specified time delay when the current amplitude exceeds the predefined threshold value. Such OC relays have two parameters that can be adjusted: the time delay and the pickup current.
3. Inverse-time OC relays: a trip signal is produced after the operating time (*T*) defined by IEEE C37.112 [15] as in Eq. (6.1) when the current amplitude exceeds the predefined threshold value.

$$T = \frac{TDS}{7} \left(\frac{\beta}{\left(I_m/I_p\right)^{\alpha} - 1} + \gamma \right) \tag{6.1}$$

where *TDS* is the relay's time delay setting; I_m is the measured current; I_p is the pick-up current of the relay; and β, α, and γ represent the relay characteristics. These parameters decide whether the relay is the moderately inverse, very inverse (VI), or extremely inverse (EI) type [15].

IEC 60255 defines a number of inverse time OC relays as standard inverse (SI), VI, EI, and DT according to the tripping time parameters (a and b) in (6.2).

$$T = TMS \left(\frac{a}{(I_m/I_p)^b - 1} \right) \tag{6.2}$$

where TMS is the time multiplier setting of the relay.

The main parameters, collectively called protection settings, that decide the operating times of a digital directional OC relay are the time multiplier-setting, the plug-setting, and curve-setting [6,16]. The selection of appropriate values for those parameters is how a coordinated protection system is achieved. The settings can be adjusted adaptively or nonadaptively. In order to address the issue of low fault currents in island mode, directional OC relays with extreme inverse characteristics are advised for the fact that they can operate quickly even for lower fault current values [6].

A protection scheme in general or a coordinated OC protection for microgrids must ensure safe operation in both grid-connected and the islanded modes of operation. The protection of the microgrid in grid-connected mode will not be meaningfully different from that of the conventional distribution network as there will be large enough fault currents due to the contribution of the host grid. This allows the employment of the conventional OC relays, though the protection coordination may be compromised or even entirely lost in some cases, due to the existence of DGs. Yet again, microgrids are supposed to operate not only in parallel with the grid, but also in island mode too. The fault currents in island mode may be significantly smaller than those of the grid-connected mode. It could be worse when the microgrid is dominated by converter-based DGs whose fault current supplying capacity is limited compared to rotating-machine-based DGs. This consequently makes conventional OC protection ineffective for microgrids operating in island mode.

OC relays are reported to be applied for protection of microgrid in different studies. A coordination of directional OC relays is investigated in [17] based on an IEC benchmark microgrid operating in different modes with a genetic algorithm applied to optimize the limits of the maximum plug setting multiplier for the OC relays coordination. Numerical IDMT OC relay is adopted in [18] for the protection of a converter-based microgrid operating in island mode.

6.4.1.1 Coordination of overcurrent protection

There are different methods for the coordination of relays in a protection system. The coordination may be achieved through discrimination based on time-grading or current-grading, or a combination of the two [19]. The basic principle in time-grading-based coordination is adjusting the time delay settings of the relays in such a way that the circuit breaker closest to location of the fault shall act first and others follow chronologically in the order of their distance to the fault current location. This is mostly achieved in a radial system with a single power source through a stepwise increment of the definite-time delay values of individual relays from the far end of the network toward the source. However, this would mean faults occurring close to the source, which are coincidently the more severe ones, being cleared after a longer time delay. On the other hand, current-grading-based coordination discriminates faults and decides the operating time of the individual breakers based on the level of fault current sensed by the respective relays. That means the relay sensing the highest fault current will operate first and others follow based on the magnitude of the fault current they sense. This type of coordination again has the disadvantage that a meaningful discrimination can be achieved only when there is significant impedance between the subsequent protection devices. It also assumes a nearly constant source fault level. Thus both techniques are not the best propositions, especially in microgrid protection where the length of lines is usually short (not large enough impedance between protection devices for proper discrimination) and the presence of variable output DG units affects the source fault level (not constant).

OC relays are one of the common elements, together with islanding detectors and reclosers, used in the protection of the point of common coupling (PCC) where microgrids are connected to the utility grid. Coordinated OC protection can also be used for the protection of cables in microgrids. However, the main issue causing difficulty in the coordination of relays is the existence of DGs. The procedure followed in designing a coordinated protection system for microgrids and the effect of DGs are illustrated in the following examples.

Example 6.1: Calculation example

Designing a coordinated OC protection for microgrid shown in Fig. 6.2 with components described in Table 6.1 and assuming line impedance of $(0.64 + j\,0.12)$ Ω/km for all cables.

1. when the DGs are not connected
2. when the DGs are connected

The source impedance can be considered neglected and the distribution transformer impedance can be assumed to be 5% with an X/R ratio of 8 (i.e., $Z_x = 0.0062 + j\,0.0496$ p.u.).

Solution:

Step 1: Calculating impedances of the connecting cables

The line impedences for each of the cables connecting the busbars and the utility grid to the PCC (line 1) are calculated from the cable parameters and the length of respective cables, as shown in Fig. 6.3. The impedance values are converted to per unit values using the base impedance Z_b, which is

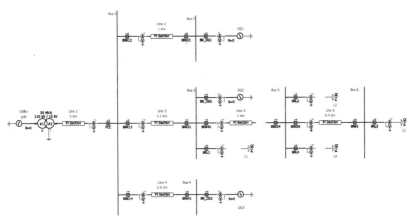

FIGURE 6.2 Microgrid test system for coordination of OC protection.

TABLE 6.1 Components of test microgrid.

Load/DG	P (MW)	Q (MVAr)
L1	2.0	0.4
L2	1.0	0.2
L3	0.5	0.1
L4	0.5	0.1
DG1	5.0	0
DG2	2.5	0
DG3	20	0

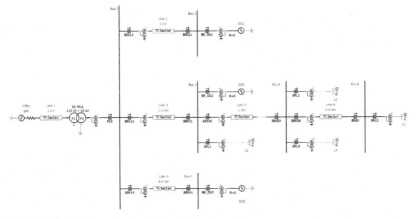

FIGURE 6.3 Line impedance of cables in the microgrid.

calculated as below for the secondary side of the main transformer connecting the microgrid to the utility grid.

$$Z_b = \frac{(10 \text{ kV})^2}{50 \text{ MVA}} = 2\Omega$$

Step 2: Fault level calculations

The next step is to calculate the fault current level at each of the protection zones corresponding to the busbars.

At Bus 1:

$$I_{f1} = \frac{S_b}{Z_{eq}} = \frac{50 \text{ MVA}}{Z_X + Z_{L1}} = 50.35 \text{ MVA} = 2.91 \text{ kA}$$

At Bus 2:

$$I_{f2} = \frac{S_b}{Z_{eq}} = \frac{50 \text{ MVA}}{Z_X + Z_{L1} + Z_{L2}} = 37.92 \text{ MVA} = 2.19 \text{ kA}$$

At Bus 3:

$$I_{f3} = \frac{S_b}{Z_{eq}} = \frac{50 \text{ MVA}}{Z_X + Z_{L1} + Z_{L3}} = 36.14 \text{ MVA} = 2.09 \text{ kA}$$

At Bus 4:

$$I_{f4} = \frac{S_b}{Z_{eq}} = \frac{50 \text{ MVA}}{Z_X + Z_{L1} + Z_{L4}} = 42.08 \text{ MVA} = 2.43 \text{ kA}$$

At Bus 5:

$$I_{f5} = \frac{S_b}{Z_{eq}} = \frac{50 \text{ MVA}}{Z_X + Z_{L1} + Z_{L3} + Z_{L5}} = 24.58 \text{ MVA} = 1.42 \text{ kA}$$

At Bus 6:

$$I_{f6} = \frac{S_b}{Z_{eq}} = \frac{50 \text{ MVA}}{Z_X + Z_{L1} + Z_{L3} + Z_{L5} + Z_{L6}} = 18.95 \text{ MVA} = 1.09 \text{ kA}$$

Step 3: Relay coordination

In order to show the straightforward customary procedure in coordinating relays, we first assume the DGs in the microgrid are not connected. This ultimately transforms the microgrid to a simple radial distribution network of Fig. 6.4.

1. *When the DGs are not connected (radial distribution network)*

Coordinating a protection system involves adjusting the two basic settings of time OC relays (ANSI code 51): time multiplier setting (*TMS*) and pickup current (I_p). For the radial section of the microgrid (Fig. 6.4), the respective setting values for each of the relays are calculated as follows.

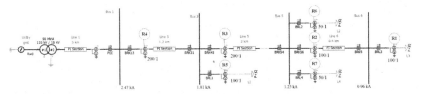

FIGURE 6.4   Radial distribution network with no DG.

OC relays with VI characteristic are considered in this example. Thus the relation between the *TMS* and operating time for the respective relays is defined by Eq. (6.2) where the constants *a* and *b* assume values of 13.5 and 1, respectively, according to IEC 60255. Eq. (6.2) will be modified to (6.3) and Eq. (6.3) can be rearranged to (6.4).

$$T = TMS\left(\frac{13.5}{PSM - 1}\right) \tag{6.3}$$

where $PSM = I_m/I_p$ is plug setting multiplier of a relay.

$$TMS = T\left(\frac{PSM - 1}{13.5}\right) \tag{6.4}$$

Now, we start calculating the setting values backwards from the relay at the far end to the one close to the source.

R1, R5, R6, and R7 (peripheral relays):

The relays R1, R5, R6, and R7 are located at the ends (periphery) of the different lines of the radial network. Hence, their operating time can be arbitrarily set at 0.1 second (considering no intentional delay as the relays serve no backup function). We will then continue calculating the *TMS* of the relays starting from R1 (which constitutes the longest possible fault path).

We assume the pickup current of the relays to be 150% of the full load current for each branch (it can be calculated from the rated load). CT selection is not addressed in this example and we simply take the CT ratio provided in the figure and we need to later check if the CT secondary current did not exceed 100 times the rated secondary current during faults. The fault current and pickup values as observed by the CT secondary (with CT ratio of 100/1) are respectively calculated as:

$$I_{f6,\ sec} = \frac{1.09\ \text{kA} \times 1}{100} = 5.47\ \text{A and } I_{P,\ sec} = \frac{1.5 \times 0.029\ \text{kA} \times 1}{100} = 0.44\ A$$

Thus $PSM = 5.47/0.44 = 12.39$ and

$$TMS_{R1} = 0.1\left(\frac{12.39 - 1}{13.5}\right) = 0.084$$

Such a *TMS* value may not be exactly available in electromechanical relays and may need to be slightly adjusted to the nearest value while digital relays allow for continuous adjustment of setting values. We continue with the value as the basic intention is to elaborate the procedure. *TMS* values of the relays R5, R6, and R7 is presented in Table 6.2.

R2:

As relay R2 shall have the function of providing backup to R1 and in order to achieve selectivity, we will apply a time interval of 0.5 second from the operating time of R1. That is considering circuit breaker delays and the relay processing time with some precautionary margin. Thus the operating time of R2 will be:

$$T_2 = T_1 + 0.5 \; second = 0.6 \; second$$

While *PSM* of R2 will be 7.09/0.44 = 16.07 and

$$TMS_{R2} = 0.6 \left(\frac{16.07 - 1}{13.5} \right) = 0.67$$

The procedure will continue in a similar fashion and the final results are summarized in Table 6.2.

2. *when the DGs are connected*

Now comes the challenge due to insertion of the DGs which will contribute to the fault currents, thus changing the fault level at the buses. Thus we need to recalculate the fault currents at the busbars. This time, the single line diagram changes back to that of Fig. 6.3. Taking an example where a fault occurs at Bus 6, in addition to the utility grid, each of the DGs will supply the fault. Fig. 6.5 shows how DG1 contributes to the fault at Bus 6. The other DGs have similar impacts. Thus when calculating the fault level at each point (or bus), we shall add the individual contribution of the DGs to the fault current with that of the utility grid (Table 6.3):

TABLE 6.2 Calculation result for relay settings.

Relay	CT ratio	T	$I_{f, \, sec}$	$I_{p, \, sec}$	PSM	TMS
R1	100/1	0.10	5.47	0.44	12.39	0.08
R2	100/1	0.60	7.09	0.44	16.07	0.67
R3	200/1	1.10	10.43	1.77	5.91	0.40
R4	200/1	1.60	14.53	1.77	8.23	0.86
R5	100/1	0.10	10.43	1.77	5.91	0.04
R6	50/1	0.10	7.09	1.77	4.02	0.02
R7	50/1	0.10	7.09	0.88	8.03	0.05

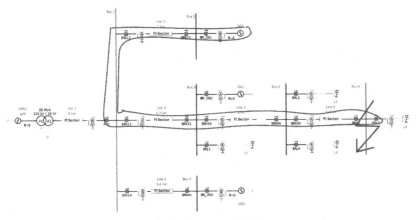

FIGURE 6.5 DG1 contributing to the fault current at Bus 6.

TABLE 6.3 Fault current calculation summary.

	I_f (kA)	I_{DG1} (kA)	I_{DG2} (kA)	I_{DG3} (kA)	I'_f (kA)
Bus 1	2.91	0.89	0.37	5.91	7.17
Bus 2	2.19	0.00	0.11	2.22	2.33
Bus 3	2.09	0.40	0.00	1.97	2.37
Bus 4	2.43	0.78	0.25	0.00	1.03
Bus 5	1.42	0.21	0.22	0.93	1.36
Bus 6	1.09	0.19	0.18	0.84	1.21

For Bus 1:
At Bus 1:

$$I'_{f1} = I_{f1} + I^1_{DG1} + I^1_{DG2} + I^1_{DG3} = I_{f1} + \frac{S_{DG1}}{Z_{L2}} + \frac{S_{DG2}}{Z_{L3}} + \frac{S_{DG3}}{Z_{L4}}$$

At Bus 2:

$$I'_{f2} = I_{f2} + I^2_{DG1} + I^2_{DG2} + I^2_{DG3} = I_{f2} + 0 + \frac{S_{DG2}}{Z_{L2} + Z_{L3}} + \frac{S_{DG3}}{Z_{L2} + Z_{L4}}$$

At Bus 3:

$$I'_{f3} = I_{f3} + I^3_{DG1} + I^3_{DG2} + I^3_{DG3} = I_{f3} + \frac{S_{DG1}}{Z_{L2} + Z_{L3}} + 0 + \frac{S_{DG3}}{Z_{L3} + Z_{L4}}$$

At Bus 4:

$$I'_{f4} = I_{f4} + I^4_{DG1} + I^4_{DG2} + I^4_{DG3} = I_{f4} + \frac{S_{DG1}}{Z_{L2} + Z_{L4}} + \frac{S_{DG2}}{Z_{L3} + Z_{L4}} + 0$$

At Bus 5:

$$I'_{f5} = I_{f5} + I^5_{DG1} + I^5_{DG2} + I^5_{DG3} = I_{f5} + \frac{S_{DG1}}{Z_{L2} + Z_{L3} + Z_{L5}} + \frac{S_{DG2}}{Z_{L5}} + \frac{S_{DG3}}{Z_{L3} + Z_{L4} + Z_{L5}}$$

At Bus 6:

$$I'_{f6} = I_{f6} + I^6_{DG1} + I^6_{DG2} + I^6_{DG3} = I_{f6} + \frac{S_{DG1}}{Z_{L2} + Z_{L3} + Z_{L5} + Z_{L6}}$$
$$+ \frac{S_{DG2}}{Z_{L5} + Z_{L6}} + \frac{S_{DG3}}{Z_{L3} + Z_{L4} + Z_{L5} + Z_{L6}}$$

Based on those new fault levels, the setting values of the relays are calculated and compared to the old values in (1) as presented in Table 6.4.

As the summarized result in Table 6.4 shows, the insertion of the DGs to the distribution network results in the need for readjustment of relay settings. In order to keep the time-grading-based discrimination between the relays effective and hence allow selective operation of the protection system, the setting of the relays (specifically the time multiplier settings) needs to be revised. However, if the *TMS* of the relays stays the same as in case (1) (Table 6.2), the operating times of the relays based on the new fault levels will be as shown in the last column of Table 6.4. Looking thoroughly at the table, we can notice that the coordination is not completely lost except that of R4. In case of a fault occurring in the region covered by R4, all the DGs will contribute to the fault and the fault level will be significantly higher. As a result, the operating time of R4 (0.6 second) has become shorter than that of R3 (0.944 second). This can be translated as relay R4 acting before R3 for a fault on line 5 which will result in unnecessary power outage for a section of the microgrid between the two relays. Therefore the relay coordination is

TABLE 6.4 Impact of DGs on OC relay coordination.

Relay	Unchanged		New				Old	
	CT ratio	$I_{p,\ sec}$	T	$I_{f,\ sec}$	PSM	TMS	TMS	T
R1	100/1	0.44	0.10	12.15	27.51	0.20	0.084	0.043
R2	100/1	0.44	0.60	13.64	30.90	1.33	0.670	0.302
R3	200/1	1.77	1.10	11.87	6.72	0.47	0.400	0.944
R4	200/1	1.77	1.60	35.84	20.29	2.29	0.857	0.600
R5	100/1	1.77	0.10	23.73	13.44	0.09	0.036	0.039
R6	50/1	1.77	0.10	27.29	15.45	0.11	0.022	0.021
R7	50/1	0.88	0.10	27.29	30.90	0.22	0.052	0.024

likely to hold if the fault injection by the DGs is small as previously outlined by [20].

Example 6.2: Simulation example using ETAP

The protection coordination and impact of DGs in microgrid protection is simulated using ETAP based on a model microgrid with similar architecture to that of Example 6.1.

1. *Microgrid where DGs are not connected* (Fig. 6.6)

Step 1: Load flow for calculation of full load current (Fig. 6.7)

Step 2: Short circuit test on each bus for calculation of fault level at each bus (Fig. 6.8)

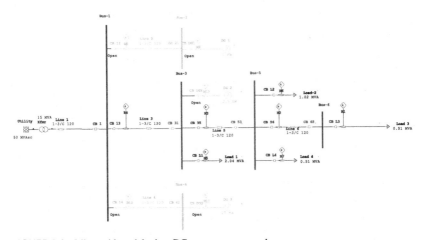

FIGURE 6.6 Microgrid model when DGs are not connected.

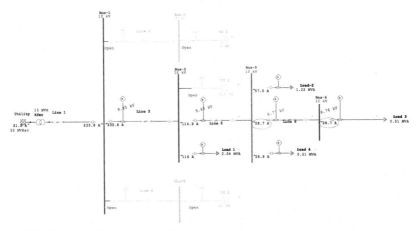

FIGURE 6.7 Load flow analysis result.

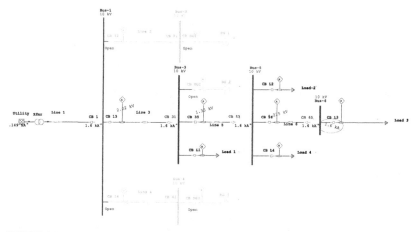

FIGURE 6.8 Short circuit test on each Bus 6.

The fault levels at the other buses are shown in Table 6.5.
Step 3: Calculation of relays' operating time and *TMS*
Step 4: Checking protection coordination (Figs. 6.9 and 6.10)

2. Microgrid with DGs connected

Step 1: Load flow for calculation of full load current
The full load current through the branches stays the same as in (1).
Step 2: Short circuit test on each bus for calculation of fault level at each bus (Fig. 6.11)
The fault levels at the buses are changed from those of (1) and they are shown in Table 6.6.
Step 3: Calculation of relays' operating time and *TMS*
Step 4: Checking protection coordination
If the setting values of the relays were left unchanged as in case (1), the relay coordination will be lost as shown in Figs. 6.12 and 6.13. Hence, we need to update the relay settings (especially the *TMS* values according to Table 6.4) and the coordination will be achieved.

6.4.2 Differential protection

Differential protection is one of the most widely used methods for power system protection and it is also the most common component of protection systems of microgrids. The basic principle of differential protection is the comparison of the difference between measurements of currents at different points to a preset threshold. The basic and simple principle of differential protection is shown in Fig. 6.14. Differential protection is mostly applied to protect busbars, transformers, or transmission lines. The protection scheme depends on the communication of measurements from two points (the input and output) of the protected

TABLE 6.5 Calculation results.

	T	CT ratio	I_{fault}	I_{FL}	I_p	I_{fs}	I_{ps}	PSM	TMS
R1	0.10	100/1	1600	28.7	43.05	16.00	0.43	37.17	0.27
R2	0.60	100/1	1630	28.7	43.05	16.30	0.43	37.86	1.64
R3	1.10	200/1	1795	114.9	172.35	8.98	0.86	10.41	0.77
R4	1.60	200/1	1905	230.8	346.2	9.53	1.73	5.50	0.53
R5	0.10	100/1	1795	116	174	17.95	1.74	10.32	0.07
R6	0.10	50/1	1630	57.5	86.25	32.60	1.73	18.90	0.13
R7	0.10	50/1	1630	28.8	43.2	32.60	0.86	37.73	0.27

Time ...	ID	If (kA)	T1 (ms)	T2 (ms)	Condition
136	R1	1.644	136		Phase - OC1 - 51
206	CB L3		70.0		Tripped by R1 Phase - OC1 - 51
323	R2	1.644	323		Phase - OC1 - 51
378	CB 65		55.0		Tripped by R2 Phase - OC1 - 51
393	CB 56		70.0		Tripped by R2 Phase - OC1 - 51
545	R3	1.644	545		Phase - OC1 - 51
600	CB 53		55.0		Tripped by R3 Phase - OC1 - 51
615	CB 35		70.0		Tripped by R3 Phase - OC1 - 51
742	R4	1.644	742		Phase - OC1 - 51
797	CB 31		55.0		Tripped by R4 Phase - OC1 - 51
812	CB 13		70.0		Tripped by R4 Phase - OC1 - 51

FIGURE 6.9 Protective device sequence-of-operation.

device or zone through means such as pilot wire, fiber optics, power line carrier, or wireless communication [14]. During normal operation or faults outside the protected zone, the sum of the currents flowing shall be essentially equal to the sum of the currents flowing out.

The relay is connected differentially, as shown in Fig. 6.14. Hence, during normal operation and external faults, the current flowing through the relay is the difference of the secondary currents from the CTs at the two ends. The current flowing through the protective relay is the difference in the exciting currents of the differentially connected CT. With I_p being the primary current entering or leaving the protected zone divided by the transformer ratio of the CT and I_e being the secondary exciting current of the CT, the secondary current $I_p - I_e$ from the two sides will be similar except for the unavoidable slight difference even with exactly the same ratio and type of CTs. Thus during normal operation there will be small nonzero current flowing through the relay due to the losses within the protected zone and the small differences between the two CTs. This relay current may be even larger when there is CT saturation or when the two CTs are of different type or transformer ratios. Hence, the pickup current setting shall consider such conditions to avoid unwanted operation of the relay during normal operation or external faults.

External faults may cause a sudden increase in current and DC offset current that could lead to significant differences in the secondary currents from the two sides, which means a considerable amount of current flowing through the differential relay. This incident is usually for a short period of time during the fault inception. Thus applying some time delay to the relay operation can help to avoid unwanted operation during external faults.

During faults occurring within the protected zone or equipment, which we call internal faults, the differential current flowing through the relay is

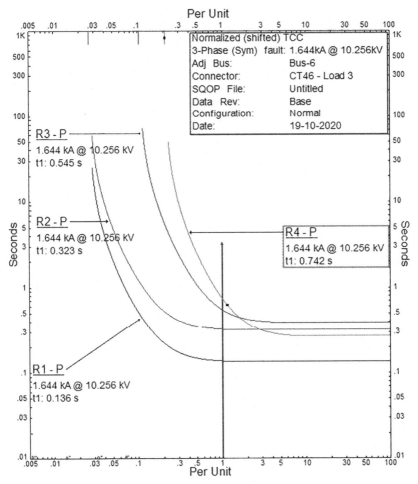

FIGURE 6.10 Normalized time−current characteristic (TCC) curves of coordinated relays.

essentially the sum of the two secondary currents as both primary currents will be flowing into the protected zone feeding the fault. Thus except in the case of high impedance faults where the fault current is low, good discrimination is achieved for the detection of faults within the protected zone. Relays operating based on percentage values rather than actual current magnitude can provide higher sensitivity toward internal faults and higher restraint for external faults. As a result, most differential relays are of percentage type with either a fixed percentage threshold of 10%−50% or with some taps to change the threshold percentage. Though it seems as simple as the comparison of two current vectors, especially in simple two-terminal differential protection systems, adjustment of the protection

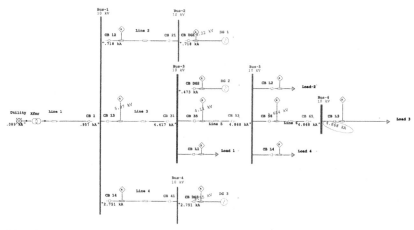

FIGURE 6.11 Short circuit current at Bus 6.

settings may be challenging for multiterminal components where a protection zone is defined by several inputs and there are continuous changes in configuration.

Differential protection has advantages, such as good selectivity, immunity to power swings and external faults, ability to operate without prior knowledge of the fault levels and convenience for network upgrading with connection of more components, sensitivity including those of high impedance faults, and simplicity [14,21]. One issue with implementation of this protection scheme is the need for duplicate measurement or instrumentation devices and the communication media which in turn result in higher cost especially for their application in smaller systems like microgrids. Differential protection is said to be more applicable if both ends of the protected device or zone are close to each other [22]. However, with the recent developments in communication technology, the scheme has become less costly.

Differential protection is widely used for the protection of single components such as transformers and busbars. In the protection of transformers, differential protection can provide the best overall protection for phase and ground faults, except in ungrounded or high impedance grounded systems. Differential protection is the common choice for the protection of transformers or transformer banks with sizes in the Mega Volt Ampere (MVA) range. However, it may also be employed in lower capacity transformers based on the importance of the transformer in the system as in microgrids.

Differential protection should also be considered and applied wherever possible for all busbars in a microgrid as primary protection. Although busbars are not frequently threatened by faults, the impact is very significant when faults happen to them. That is why fast and sensitive protection is needed for the protection of busbars. Differential protection is the usual choice for the primary protection of busbars in microgrids while protection

TABLE 6.6 Calculation results when DGs are connected.

	T	CT ratio	I_f	I_FL	I_p	I_{fs}	I_{ps}	PSM	TMS
R1	0.10	100/1	4868	28.7	43.05	48.68	0.43	113.08	0.83
R2	0.60	100/1	5193	28.7	43.05	51.93	0.43	120.63	5.32
R3	1.10	200/1	6881	114.9	172.35	34.41	0.86	39.92	3.17
R4	1.60	200/1	1905	230.8	346.2	9.53	1.73	5.50	0.53
R5	0.10	100/1	6881	116	174	68.81	1.74	39.55	0.29
R6	0.10	50/1	5193	57.5	86.25	103.86	1.73	60.21	0.44
R7	0.10	50/1	5193	28.8	43.2	103.86	0.86	120.21	0.88

Time ...	ID	If (kA)	T1 (ms)	T2 (ms)	Condition
136	R1	5.226	136		Phase - OC1 - 51
206	CB L3		70.0		Tripped by R1 Phase - OC1 - E
316	R4	4.746	316		Phase - OC1 - 51
322	R2	5.226	322		Phase - OC1 - 51
371	CB 31		55.0		Tripped by R4 Phase - OC1 - E
377	CB 65		55.0		Tripped by R2 Phase - OC1 - E
386	CB 13		70.0		Tripped by R4 Phase - OC1 - E
392	CB 56		70.0		Tripped by R2 Phase - OC1 - E
394	R3	5.226	394		Phase - OC1 - 51
449	CB 53		55.0		Tripped by R3 Phase - OC1 - E
464	CB 35		70.0		Tripped by R3 Phase - OC1 - E

FIGURE 6.12 Protective device sequence-of-operation.

of the circuits connected to the bus generally provides backup protection for the busbar (Fig. 6.15).

Differential protection is applied for protection of feeders in microgrids in [23]. A current differential protection is said to be chosen for being operable under conditions of bidirectional power flow, changing fault current levels, and the plug-and-play of DGs, while it can function well for both grid-connected and islanded modes. A relay composed of five elements (three phase elements for each phase, a negative sequence current element, and zero sequence current element) is used in the study. While the phase differential elements are intended to provide protection for faults causing significantly large fault currents the negative and zero sequence elements will provide more sensitive earth fault protection for unbalanced faults of small fault current level such as high impedance ground faults in a feeder.

Differential protection is discussed in [23] and other more studies as suitable for protection of feeders where the fault current could be of varying magnitude with DT and IDMT characteristics with a current−time characteristic curve shown in Fig. 6.16. As it can be seen from the figure, the characteristic curve of the relay has two slopes for the low fault current stage while the higher fault current stage is a horizontal line (constant). The two slopes for the low current range are defined by the percentage bias settings K_1 and K_2. This dual slope characteristic ensures higher sensitivity during low fault currents and improved security for large fault currents where CTs could be subject to saturation. A trip command will be issued by a relay when one of the conditions in (6.5) or (6.6) is satisfied.

$$I_b < I_{b1} \text{ and } I_{dif} > K_1 \cdot I_b + I_{dif1} \tag{6.5}$$

$$I_b > = I_{b1} \text{ and } I_{dif} > K_2 I_b - (K_2 - K_1)I_{b1} + I_{dif1} \tag{6.6}$$

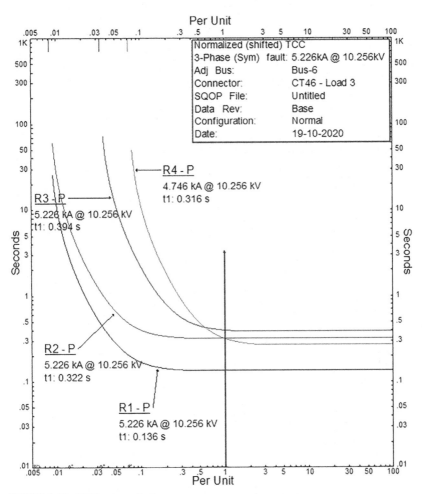

FIGURE 6.13 TCC curves of relays.

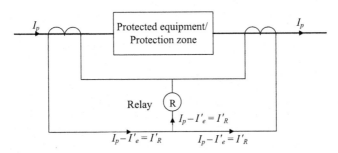

FIGURE 6.14 Principle of basic differential protection.

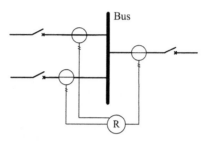

FIGURE 6.15 Differential bus protection.

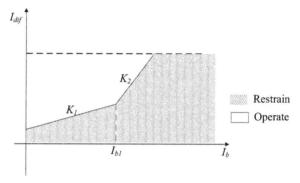

FIGURE 6.16 Characteristic curve of dual slope differential relay.

where I_b and I_{dif} are the bias and differential currents defined by Eqs. (6.7) and (6.8), respectively.

$$I_{dif} = |I_1 + I_2| \qquad (6.7)$$

$$I_b = \frac{|I_1| + |I_2|}{2} \qquad (6.8)$$

The setting of differential relays (such as the pickup currents I_{dif1}, I_{dif2}, and I_{b1}, and the slopes K_1 and K_2) should be defined by detailed calculations based on the feeder and system parameters.

Aside from its cost, the other drawback from application of differential protection in the protection of microgrid feeders is related to the communication channel required to transfer the current information between the far and local end relays. Based on the type of protection media chosen, there could be a varying level of risk of communication failure and time delay that can be seen as a phase shift between the two currents. The latter issue can be addressed with proper time synchronization of current phasors during calculations, while the former issue requires the need for a backup protection that does not depend on the same communication network.

6.4.3   Distance protection

We have earlier seen how complicated and challenging it could be to ensure coordination between OC relays in the protection of microgrids based on fault current magnitude. Another option that has been implemented to the traditional power system for a long time and yet has been found to be useful for microgrids is distance protection that functions based on calculation and comparison of fault impedance values seen by the relays.

The basic operating principle of distance protection relies on the increase in current and decrease in voltage as a result of faults so that the impedance at the fundamental frequency between the fault point and the relay location, which is calculated from the voltage and current at the relay point, changes accordingly. This calculated impedance value is used to identify whether a fault has occurred or not within a protection zone. The impedance at the relay point is calculated by dividing the voltage measured at the point by the current flowing through the protected branch and will be essentially equal with the load impedance under normal operation. However, when a fault happens somewhere in the protection zone the calculated impedance will drop to be close to the line impedance. Thus comparing the calculated impedance to setting values is the way a distance protection detects, locates, and isolates a fault. Digital relays allow computation of impedance from measured voltage and current values. The electromechanical type distance relays perform the impedance calculation through interaction of their two coils, one of which is energized by voltage and the other by the current that produces a positive or pickup torque and a negative or reset torque respectively.

Distance relays can have different principal types based on their internal characteristics, such as impedance, reactance, admittance (mho), ohm, and offset mho type relays. The relay characteristics are obtained by an amplitude comparator or phase comparator as shown in Fig. 6.17.

Mho type relays are the basic and most commonly applied element used in the industrial distance protection. As shown in Fig. 6.17B its characteristic curve passes through the origin of the $R-X$ plane which makes the relay directional. Coordination of such relays can be achieved as shown in Fig. 6.17 with each protection zone associated with a respective circular characteristic curve. The decision on the occurrence of a fault in a protected zone is decided based on a comparison of calculated Z to the setting impedance Z_{set} such that $Z < Z_{set,i}$ implies a fault in zone i, while $Z > Z_{set,i}$ implies either the fault is outside zone i or there is no fault at all. The rule of thumb procedure to decide the setting values for Mho distance relay for a three-zone system of Fig. 6.18 is:

$$Z_{set,1} = K_1 Z_L \qquad (6.9)$$

$$Z_{set,2} = K_2(Z_L + Z_{set,1}) \qquad (6.10)$$

$$Z_{set,3} = K_3(Z_L + Z_{set,2}) \qquad (6.11)$$

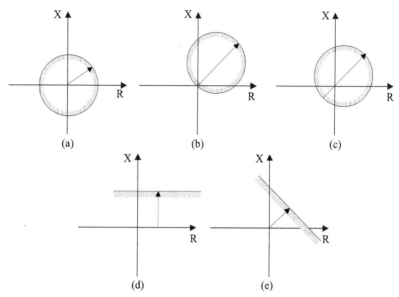

FIGURE 6.17 Characteristic (R−X) curves of different types of distance relays (A) impedance relay (B) mho relay (C) Offset mho relay (D) reactance relay (E) Ohm relay.

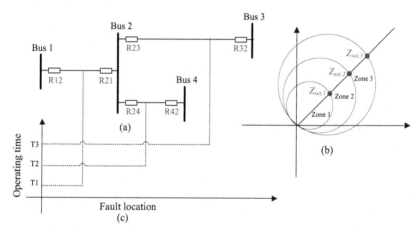

FIGURE 6.18 Mho distance protection for (A) a four bus system with (B) three zones of protection and (C) corresponding operating time of relay.

where $Z_{set,i}$, is the setting impedance for protected zone i, K_i is the reliability coefficient of zone i (usually assumes a value between 0.8 and 0.85), and Z_L is impedance of the primary protected line. If the fault occurs in the section between Bus 1 and Bus 2, the relay R12 will operate instantaneously (without an intentional time delay). However, there will be an intrinsic delay of T1 as shown in Fig. 6.18C. The time difference between T2 and T1, and T3 and T2 are the coordination time intervals intended to separate the primary protection of a protection zone from backup protections so that the primary protection relay will be given the first opportunity to clear the fault in the primary protection zone. As shown in equation (6.7), the instantaneous primary protection ($Z_{set,1}$) is set below 100% line impedance (usually 80% or 85% as decided by the reliability coefficient) which implies zone 1 is set to cover 80% or 85% of the primary line length. That is to avoid selectivity problems due to errors (reaching up to 15% or 20%) in measurements or calculations of fault impedance.

Fig. 6.18 and the above discussion assume the network to have a radial structure where the power source is connected to Bus 1 and power flows only toward the left. However, today's distribution network, especially microgrids, are characteristically composed of power sources distributed across the different buses in the network. Such a scenario would bring about an issue called the *infeed current influence* or the *infeed effect* on the impedance measurement or calculation and as a result the operation of distance relays. The microgrid shown in Fig. 6.19 demonstrates this situation. DG1, which is now connected to Bus 2, will feed the fault current together with the grid when a fault occurs on Line23. This infeed will cause an equivalent increase in the impedance seen by the relay R23. The larger sensed impedance by relay R23 means the fault will be misinterpreted as occurring in the subsequent zones while it actually occurred within zone 1. This creates selectivity issue.

To offset the infeed current effect due to DG1, a branch coefficient K_b, defined by Eq. (6.12), can be adopted as suggested in [24] for the setting impedance of the relay. This leads to the settings for the relay for zone 2 and 3 modified as Eqs. (6.13) and (6.14) while $Z_{set,1}$ remains unchanged.

$$K_b = \frac{I_R + I_F}{I_R} \qquad (6.12)$$

where I_R and I_F are the current measured by the relay and current fed by DG on the other side of the line, respectively.

$$Z_{set,2} = K_2(Z_L + K_b Z_{set,1}) \qquad (6.13)$$

$$Z_{set,3} = K_3(Z_L + K_b Z_{set,2}) \qquad (6.14)$$

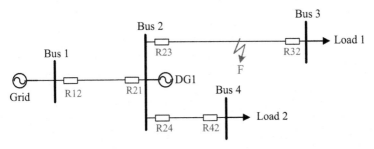

FIGURE 6.19 Infeed effect of DG on distance protection for microgrid.

The most common factors that affect the proper impedance measurement and hence operation of distance protection are:

- Power system oscillation
- Existence of a branch circuit between the protection unit and the fault point
- Short-circuit resistance
- Measurement (CT or VT) error
- Secondary circuit of VT being disconnected
- Insertion of compensation capacitor in series

Out of those factors, the involvement of multiple DGs, change of operating modes, and complexity in configuration in microgrids mean there is a greater possibility of the first two issues.

The practical faults happening in microgrids are generally nonmetallic, which implies that the fault resistance causes a change to the measured impedance by the relay. This would lead to the protection range being shortened and hence faults left undetected or misoperation in the opposite direction. The impact of fault resistance is linked to the fault point location and the relay setting value. The closer the short-circuit point is to the protection installation location, the greater the impact, and vice versa. The smaller the setting value of the protection device, the greater the relative impact of the fault resistance.

The suggested techniques to address the impact of fault resistance are:

- using directional impedance relays with a larger area occupied by the action characteristic curve in the + R axis direction, and
- using an instantaneous measuring device in order to fix the initial operating state of the distance element through the action of a starting element.

The other issue relatively prevalent in microgrids is oscillations, during which the phase angle difference between the potentials of the generators in the system changes periodically with time. In such cases the voltage, line current, and the measured impedance of the distance protection at each point in the system will also change periodically, which may lead to malfunctioning of distance protection.

Distance protection has the economic and reliability advantages that no communication network is involved and the operation depends on measurements at the relay location. This scheme is principally used at high voltage levels in the conventional power system. However, there have been attempts to apply it in the protection of microgrids with converter-based DGs as reported in [25].

Distance protection can be used for the protection of phase faults and ground faults. It is also independent of changes in the fault current magnitude, and hence not much affected by changes in generation capacity and system configuration. This makes the scheme suitable for application in microgrids.

Though distance protection may have the mentioned and other advantages, it has been reported to have issues with the impedance seen by the relays being affected by the fault current limiting nature of the converted-based DGs. Another issue about the scheme is a possible overreach problem in case of squirrel cage induction machine-based DGs that may absorb reactive power at times, causing the line current to lead the voltage. The fact that conventional distance relays assume zero fault impedance while measuring or calculating the impedance also leads to high impedance faults being sensed as out of the actual protection zones or completely unrecognized [14]. Applicability of distance protection for a radial microgrid in Aalborg, Denmark is investigated in [24] with satisfactory performance reported.

6.4.4   Voltage-based protection

A voltage-based scheme of protection based on the conversion of the measured three-phase voltage signal from the abc frame to dq frame is proposed in [26]. The dq stationary frame representation is reported to have the advantage that the line frequency components can be converted to dc values which can be digitally filtered and then processed by a dual hysteresis comparator for fault detection.

$$\begin{bmatrix} U_\alpha \\ U_\beta \\ U_0 \end{bmatrix} = \frac{2}{3} \begin{bmatrix} 1 & -1/2 & -1/2 \\ 0 & -\sqrt{3}/2 & \sqrt{3}/2 \\ 1/2 & 1/2 & 1/2 \end{bmatrix} \begin{bmatrix} U_a \\ U_b \\ U_c \end{bmatrix} \tag{6.15}$$

$$\begin{bmatrix} U_d \\ U_q \end{bmatrix} = \frac{2}{3} \begin{bmatrix} \cos(wt) & -\sin(wt) \\ \sin(wt) & \cos(wt) \end{bmatrix} \begin{bmatrix} U_\alpha \\ U_\beta \end{bmatrix} \tag{6.16}$$

$$U_{d,err} = U_{d,ref} - U_d \tag{6.17}$$

$$U_{q,err} = U_{q,ref} - U_q \tag{6.18}$$

6.4.5   Adaptive protection

It is discussed that one important issue in the protection of microgrids is the low fault current level, especially in island mode. One of the suggested ways to address this problem is installing a source of high short-circuit current. That can be a flywheel or a supercapacitor. However, this may require the use of power electronic interfaces with high short-circuit current handling capability. Production of inverters with a superior capacity to generate high short-circuit currents is still a work in progress and leads to extra investment cost even in the currently applicable context. Thus it is better to find solutions that the current technology already allows, and that are economically viable. The directional OC protection function can be sufficient to protect a microgrid provided that its settings take into account the network topology and changes in generation type and amount. These conditions must be checked at all times to ensure that the protection settings are appropriate for each circumstance, which leads us to idea of adaptive microgrid protection systems.

Adaptive protection is defined by [27] as

[A]n online activity that modifies the preferred protective response to a change in system conditions or requirements. It is usually automatic, but can include timely human intervention [27].

Another definition by [28] is:

[A] protection philosophy which permits and seeks to make adjustments automatically in various protection functions in order to make them more attuned to prevailing system conditions [28].

Adaptive protection is applied through use of adaptive relays which are defined by [27] as relays whose settings, characteristics or logic functions can be changed online in a timely manner by means of externally generated signals or control action.

Adaptive protection involves protective relays whose setting can be dynamically adjusted as well as a fast and effective communication infrastructure. Fuses and electromechanical relays are not suited for this purpose. The relays for effective implementation of such a protection should be relays with various "setting tables" and/or "multiple instances" of the OC units, so that they can be activated and deactivated locally or remotely, automatically or manually. Numerical directional OC relays are mostly utilized for such a protection. The installation of intelligent electronic devices (IEDs) at required points in the microgrid help to gather the information required to make the adjustment of protective settings. The adjustments are made based on power flow and short circuit studies carried out on each possible topological configuration. The decisions based on these computations are communicated to the numerical relays through a communications system which can be of different protocols and physical media. The two options, architecture-wise, are a centralized or decentralized system. An adaptive system usually requires a higher investment than a "traditional" low-voltage system based on fuses and circuit breakers, though it is worth considering the benefits that a reliably protected microgrid can provide.

One aspect of adaptive microgrid protection that needs further discussion is the technique used in order to adjust the protective settings according to the dynamic condition of the microgrid. Generally, there are two options:

1. *Online calculation*: this technique involves calculating the protection settings online based on the electrical model of the microgrid at a specific time step. This can be done on a personal computer. However, due to the computing power required, it is unthinkable that a distribution protection does it today. It could also be applied with a centralized communications system in which a central computer unit with sufficient computing power

and all the information on the status of the microgrid. The system can be implemented though a calculation process that is either run periodically or is triggered by an event such as circuit breaker tripping, protection alarm, etc. [29]. In case of the later one, every time a change in the microgrid was detected, the settings for each component of the protection system would have to be recalculated, and once the process had finished, they would have to be loaded using the communications system. This option is a complex system, which allows simple relays to be used (a single setting table and one instance of each protection unit), but which makes its own decisions about the settings to be loaded, and which will require a calculation each time that there is any change in the microgrid.

An adaptive protection scheme is used in [30] for microgrids based on microprocessor-based overcurrent relay and reported to improve reliability of the microgrid by enabling autoreclosure for temporary faults and proved effective in detecting low impedance faults. An adaptive OC protection strategy that can adjust the current and time settings based on changes in the mode of operation is also devised in [31]. The paper tested the proposed scheme on an RTDS simulation of a modified CIGRE test microgrid and reported a fast and accurate detection of different types of faults in both island and grid-connected modes. Another study by [32] proposed an online adaptive protection scheme that uses phasor measurement data from microphasor measurement units. The method computes fault index and abnormality coefficients for each feeder of a microgrid to detect the location of a fault and the abnormality case.

One challenge with online adaptive protection schemes is the risk of failure of communication network and cyber security threats and it is suggested to have a contingency in the event that a communication failure happens [33].

2. *Precalculated protection settings:* The protection setting is decided by referring to a table of precalculated values based on possible expected cases. This system first requires identifying the number of possible configurations of the microgrid, basically a function of the number of switches and the feeding-in states of DERs. This gives rise to a set of possible cases, called an event table, which can be simplified by eliminating those that no longer make sense when considering different priority levels between switches; for example, the switch that connects the microgrid with the medium-voltage network is the one with the highest priority, and so on toward lower levels. Each of these cases and the possible faults (locations and types) that may occur need to be simulated and fault currents passing through all monitored circuit breakers need to be estimated. IEC 60909 [34], which is applicable to calculation of short-circuit currents in low-voltage (LV) AC systems, may be referred to when deciding the timing of the simulated faults. As a result of all these simulations, a list of actions will be obtained for each protection equipment. That is presented as an action table which has the same dimension as the event table.

This system requires protection equipment with multiple tables and instanced protection units. The complexity remains in the configuration phase, since neither the relays nor any external system have to make complex online calculations later on. It has the advantage that it allows a prior knowledge of what the settings to be used in any of the protections can be, so that it can be tested before starting the installation. Furthermore, this system allows the use of both centralized and decentralized communication architectures. The biggest problem with this system is that the maximum number of possible settings is twice the number of microgrid switches. Therefore in very large microgrids, an unmanageable number of cases may emerge. One solution to this problem is to divide the microgrid into parts, and create what is called a multimicrogrid; in this way it is possible to work with smaller microgrids.

Such an automatic adaptive protection strategy for microgrids based on precalculated settings of OC protection relays is devised and verified in [35]. The proposed method involves automatic adjustment of settings of the OC relays so that the coordination time intervals between relays stay similar while the microgrid experiences new situations such as reconnection of a DG.

6.4.6 Machine learning-based protection schemes

With the complexity of the nature of fault detection and location in microgrids, the coordination of conventional protection relays has become a tedious, and at times less effective, approach. However, innovation and deployment of digital relays has opened the lane for the intelligent protection relays which use a combination of machine learning tools and digital signal processing methods. Such methods are gaining significant attention for the detection and location of faults in microgrids in recent years. Though the methods are mostly under research and development stages, there are encouraging reports that would allow for predicting a wider application of those intelligent relays in the near future. Some of the attempts in this regard are briefly discussed in this section.

A study by [36] presented a rule-based adaptive microgrid protection scheme that makes use of machine learning methodology. The procedure involves quantitative analysis of the uncertain elements in a microgrid through Pearson correlation coefficients followed by a state estimation procedure. The latter procedure is implemented through an artificial neural network (ANN)-support vector machine (SVM) model, which is a combination of two of the common machine learning tools: ANN and SVM. The state recognition algorithm is used to assist the decision-making in adjusting the adaptive protective settings based on the network topology and mode of operation so as to ensure the reliable protection of the microgrid. The method allows for utilization of the massive data streams which are becoming easily available in smart grids.

An approach closely similar to the one in the earlier study was applied in [37], with approximate component coefficients resulting from discrete wavelet transform (DWT) on the voltage and current signals being used this time as inputs to the SVM-based classifier for fault detection in both island and grid-connected modes of a microgrid. A protection method based on combination of multiagent system and machine learning is proposed for protection coordination and updating the relay settings in [38].

Example 6.3: Cognitive edge computing-based protection strategy

This example proposes a fault detection and protection strategy based on cognitive edge computing in an effort to harvest the benefits of cognitive edge computing and address the special needs of microgrids. In the proposed strategy, a microgrid smart gateway is used to accumulate data into a central repository where it will be processed and analyzed (Fig. 6.20). The intermediary smart gateway includes a protection unit where the fault detection, location, and isolation are accomplished through a combination of virtual mode decomposition (VMD), SVM and long short-term memory (LSTM)-type deep machine learning tools. The procedure used in the protection strategy is summarized in Fig. 6.21. The local measurements of branch currents and bus voltages (Fig. 6.22) are processed through VMD and the informative decomposed components are provided as inputs to the SVM-based fault detection unit and LSTM-based fault location unit. The smart digital relay (Fig. 6.22) passes trip commands to the respective circuit breaker/s and submits compiled data regarding history of faults and protection actions to the upper level units. The findings from simulation results demonstrate the effectiveness of the proposed strategy to provide fast and accurate fault detection and protection against all types of faults and locations in the microgrid.

The data for training and testing the devised method is generated from recording (with a sampling frequency of 1 kHz) the voltage at each busbar

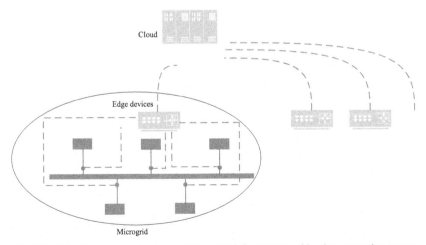

FIGURE 6.20   Structure of cloud-based microgrid infrastructure with edge computing protection devices.

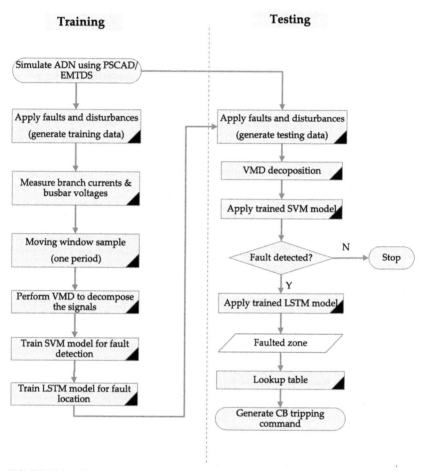

FIGURE 6.21 Flowchart of procedure followed.

and the current in each branch of the circuit shown in Fig. 6.23 after sets of disturbances and faults are applied to the system.

In order to evaluate the performance of the devised fault detection scheme in this study, the confusion matrix and accuracy measures of f-score, recall, and precision parameters are used, as defined in [39]. The parameters are expressed in percentage forms in the following equations:

$$\text{Precision} = \frac{TP}{TP + FP} * 100\% \qquad (6.19)$$

$$\text{Recall} = \frac{TP}{TP + FN} * 100\% \qquad (6.20)$$

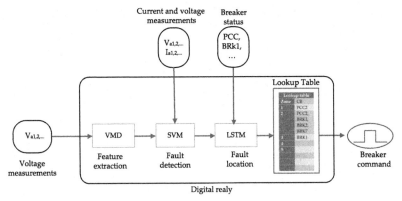

FIGURE 6.22   Structure of designed digital protection relay.

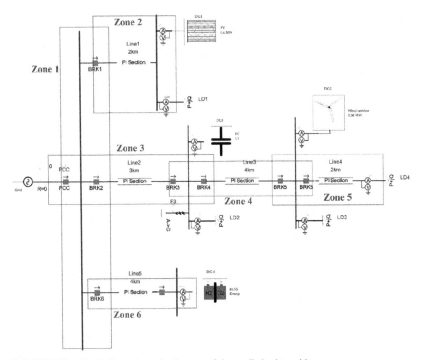

FIGURE 6.23   Single-line schematic diagram of the studied microgrid.

$$F\text{-Score} = 2* \frac{\text{Precision} * \text{Recall}}{\text{Precision} + \text{Recall}'} \qquad (6.21)$$

where TP = true positive, FP = false positive, TN = true negative, and FN = false negative.

Accuracy of the fault location model is also evaluated using a confusion chart and respective accuracies of the classification for each protection zone Eq. (6.22):

$$\text{Accuracy} = \frac{\text{Count}(Pred\_Loc = Act\_Loc)}{\text{Count}(Act\_Loc)} * 100\% \qquad (6.22)$$

where $Pred\_Loc$ = predicted location (zone), Act_Loc = actual location (zone).

After the faults are detected and located through the fault detection and location units, the appropriate fault clearing action will be carried out through a breaker opening command that can be generated based on a lookup table that associates the protection zones with the circuit breakers.

The effectiveness of the devised fault detection and location techniques are verified based on a test data generated using the test microgrid model simulated in PSCAD-EMTDS. The proposed methods have proved to be effective in addressing the challenges specific to the protection of microgrids and general protection issues related to high impedance faults. The procedure involves VMD decomposition of the branch current (Fig. 6.24 (A)) and bus voltage (Fig. 6.24 (B)). The fault detection unit of the relay was able to identify diverse fault types based on the features of the voltage extracted by VMD decomposition (Fig. 6.25). The LSTM-based fault location unit uses the features extracted from the bus voltage and branch currents as well as status of the breakers and the output of the fault detection unit as inputs. It also performed well with close to 100% accuracy of locating all types of faults in the microgrid (Figs. 6.26 and 6.27; Table 6.7). The technique has the advantages that adaptive adjustment of protection relay settings is not needed, the complex fault conditions related to the plug-and-play features of a microgrid and the generation variation from renewable DGs are easily learnt by the LSTM and SVM models, and the relay was able to respond properly under varying conditions of generation and loading.

6.5 Centralized protection for microgrids

Centralized protection analyzes comprehensive information of the entire network, so as to accurately and quickly determine the location and cause of a fault. A generic centralized microgrid protection scheme may have the structure shown in Fig. 6.28.

It is mainly composed of a smart local protection device and a central protection system (or sometimes central protection and control system). The local protection devices include relays and measurement units. The main functions of these devices include collection and calculation of electric energy information, analysis of the collected data according to the protection algorithm, sending the calculation results to the upper protection system, receiving the control signal of the centralized protection center, and executing the instructions and monitoring of the upper control center device status.

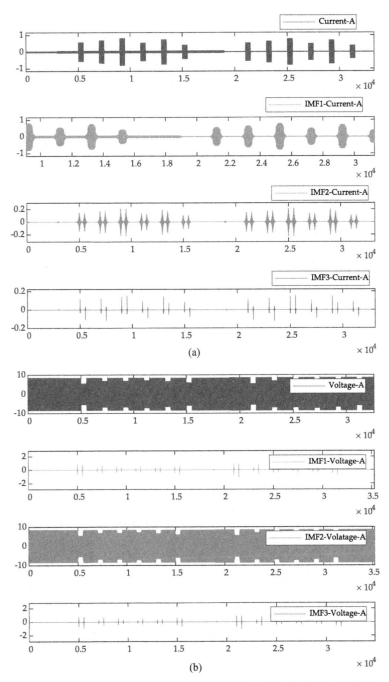

FIGURE 6.24 VMD decomposition of Phase A (A) current (Branch 1) (B) voltage (Bus 1).

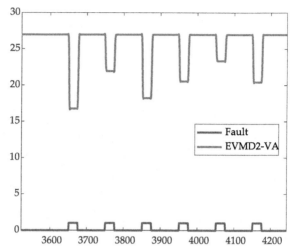

FIGURE 6.25 Energy of second intrinsic mode function (IMF) of voltage signal serving as input for SVM-based fault detection.

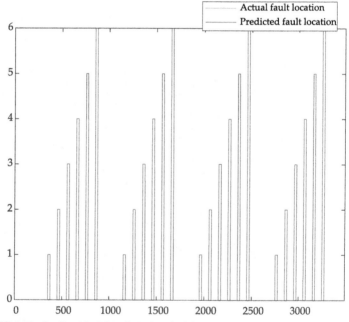

FIGURE 6.26 Predicted fault location using the LSTM model (Phase A).

The centralized protection system conducts fault analysis based on the data from the local protection equipment to determine whether the microgrid is in a normal operating state or not. Once a fault happens, corresponding protective measures shall be taken according to the fault situation.

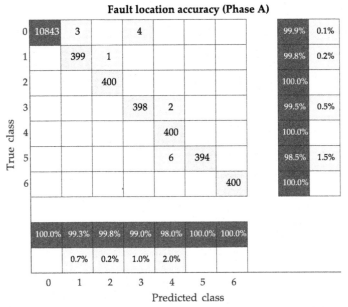

Fault location accuracy (Phase A)

FIGURE 6.27   Confusion chart for locating faults involving Phase A.

TABLE 6.7 Accuracy of fault detection using SVM-based model.

	TP	FP	TN	FN	Precision	Recall	F-Score
Phase A	92	0	434	4	100	95.83	97.87
Phase B	62	0	466	2	100	96.88	98.41
Phase C	31	0	498	1	100	96.88	98.41

The centralized protection system can analyze and compare the information of on-site monitoring equipment to determine the location of the fault.

This kind of protection system usually utilizes extensive communication to monitor the microgrid and update relay settings according to the changes in the system. This high dependence on extensive communication can be eased by limiting the need for communication only to when there is a change in the microgrid. Once the setting values are estimated and the operating currents are reported to relays, the relays can operate individually without any communication process.

When there is a problem in the microgrid protection communication system, the local protection device will not be able to receive the action command from the centralized protection unit. Thus in such events, the local protection equipment should have the ability to provide protection according

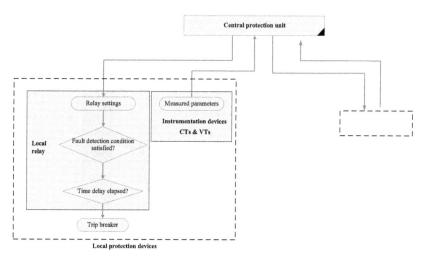

FIGURE 6.28 Central protection system and local decision-making.

to the local setting, and detect and judge the fault according to the local information to complete the protection action.

Though a considerable number of research works are available indicating the level of attention being drawn to centralized protection in recent years, a modest exploration of websites of known protection device manufacturers and online shopping sites showed the market penetration of such protection systems is still quite low. However, looking at the amount of studies and reports coming up regarding the technology and the growing demand for microgrids and reliable protection of them, it is inevitable that centralized protection will be applied more widely sooner than later. Some of the attempts and research findings regarding the approach are reviewed hereunder.

A microgrid protection scheme that employs a central protection unit and phasor measurement units is proposed in [40]. The operating algorithm of the protection system is based on positive-sequence impedance components. The method is reported to be applicable for microgrids with both radial and looped architecture while the central protection unit has the capability of updating its pickup values (which actually are the upstream and downstream equivalent positive-sequence impedances of each line) when there is a change in configuration of the microgrid or mode transfer between grid-connected and island modes. While the findings of the study are interesting, the proposed schemed is verified only through simulation tools.

A centralized protection system capable of responding to dynamic changes in microgrid such as connection and disconnection of DGs is also developed in [41]. The protection scheme is yet again dependent on extensive communication based on IEC61850 to monitor the status of the microgrid and update relay settings. The Transmission Control Protocol/Internet

Protocol (TCP/IP)-based Ethernet network is chosen as a communication medium for its cost-effectiveness and availability. The operating fault current and time delay of each local relay are adjusted based on the microgrid's operating mode and the status of the DGs. The operating fault current or pickup current setting of each relay is calculated as Eq. (6.23) [41].

$$I_{set} = (I_{fG}x \ mode) + \sum_{i=1}^{n}(k_i x I_{fDG,i} x DG_i) \tag{6.23}$$

where I_{set} is the relay pick-up current; I_{fG} is the fault current from grid calculated using Thevenin equivalent; $mode$ is the operating mode of the microgrid (0 for island and 1 for grid-connected); k_i is impact factor of the ith DG on the fault current sensed by the relay (takes a value between 0 and 1 and represents the effect of line impedance on the fault current); n is the number of DGs in the microgrid; $I_{fDG,i}$ is the fault current contribution of the ith DG; and DG_i is the status of the ith DG.

A study by [42] took the concept of centralized protection to an even higher level with the proposition of the concept "*i-protection.*" The protection scheme uses informatics to accurately detect faults in multitapped complex configurations such as microgrids through IEDs and Internet of Energy (IOE) using wide-area wireless networks such as IEEE 802.16 wireless protocol. The method considers a configuration as one zone rather than multiple zones. It consists of a base station, a switching center, and IEDs. The protection scheme solves the coordination issues of conventional protection schemes.

6.6   Protection of looped microgrids

The impact of bidirectional power flow on the effectiveness and coordination of protection systems is more evident in looped or meshed networks. The difficulty increases considerably when multiple DG units are involved, as in most microgrids, due to the increased complexity and the need for frequent revisions of relay settings to fit to the continuous changes [43]. Hence, investigating a reliable protection concept for application to such systems is a hot topic of research and development.

Applying directional OC relays is one of the suggested approaches to the protection of meshed distribution networks with DGs, although that leaves us with complicated tasks of deciding optimal settings and protection coordination. A loop network is a particularly common arrangement within distribution for the sake of ensuring power supply availability in the case of faults occurring on interconnecting feeders. Applying directional OC relays and coordinating them based on a grading procedure that starts with opening the ring at the supply point and grading the relays first clockwise and then anticlockwise is a common practice when the network involves a single source. That would comprise determining the short-circuit impedance by network reduction (series connection, parallel connection, and delta-star transformation, etc.) using the positive-sequence short-circuit impedances of network components. The more challenging aspect is when there is more than one source in the looped network which makes it difficult to apply and ensure discrimination through time-graded OC protection. One suggested approach that still allows use of directional OC relays is to open the loop at one of the supply points using a suitable high set instantaneous OC relay and grading the rest of the relays in a similar approach as in the case of a single infeed. Differential protection is also another of the popular protection schemes that is considered for looped microgrids. An improved version of differential relay, variable tripping time differential protection scheme, is applied for the protection of microgrids composed of a loop in [44]. The variable tripping time is intended to address the protection coordination with no need for additional measurement and communication. A study by [45] proposes a concept of single-strategy and dual strategy protection to protect loop-based microgrids divided into four protection levels as the load, loop, feeder, and microgrid level protections. Applicability of the conventional protection systems for closed loop distribution networks with DGs was analyzed and a pilot wire instantaneous protection scheme was proposed for better effectiveness for such networks in [46].

The latest attempts on addressing the protection of loop distribution networks include application of signal processing tools and machine learning or artificial network-based techniques. One such study by Deng et al. [47] proposes a fault location technique based on DWT and SVM for noneffectively grounded loop distribution networks.

6.7 Earthing system in protection of microgrids

An earthing system may simply be defined as an arrangement by which an electrical installation is connected to a means of earthing for safety and functional purposes. The earthing system of a microgrid plays a very essential role in the fault current level and the choice and effectiveness of a protection system for microgrids, especially in island mode of operation. The suggestion regarding the earthing system of LV microgrids in [48] is to use TN-C-S[1] or TT earthing systems (shown in Fig. 6.29) for neutral earthing of LV microgrids while neutral earthing may not be necessarily needed for the DG units.

TN-C-S (sometimes called as known as protective multiple earthing (PME)) is the most common earthing configuration in many countries. The configuration provides a low-voltage supply with reliable and safe earthing while allowing multiple users to utilize one supply cable. This may lead to a voltage rise in the protective earthed neutral requiring multiple connections to earth, though it may not be a big concern in microgrids where the line length and users of the single supply line are limited. In a conventional

---

1. T = terre (earth), N = neutral, C = combined, S = separate.

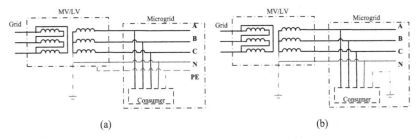

FIGURE 6.29 Circuit diagram of (A) TN-C-S earthing system (B) TT Earthing system.

distribution network, the neutral is earthed close to the source of supply which is at the intake of the installation. This logic can be transcended to microgrids so that the neutral is earthed at the PCC while there could be more earthing points throughout the network when necessary.

The TN-C-S earthing system has a drawback that an open circuited neutral conductor could be too hazardous as there will not be a return current path. Thus it may not be allowed or advised for application in some risky installations such as petrol stations. It is in such installations that a TT configuration may be used as an alternative. In a TT earthing system consumers of a microgrid are expected to supply their own earth by burying rods underground or other means. TT is also a choice for microgrids in rural areas with overhead lines. The earth fault loop impedance is higher in TT configuration compared to that of TN-C-S.

One of the issues in protection of nonisolated microgrids operating in island mode is the loss of neutral connection of medium voltage (MV)/LV transformer when the microgrid changes to island mode. This case happens when the protection device at the microgrid's PCC is located downstream from the MV/LV transformer [49]. Thus the suggested approach is to connect the microgrid to the utility grid through delta-wye grounded transformer with the microgrid side of this transformer being directly earthed [49]. That ensures there will be a path for neutral current and the earth fault currents will be of significantly high magnitude that the protection units can detect them. A similar approach (delta-wye grounded connection transformers) is advised for the connection of DGs to the microgrid as well. It is worth noting that it is a customary practice in many countries to use isolation transformers for the integration of DGs to a network.

References

[1] IEC, International Electrotechnical Vocabulary (IEV), IEV Definitions. <http://www.electropedia.org/>.

[2] C. Russell Mason, The Art and Science of Protective Relaying, Wiley, 1956.

[3] E. Csanyi, 8 essential relay operating principles of catching faults, Electrical Engineering Portal. <https://electrical-engineering-portal.com/relay-operating-principles-catching-faults>.

[4] K. Keller, Electrical system grounding and bonding, Electrical Safety Code Manual, 2010.

[5] L. Fusheng, L. Ruisheng, Z. Fengquan, Protection of the microgrid, Microgrid Technology and Engineering Application, 2016.

[6] M.N. Alam, Overcurrent protection of AC microgrids using mixed characteristic curves of relays, Comput. Electr. Eng. (2019). Available from: https://doi.org/10.1016/j.compeleceng.2019.01.003.

[7] E.J. Coster, J.M.A. Myrzik, B. Kruimer, W.L. Kling, Integration issues of distributed generation in distribution grids, Proc. IEEE (2011). Available from: https://doi.org/10.1109/JPROC.2010.2052776.

[8] D. Jones, J.J. Kumm, Future distribution feeder protection using directional overcurrent elements, IEEE Trans. Ind. Appl. (2014). Available from: https://doi.org/10.1109/TIA.2013.2283237.

[9] H.H. Zeineldin, H.M. Sharaf, D.K. Ibrahim, E.E.D.A. El-Zahab, Optimal protection coordination for meshed distribution systems with DG using dual setting directional overcurrent relays, IEEE Trans. Smart Grid (2015). Available from: https://doi.org/10.1109/TSG.2014.2357813.

[10] A. Yazdaninejadi, S. Golshannavaz, D. Nazarpour, S. Teimourzadeh, F. Aminifar, Dual-setting directional overcurrent relays for protecting automated distribution networks, IEEE Trans. Ind. Inf. (2019). Available from: https://doi.org/10.1109/TII.2018.2821175.

[11] P. Thararak, P. Jirapong, Implementation of optimal protection coordination for microgrids with distributed generations using quaternary protection scheme, J. Electr. Comput. Eng. (2020). Available from: https://doi.org/10.1155/2020/2568652.

[12] P.M. Anderson, Protective device characteristics, Power System Protection, 2010.

[13] C.A. Kramer, W.A. Elmore, Flexible inverse overcurrent relaying using a microprocessor, IEEE Trans. Power Deliv. (1990). Available from: https://doi.org/10.1109/61.53102.

[14] S. Beheshtaein, R. Cuzner, M. Savaghebi, J.M. Guerrero, Review on microgrids protection, IET Generation, Transm. Distrib. (2019). Available from: https://doi.org/10.1049/iet-gtd.2018.5212.

[15] G. Benmouyal, et al., IEEE standard inverse-time characteristic equations for overcurrent relays, IEEE Trans. Power Deliv. (1999). Available from: https://doi.org/10.1109/61.772326.

[16] M.Y. Shih, C.A. Castillo Salazar, A. Conde Enríquez, Adaptive directional overcurrent relay coordination using ant colony optimisation, IET Gener. Transm. Distrib. (2015). Available from: https://doi.org/10.1049/iet-gtd.2015.0394.

[17] S.D. Saldarriaga-Zuluaga, J.M. López-Lezama, N. Muñoz-Galeano, Optimal coordination of overcurrent relays in microgrids considering a non-standard characteristic, Energies (2020). Available from: https://doi.org/10.3390/en13040922.

[18] S. Kannuppaiyan, V. Chenniappan, Numerical inverse definite minimum time overcurrent relay for microgrid power system protection, IEEJ Trans. Electr. Electron. Eng. (2015). Available from: https://doi.org/10.1002/tee.22066.

[19] Alstom, Chapter 9: Overcurrent protection for phase and earth faults, Network Protection & Automation Guide Protective Relays, Measurement & Control, 2011.

[20] A. Girgis, S. Brahma, Effect of distributed generation on protective device coordination in distribution system, in: LESCOPE 2001 - 2001 Large Engineering Systems Conference on Power Engineering: Powering Beyond 2001, Conference Proceedings, 2001. Available from: https://doi.org/10.1109/LESCPE.2001.941636.

[21] T.S. Ustun, C. Ozansoy, A. Zayegh, Differential protection of microgrids with central protection unit support, in: IEEE 2013 Tencon - Spring, TENCONSpring 2013 - Conference Proceedings, 2013. Available from: https://doi.org/10.1109/TENCONSpring.2013.6584408.

[22] D. Bejmert, W. Rebizant, L. Schiel, Differential protection restraining procedures for objects with more than two supply ends, in: Proceedings of the Universities Power Engineering Conference, 2010.

[23] M. Dewadasa, A. Ghosh, G. Ledwich, Protection of microgrids using differential relays, in: 2011 21st Australasian Universities Power Engineering Conference, AUPEC 2011, 2011.

[24] H. Lin, C. Liu, J.M. Guerrero, J.C. Vasquez, Distance protection for microgrids in distribution system, in: IECON 2015 - 41st Annual Conference of the IEEE Industrial Electronics Society, 2015. Available from: https://doi.org/10.1109/IECON.2015.7392186.

[25] J.M. Dewadasa, A. Ghosh, G. Ledwich, Distance protection solution for a converter controlled microgrid, Power (2008). Available from: https://doi.org/10.1016/j.socscimed.2015.03.020.

[26] H. Al-Nasseri, M.A. Redfern, F. Li, A voltage based protection for micro-grids containing power electronic converters, in: 2006 IEEE Power Engineering Society General Meeting, PES, 2006. Available from: https://doi.org/10.1109/pes.2006.1709423.

[27] G.D. Rockefeller, C.L. Wagner, J.R. Linders, K.L. Hicks, D.T. Rizy, Adaptive transmission relaying concepts for improved performance, IEEE Trans. Power Deliv. (1988). Available from: https://doi.org/10.1109/61.193943.

[28] S. Horowitz, D. Novosel, V. Madani, M. Adamiak, System-wide protection, IEEE Power Energy Mag. 6 (5) (2008) 34–42. Available from: https://doi.org/10.1109/mpe.2008.927473.

[29] A. Oudalov, A. Fidigatti, Adaptive network protection in MIcrogrids, ABB Int. J. Distrib. Energy Resour. (2009). A76C7058B54F5280C12578DF00001585.

[30] R. Sitharthan, M. Geethanjali, T. Karpaga Senthil Pandy, Adaptive protection scheme for smart microgrid with electronically coupled distributed generations, Alex. Eng. J. (2016). Available from: https://doi.org/10.1016/j.aej.2016.06.025.

[31] Z. Kailun, D.S. Kumar, D. Srinivasan, A. Sharma, An adaptive overcurrent protection scheme for microgrids based on real time digital simulation, 2017 IEEE Innovative Smart Grid Technologies - Asia: Smart Grid for Smart Community, ISGT-Asia 2017, 2018. Available from: http://doi.org/10.1109/ISGT-Asia.2017.8378368.

[32] M.S. Elbana, N. Abbasy, A. Meghed, N. Shaker, μPMU-based smart adaptive protection scheme for microgrids, J. Mod. Power Syst. Clean. Energy (2019). Available from: https://doi.org/10.1007/s40565-019-0533-6.

[33] H.F. Habib, C.R. Lashway, O.A. Mohammed, A review of communication failure impacts on adaptive microgrid protection schemes and the use of energy storage as a contingency, IEEE Transactions on Industry Applications, 2018. Available from: http://doi.org/10.1109/TIA.2017.2776858.

[34] IEC 60909, Short-circuit currents in three-phase a.c. systems - calculation of currents, IEC, 2016.

[35] K. Sedghisigarchi, K.T. Sardari, An adaptive protection strategy for reliable operation of microgrids, in: 2018 IEEE International Energy Conference, ENERGYCON 2018, 2018. Available from: https://doi.org/10.1109/ENERGYCON.2018.8398779.

[36] H. Lin, K. Sun, Z.H. Tan, C. Liu, J.M. Guerrero, J.C. Vasquez, Adaptive protection combined with machine learning for microgrids, IET Gener. Transm. Distrib. (2019). Available from: https://doi.org/10.1049/iet-gtd.2018.6230.

[37] M. Manohar, E. Koley, SVM based protection scheme for microgrid, in: 2017 International Conference on Intelligent Computing, Instrumentation and Control Technologies, ICICICT 2017, 2018. Available from: https://doi.org/10.1109/ICICICT1.2017.8342601.

[38] M. Uzair, L. Li, J.G. Zhu, M. Eskandari, A protection scheme for AC microgrids based on multi-agent system combined with machine learning, in: 2019 29th Australasian Universities Power Engineering Conference, AUPEC 2019, 2019. Available from: https://doi.org/10.1109/AUPEC48547.2019.211845.

[39] W. Fei, P. Moses, Fault current tracing and identification via machine learning considering distributed energy resources in distribution networks, Energies (2019). Available from: https://doi.org/10.3390/en12224333.

[40] S. Mirsaeidi, D.M. Said, M.W. Mustafa, M.H. Habibuddin, K. Ghaffari, Design and testing of a centralized protection scheme for micro-grids, J. Cent. South. Univ. (2015). Available from: https://doi.org/10.1007/s11771-015-2932-9.

[41] T.S. Ustun, C. Ozansoy, A. Zayegh, Modeling of a centralized microgrid protection system and distributed energy resources according to IEC 61850-7-420, IEEE Trans. Power Syst. (2012). Available from: https://doi.org/10.1109/TPWRS.2012.2185072.

[42] M.M. Eissa, M.H.A. Awadalla, Centralized protection scheme for smart grid integrated with multiple renewable resources using Internet of Energy, Glob. Transit. (2019). Available from: https://doi.org/10.1016/j.glt.2019.01.002.

[43] A.M. Tsimtsios, V.C. Nikolaidis, Toward plug-and-play protection for meshed distribution systems with DG, IEEE Trans. Smart Grid (2020). Available from: https://doi.org/10.1109/TSG.2019.2945694.

[44] T.S. Aghdam, H. Kazemi Karegar, H.H. Zeineldin, Variable tripping time differential protection for microgrids considering DG stability, IEEE Trans. Smart Grid (2019). Available from: https://doi.org/10.1109/TSG.2018.2797367.

[45] X. Liu, M. Shahidehpour, Z. Li, X. Liu, Y. Cao, W. Tian, Protection scheme for loop-based microgrids, IEEE Trans. Smart Grid (2017). Available from: https://doi.org/10.1109/TSG.2016.2626791.

[46] B. Li, X. Yu, Z. Bo, A. Klimek, Investigation of protection schemes for closed loop distribution network with distributed generators, Dianli Xitong Zidonghua/Automation Electr. Power Syst. (2010).

[47] X. Deng, R. Yuan, Z. Xiao, T. Li, K.L.L. Wang, Fault location in loop distribution network using SVM technology, Int. J. Electr. Power Energy Syst. (2015). Available from: https://doi.org/10.1016/j.ijepes.2014.10.010.

[48] N. Jayawarna, N. Jenkins, M. Barnes, M. Lorentzou, S. Papthanassiou, N. Hatziagyriou, Safety analysis of a MicroGrid, in: 2005 International Conference on Future Power Systems, 2005. Available from: https://doi.org/10.1109/fps.2005.204228.

[49] H.J. Laaksonen, Protection principles for future microgrids, IEEE Trans. Power Electron. (2010). Available from: https://doi.org/10.1109/TPEL.2010.2066990.

Chapter 7

Dynamic control of microgrids

7.1 Introduction

Electrical energy in conventional interconnected power systems is mainly generated by synchronous generators and the stability and control analysis in such systems is based on the theory of synchronous generator. As such, the stability problems of power angle stability, frequency stability, voltage stability, and their corresponding small disturbance dynamics and transient stability, short-term and long-term stabilities are closely related to the dynamic characteristics of synchronous generators. However, today's power systems are commonly featuring microgrids that integrate distributed generation (DG) of different characteristics (including synchronous and asynchronous generators, power electronic converter interfaced, and other types of DGs) in the medium- and low-voltage distribution networks. Meanwhile, the nature of the loads has also shown significant changes with the incorporation of largely diverse and nonlinear loads, such as motors and power electronic interfaced loads, which are also commonly present in microgrids. The dynamic characteristics of multiple types of DG and loads are coupled and influence each other in microgrids. In addition, the most common type of DGs in microgrids interface with the network through power electronic converters. Power electronic interfaced generators, commonly known as converter-based generators (CBGs), greatly differ from conventional synchronous generators in terms of power transformation, control strategy, and dynamic characteristics. The diversity of control methods and higher penetration of power electronic interface generators will bring severe challenges to the coordination control and stable operation of microgrids. Due to the above-discussed special nature of microgrids, the traditional power system stability analysis and control methods may not meet the requirements of microgrids, which have become one of the basic theoretical and technical bottlenecks for the popularization of microgrids.

Furthermore, due to their hardly predictable and strongly intermittent generation, renewable DG units such as wind and photovoltaic power systems have a great impact on the dynamic characteristics and stable operation of the microgrid system. The other common phenomenon related with dynamic stability of power systems is the low-voltage ride-through capability of DGs in microgrids operating in grid-connected mode in the case of short-circuit faults in the grid side. Microgrids are expected to change to island

Microgrid Protection and Control. DOI: https://doi.org/10.1016/B978-0-12-821189-2.00002-4
169

mode in such events which would lead to insufficient support for the recovery of system stability and hence cause dynamic instability and possibly collapse of the microgrid system.

In recent years, the stability of microgrids has been one of the extensively studied topics. Special characteristics of microgrids which differ from the traditional power system in terms of system capacity, voltage level, structure, control architecture, operation mode, and involvement of power electronic interfaces are the concerns of those studies on the dynamic stability of microgrids. Unlike the considerable level of recent research work on the issue, there still is a lack of comprehensive understanding and proven solutions to the dynamic stability issues, specially those related to higher penetration of renewable DG, post fault events, nonlinear loads, communication delay, interaction between dynamic characteristics of multiple DGs and loads, and interaction of multiple microgrids. Thus the dynamic disturbance and control of microgrids still requires in-depth research.

7.2 Dynamic characteristic of microgrids

DG units in microgrids can be either alternating current (AC) rotating motor type or power electronic converter interfaced type based on the type of primary energy source and the employed energy conversion mechanism.

AC rotating motor type DGs are those such as diesel generators, small hydro, and fixed speed wind turbines. Those DGs are usually characterized by their large inertia time constants with the response time of their mechanical rotor being greater than 500 ms. As those DGs are connected to the microgrid without interfacing device, mechanical dynamics of the primary energy sources is directly transferred to the electric network.

Power electronic interfaced DGs include variable speed and constant frequency wind turbines, gas turbines, PV systems, energy storage systems, etc. They can be connected to the microgrid through single-stage direct current (DC)-AC, or two-stage AC-DC-AC, DC-DC-AC, etc. The outstanding characteristic of power electronic interfaced DGs is the inherent fast dynamic response capability through the power electronic transformation interface. It should be pointed out that the control loops applied in the power electronic converter interfaces are either in the DC-link (capacitance) voltage control timescale (which is in the range of 100 ms) or AC (inductor) current control timescale (about 10 ms) [1]. The involvement of different types of DGs makes microgrids strong nonlinear systems with the coupling of primary energy sources, power electronic converters, and control systems, with their dynamic characteristics being the superposition of dynamic characteristics of the constituting units on multiple timescales. Thus dynamic characteristic of microgrids is different from that of traditional grid.

Microgrids are characterized by the following features:

1. Microgrids of nonisolated type are able to operate in two modes: grid-connected mode and island mode. The grid-connected mode involves bidirectional power flow at the point of connection. The most important characteristic of microgrids is the island mode in which the point of connection must be disconnected as occasions require under abnormal conditions (mostly in the grid side). With no support from utility grid, the stability of voltage and frequency of isolated microgrid and nonisolated microgrid in island mode becomes a challenging aspect. A nonisolated microgrid operating in island mode due to temporary disturbance need be synchronized and reconnected with the utility grid after disturbance clearance. In addition, the switching between grid-connected mode and island mode may cause the stability issues. This needs to be addressed through a control strategy that ensures smooth switching and uninterrupted electricity supply.

2. Microgrids, especially those dominated by CBGs, are characteristically vulnerable to stability issues due to lack (or low level) of system inertia. It is known that DG units, such as wind turbines and gas turbines, have a certain level of inertia. However, their equivalent rotating inertia is reduced, even some or all is lost, due to their converters that effectively decouple the respective rotor speed from electric frequency. In the transient progress, the DC-side capacitance of the converters in CBGs may release energy, which is similar with synchronous motor releasing rotational kinetic energy. However, the storage capacity of the DC-side capacitance is much less than the rotational kinetic energy of conventional rotating machine type generators. Thus energy storage systems with proper control system are necessary for microgrids to increase the equivalent inertia.

In the case of conventional rotating machine type generators, the rotor speed of the generator which determines the frequency of the voltage output, is adjusted by the speed governor. The source end voltage magnitude is also controlled by the excitation system of the generator while the voltage at the load end could be adjusted through local reactive power compensation or transformer tapping of the distribution network. However, the above-stated options of the speed governor and excitation system are not available in CBG-based microgrids.

There are several manifestations that microgrids exhibit when they suffer dynamic disturbance. The first and most important is the system frequency and voltage amplitude deviating from the normal values. Secondly, with regard to power quality, voltage and current harmonics will increase sharply during the dynamic disturbances. The odd (3rd to 13th) harmonics of the voltage waveform make up 95% of total harmonic distortion (THD).

Virtual synchronous machine (VSM) control that mimics the characteristics of mechanical rotating machines is one of the suggested techniques for ensuring stability in CBG-based microgrids. This is implemented through the use of energy storage systems. Dynamic characteristics of VSM type control are defined by droop gains, output impedances, operating points, and inner control loops. Changes in system parameters, small-signal interferences, topology changes, and the like may lead to large oscillations or even instability. One of the forms of dynamic instability exhibited in such microgrids involving VSM is oscillation of different frequencies such as "synchronous oscillation" (that occurs near synchronous frequency) and subsynchronous oscillation (occurring lower than synchronous frequency) [2].

Here it is important to note that the time duration involved in dynamic disturbances is different from that of transient disturbances and is one of the key features to differentiate the two types of disturbances. As defined in the IEC 62898-3-1, transient disturbances are characterized by a shorter duration which is in the range of milliseconds (up to 50 ms), while dynamic disturbance is defined by time duration in the range of 50 ms$-$2 s [3]. In addition, during dynamic disturbances, the frequency and voltage deviation may not be as large as in transient disturbance though it may also cause system failure gradually if the system's damping method is not strong enough.

7.3 Modeling of dynamic disturbance system for microgrid

Most DGs in microgrids are designed based on power electronic interferences, which means they are connected to the system through a converter. In principle, parallel operating DGs should be coordinated and generate active power and reactive power which are decided by loads and set by the central control system. There are two architectures that may be applied in control of microgrids. The first is master$-$slave architecture which is formed by a DG unit that works as a grid-forming source while the other DGs operate as grid-following sources. The grid-forming source should make sure that the network has a constant voltage and frequency at standard values. Thus the grid-forming source is chosen to be the DG with a stable primary power resource, such as energy storage, gas turbine, diesel genset, and so forth. The grid-following sources are intended to generate as much active power and reactive power as possible. Such sources may have intermittent or unstable primary sources as in wind turbines, photovoltaic systems, and other renewable sources.

The other architecture is droop control which is a prevalent method in a microgrid. Droop control allows multiple grid-forming resources to coexist in a microgrid. The grid-forming resources can be usually designed as virtual synchronous generators.

An internal model control consisting of current and voltage controllers is derived in [4] for an inverter-based microgrid and verified using a small-signal model.

A control strategy that utilizes robust servomechanism problem control theory for three-phase voltage inverters in an isolated AC microgrid is proposed. The devised control scheme is experimentally tested to verify the performance of the method for fast recovery on load variation and fast current limiting under overload or short-circuit conditions.

A typical converter-based grid-forming DG (such as the one shown in Fig. 7.1) consists of a DC part, converter, and AC part. DC voltage sources are used to model the DC part of the grid-forming source. The AC part

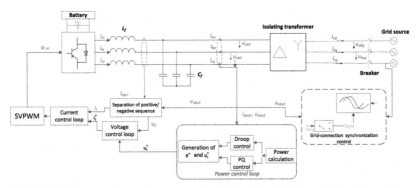

FIGURE 7.1 The overall control structure and circuitry of a grid-forming DG in microgrid.

contains isolation transformers and impedances of the lines. The complex part of the system design lies in the converter which includes parameter measurement, selection of filter parameters, phase angle generation, proportional and integral (PI) control gains for voltage and current control loops, and space vector pulse width modulation (SVPWM) modeling. Fig. 7.1 presents the main control structure of a grid-forming source. The small-signal stability model of this chapter is developed considering this topology and based on Ref. [4]. Thus some details can be referred in this work.

7.3.1 Modeling of power control loop

The power control is designed with components including average power calculation, filter, and droop controller (shown in Fig. 7.2). In the proceeding steps of the control system design, the horizontal (d) axis is taken as a reference axis (thus $u_{oq}^* = 0$).

The average power supplied by the inverter is determined by the output of the filter and defined as:

$$P = \frac{\omega_c}{s + \omega_c} p \Rightarrow \frac{d(P)}{dt} = -P\omega_c + 1.5\omega_c(u_{d0}i_{d0} + u_{q0}i_{q0})$$

$$Q = \frac{\omega_c}{s + \omega_c} q \Rightarrow \frac{d(Q)}{dt} = -Q\omega_c + 1.5\omega_c(u_{d0}i_{q0} - u_{q0}i_{d0})$$

(7.1)

This external power control loop basically sets the magnitude and frequency for the fundamental component of the inverter output voltage based on the active and reactive powers droop characteristics.

7.3.1.1 Modeling of phase angle generation

Controlling active power is controlling frequency. Phase angle generation can be performed in a similar way as a phase locked loop. This angle is used

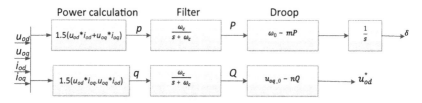

FIGURE 7.2 Block diagram of power control loop.

as input to the Park transformation, Clark transformation, and SVPWM. This angle is basically the integral of angular frequency, and expressed by:

$$\dot{\delta} = \omega \qquad (7.2)$$

The angular frequency ω is affected by change of active power flow which can be treated by two distinct operations named as droop control and synchronization. It is defined by (7.3).

$$\omega = \omega_0 + \omega_p + \omega_{syn} \qquad (7.3)$$

where ω_0 is the nominal value, ω_p is angular frequency deviation when different operation incidences which cause power imbalance happened and treated by the droop control, and ω_{syn} is the angular frequency feedback which is used to perform fast synchronization between microgrid and the distributed energy resource (DER).

To convert the variables from an individual inverter reference frame onto a common reference frame, we will define an angle for each inverter. It should be noted that δ represents the angle between an individual inverter reference frame and the common reference frame.

$$\delta = \int (\omega - \omega_{com})dt = \int (\omega_0 + \omega_p + \omega_{syn} - \omega_{com})dt \qquad (7.4)$$

Differentiating (7.4) with respect to time,

$$\frac{d(\delta)}{dt} = \omega_0 + \omega_p + \omega_{syn} - \omega_{com} \qquad (7.5)$$

Hence, the droop equation of active power control is expressed by Eq. (7.6).

$$\omega = \omega_0 - 2\pi m(P_{ref} - P) + \omega_{syn} - \omega_{com} \qquad (7.6)$$

where P_{ref} is the setting value according to the specific requirements of the microgrid, P is the active power of the grid-forming source, and m is the slope parameter of the droop curve.

7.3.1.2 Modeling of reactive power control (voltage amplitude generation)

Grid-forming sources should maintain the system frequency and voltage amplitude at their nominal values. To get the proper voltage amplitude, reactive power feedback droop control is introduced. The power control block diagram is shown in Fig. 7.2.

The generic mathematical equation for reactive power droop control model is

$$u_{d0}^* = u_{q0\_n} - n(Q_{ref} - Q) \tag{7.7}$$

where u_{q0_n} is the nominal setting value for voltage amplitude. Q_{ref} is always set to 0. Q is the generated reactive power and fed back to the power control loop.

7.3.1.3 Modeling of double loop control (voltage and current control loops)

Active and reactive power decouple controller is one of the prevalent methods for the converters. This control structure is illustrated in Fig. 7.3. It contains two control loops (i.e., outer voltage control loop and inner current control loop), and active damping and impedance reconstruction circuitries. First, the basic voltage and current control loops are designed and analyzed. Then, the concepts, mathematical models, and advantages of the active damping and impedance reconstruction blocks follow with verified results.

The voltage controller is designed to reject high-frequency disturbances and provide sufficient damping for the output inductance-capacitance (LC) filter [5]. The voltage controller block diagram includes all feedback and feedforward terms. Output voltage control is achieved with a standard PI controller (Fig. 7.4).

Accordingly, the state equations are given by Eq. (7.8) and expanded as (7.9).

$$\frac{d(\varphi_d)}{dt} = u_{od}^* - u_{od}$$
$$\frac{d(\varphi_q)}{dt} = u_{oq}^* - u_{oq} \tag{7.8}$$

$$i_{Ld}^* = Fi_{od} - \omega_0 C_f u_{oq} + k_{ui}\varphi_d + k_{up}(u_{od}^* - u_{od})$$
$$i_{Lq}^* = Fi_{oq} + \omega_0 C_f u_{od} + k_{ui}\varphi_q + k_{up}(u_{oq}^* - u_{oq}) \tag{7.9}$$

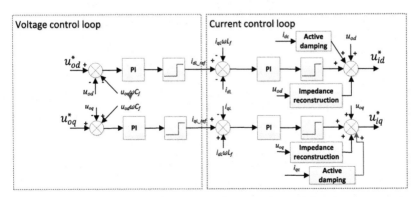

FIGURE 7.3   Block diagram of double loop control structure along with impedance reconstruction block.

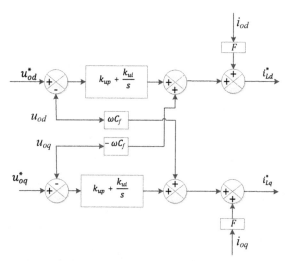

FIGURE 7.4 Voltage control structure.

where u_{od} and u_{oq} are the d and q axis voltage components which are calculated through Park transformation of output phase voltage before the isolating transformer. To get proper and fast control, the maximum and minimum limits are set for the PI regulator parameters. For the purpose of achieving decoupled control, $\omega_0 C_f u_{oq}$ and $\omega_0 C_f u_{od}$ are used as the feedback signals, and C_f is the filter capacitor.

Fig. 7.5 illustrates the current control structure which is the most important control loop of the whole control system. It contains feedback and feedforward terms. The current control should provide sufficient active damping for the output LC resonance. The output current control is achieved with a standard PI controller.

In Fig. 7.5, i_{Ld}^* is the reference of the d axis component which is the regulated output current of the voltage control loop; i_{Ld} is the d axis component of output current after the LC filter. To get proper and fast control, the maximum and minimum limits are set for the PI regulator; for the purpose of achieving decoupled control, $\omega_0 L_f i_{Lq}$ is used as the feedback; L_f is the filter inductance. On the other hand, i_{Lq}^* is the reference of the q axis component which is also the output of voltage control loop; i_{Lq} is the q axis component of output current after LC filter.

Accordingly, the state equations are given by Eq. (7.10) and expanded as (7.11).

$$\frac{d(\gamma_d)}{dt} = i_{Ld}^* - i_{Ld}$$

$$\frac{d(\gamma_q)}{dt} = i_{Lq}^* - i_{Lq} \tag{7.10}$$

$$u_d^* = -\omega_0 L_f i_{Lq} + k_{ii}\gamma_d + k_{ip}(i_{Ld}^* - i_{Ld}) + u_{d0}$$
$$u_q^* = \omega_0 L_f i_{Ld} + k_{ii}\gamma_q + k_{ip}(i_{Lq}^* - i_{Lq}) + u_{q0} \tag{7.11}$$

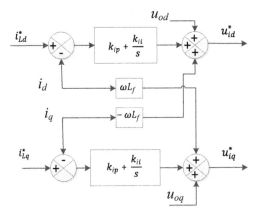

FIGURE 7.5 Current controller.

7.3.1.4 Modeling of low-pass filter

The electric network in the block diagram of Fig. 7.1 consists of a low-pass filter (LPF) and an isolating transformer. The isolating transformer can be represented as coupling inductance which is lumped with the LPF and forms an inductance-capacitance-inductance (LCL) LPF. The LPF in this design example is shown in Fig. 7.6.

In Fig. 7.6, R_d represents the resonant damping resistor.

The differential state equation of the converter electric network (or the filter) is shown as follows [6].

$$\frac{d(i_{dx})}{dt} = \frac{-r_f}{L_f}i_{dx} + \frac{1}{L_f}(u_d - u_{d0}) + \omega i_{qx}$$

$$\frac{d(i_{qx})}{dt} = \frac{-r_f}{L_f}i_{qx} + \frac{1}{L_f}\left(u_q - u_{q0}\right) - \omega i_{dx}$$
(7.12)

$$\frac{d(u_{d0})}{dt} = \frac{1}{C_f}(i_{dx} - i_{d0}) + \omega u_{q0}$$

$$\frac{d(u_{q0})}{dt} = \frac{1}{C_f}\left(i_{qx} - i_{q0}\right) - \omega u_{d0}$$
(7.13)

The differential equations of the interface circuit between microgrid and converter are expressed as:

$$\frac{d(i_{d0})}{dt} = \frac{-r_t}{L_t}i_{d0} + \frac{1}{L_t}(u_{d0} - u_{dg}) + \omega i_{q0}$$

$$\frac{d(i_{q0})}{dt} = \frac{-r_t}{L_t}i_{q0} + \frac{1}{L_t}(u_{q0} - u_{qg}) - \omega i_{d0}$$
(7.14)

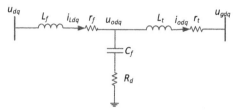

FIGURE 7.6 Low-pass filter.

7.3.1.5 Modeling the distribution (microgrid) network

A network model can be generally developed in such a way that n lines and m nodes with s inverters and p load points are built. On a common reference frame, the state equations of line current of ith line connected between buses j (inverter bus) and k (load bus) can be considered. However, for this chapter, two inverters, one connecting line, and one load are taken into consideration. In the general case of a distribution network, the dynamics equations of a line in this network are defined as

$$
\begin{aligned}
\frac{d(i_{lineD})}{dt} &= \frac{-R_{line}}{L_{line}} i_{lineD} + \omega i_{lineQ} + \frac{1}{L_{line}} u_{gDj} - \frac{1}{L_{line}} u_{gDk} \\
\frac{d(i_{lineQ})}{dt} &= \frac{-R_{line}}{L_{line}} i_{lineQ} - \omega i_{lineD} + \frac{1}{L_{line}} u_{gQj} - \frac{1}{L_{line}} u_{gQk}
\end{aligned}
\tag{7.15}
$$

Resistance R_{line} represents the copper loss component of the line. Inductance L_{line} is considered as the lumped inductance resulting from the line (cable).

7.3.1.6 Modeling the load

Loads in microgrids are usually inductive and/or resistive loads. This book also takes these two kinds of loads into consideration using a typical RL load connected to the microgrid bus (as in Fig. 7.7). To perform the dynamic stability analysis, load perturbation will be realized by changing the value of resistor and inductor. Differential equations for the load can be expressed in Eq. (7.16).

$$
\begin{aligned}
\frac{d(i_{LoD})}{dt} &= \frac{1}{L_{Lo}} (-R_{Lo} i_{LoD} + u_{gDk}) + \omega i_{LoQ} \\
\frac{d(i_{LoQ})}{dt} &= \frac{1}{L_{Lo}} (-R_{Lo} i_{LoQ} + u_{gQk}) - \omega i_{LoD}
\end{aligned}
\tag{7.16}
$$

7.3.1.7 Approximated linear model

The nonlinear equations above are derived to express the proposed system in this chapter which reflect the system dynamics. To get better analysis of the dynamic characteristic of microgrid, the prevalent method is to approximate the nonlinear system. The approximated linear design can establish the dynamic

FIGURE 7.7 Load model.

nonlinear model without the consideration of variable parameters and uncertainty. Thus the nonlinear equations are linearized around steady-state operating points to develop a state-space model of the microgrid. Then, an eigenvalue analysis is done using the linearized model to determine the small-signal stability of the system.

A general nonlinear binary function can be expressed as:

$$y = f(x_1 + x_2) \tag{7.17}$$

Thus the approximate linearity can be realized through the Taylor formula which can be expressed as:

$$
\begin{aligned}
y &= f(x_1 + x_2) = f(x_1 + \Delta x_1 + x_2 + \Delta x_2) \\
&= f(x_1 + x_2) + \frac{\partial f}{\partial x_1} \Delta x_1 + \frac{\partial f}{\partial x_1} \Delta x_2 + \frac{1}{2} \left(\frac{\partial f}{\partial x_1} \Delta x_1 + \frac{\partial f}{\partial x_1} \Delta x_2 \right)^2
\end{aligned} \tag{7.18}
$$

Ignoring the second-order partial derivative (as it has insignificant impact), the approximate binary function could be expressed as in (7.19).

$$y = f(x_1 + x_2) + \frac{\partial f}{\partial x_1} \Delta x_1 + \frac{\partial f}{\partial x_1} \Delta x_2 \tag{7.19}$$

The approximate linearity, which is also called the *Lyapunov* linearity equation, can be defined as:

$$y - f(x_1 + x_2) = \Delta y = \frac{\partial f}{\partial x_1} \Delta x_1 + \frac{\partial f}{\partial x_1} \Delta x_2 \tag{7.20}$$

As the power system could be expressed by differential-algebra equations, the solution of the equilibrium point will be the first step. Supposing x is the differential variable of the power system, y is the algebraic variable of the power system, and μ is the control variable of the power system, the mathematical model of the power system could be described as:

$$
\begin{cases}
\dot{x} = f(x, y, \mu) \\
0 = g(x, y, \mu)
\end{cases} \tag{7.21}
$$

In the mathematical model established in (7.21), f is the differential equation, and g is the algebra equation which is also the power flow equation. Thus if $\exists(x_0, y_0, \mu_0)$ exists, the equations in (7.22) are satisfied.

$$\begin{cases} 0 = f(x_0, y_0, \mu_0) \\ 0 = g(x_0, y_0, \mu_0) \end{cases} \tag{7.22}$$

All the solutions which satisfy Eq. (7.22) are known as equilibrium solutions. Such solutions can be described as in (7.23).

$$M = \{(x, y, \mu)/f(x, y, \mu) = 0, g(x, y, \mu) = 0\} \tag{7.23}$$

For the dynamic stability analysis, *Lyapunov* linearity is performed based on Eq. (7.22).

$$\begin{cases} \dfrac{dx}{dt} = D_x f(x_0, y_0)dx + D_y f(x_0, y_0)dx \\ 0 = D_x g(x_0, y_0)dx + D_y g(x_0, y_0)dx \end{cases} \tag{7.24}$$

where $D_x f(x_0, y_0)$ and $D_y f(x_0, y_0)$ are the partial differential equations of $f(x, y, \mu)$; and $D_x g(x_0, y_0)$ and $D_y g(x_0, y_0)$ are the partial differential equations of $g(x, y, \mu)$.

Simultaneously equating (7.24), Eq. (7.25) is obtained and can be simplified as in (7.26).

$$\frac{dx}{dt} = (D_x f(x_0, y_0) - D_y f(x_0, y_0)[D_y g(x_0, y_0)]^{-1} D_x f(x_0, y_0))dx \tag{7.25}$$

$$\dot{x} = (A - BD^{-1}C)x \tag{7.26}$$

where $A = D_x f(x_0, y_0)$, $B = D_y f(x_0, y_0)$, $C = D_x g(x_0, y_0)$, and $D = D_y g(x_0, y_0)$.

According to the *Lyapunov* stability method, the necessary and sufficient conditions of the stability of linear system are that there is no positive real part eigenvalue of matrix $(A - BD^{-1}C)$. Thus the state-space model is designed and verified with the eigenvalue analysis approach, and simulated results are presented in the following sections.

7.4 State-space model and analysis of dynamic disturbance stability

Eqs. (7.4)–(7.26) describes the mathematical model of grid-forming sources. A grid-forming source-based microgrid topology is depicted in Fig. 7.8.

As per the figure, two DERs are taken into consideration. DER1 is a lead−carbon battery with converter, while DER2 is a supercapacitor battery

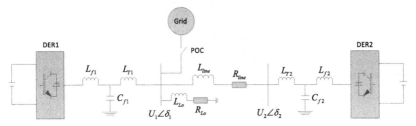

FIGURE 7.8 Topology of microgrid with two DERs.

with converter. L_{f1} represents the LPF inductance and L_{T1} is the equivalent inductance of isolating transformer of DER1. C_{f1} represents low-pass filter capacitance of DER1. POC is the point of connection between microgrid and utility grid. If the circuit breaker of POC is open, the microgrid will operate in island mode and the DERs become grid-forming sources. On the other hand, if the circuit breaker of POC is closed, the microgrid will operate in grid model and DERs are grid-following sources. L_{line} represents inductance of transmission line, R_{line} represents resistance of transmission line. L_{f2} represents the LPF inductance, and L_{T2} is the equivalent inductance value of the isolating transformer of DER2. C_{f2} represents LPF capacitance of DER2.

What should be emphasized here is that the nominal system voltage level of the microgrid bus is 10 kV and the voltage level of the primary side of the isolating transformer is 315 V. For the convenience of analysis, system voltage level is equivalent to 400 V. Phase angles of output voltage of DER1 and DER2 are set as δ_1 and δ_2, respectively.

The modeling approach presented in this chapter divides the whole system into three major submodules: inverter, network, and load (Fig. 7.9). Each inverter is modeled on its individual reference frame whose rotational frequency is set by its local power sharing controller. The inverter model includes the power sharing controller dynamics, output filter dynamics, coupling inductor dynamics, and voltage and current controller dynamics. These last two elements introduce high frequency dynamics which are apparent at peak and light load conditions and during large changes in load. The small-signal flow among the submodules shown in Fig. 7.9 will be explained in the following sections.

Network dynamics are generally neglected in small-signal modeling of conventional power systems. The reason behind this is that the time constants of rotating machines and their controls are much larger than those of the network. In the case of microgrids, the microsources are connected through inverters whose response times are very small and network dynamics would influence the system stability.

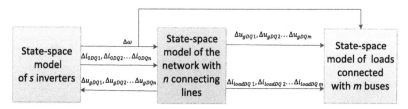

FIGURE 7.9 Block diagram of complete small-signal state-space model of a microgrid.

7.4.1 Designing small-signal stability model of microgrid

7.4.1.1 Power controller

Taking the differential Eqs. (7.1)–(7.7) into consideration and linearizing them, the small-signal power controller model can be written in a state-space form as:

$$
\begin{bmatrix} \Delta\dot\delta \\ \Delta\dot P \\ \Delta\dot Q \end{bmatrix} = A_p \begin{bmatrix} \Delta\delta \\ \Delta P \\ \Delta Q \end{bmatrix} + B_p \begin{bmatrix} \Delta i_{Ldq} \\ \Delta u_{odq} \\ \Delta i_{odq} \end{bmatrix} + B_{p\omega com}[\Delta\omega_{com}] \tag{7.27}
$$

$$
\begin{bmatrix} \Delta\omega \\ \Delta u^*_{odq} \end{bmatrix} = \begin{bmatrix} C_{p\omega} \\ C_{pu} \end{bmatrix} \begin{bmatrix} \Delta\delta \\ \Delta P \\ \Delta Q \end{bmatrix} \tag{7.28}
$$

where the matrices of the coefficients of state variables are

$$
A_p = \begin{bmatrix} 0 & -m & 0 \\ 0 & -\omega_c & 0 \\ 0 & 0 & -\omega_c \end{bmatrix}
$$

$$
B_p = \begin{bmatrix} 0 & 0 & 0 & 0 & 0 & 0 \\ 0 & 0 & 1.5\ \omega_c i_{od} & 1.5\ \omega_c i_{oq} & 1.5\ \omega_c u_{od} & 1.5\ \omega_c u_{oq} \\ 0 & 0 & 1.5\ \omega_c i_{oq} & -1.5\ \omega_c i_{od} & -1.5\ \omega_c u_{oq} & 1.5\ \omega_c u_{od} \end{bmatrix}
$$

$$
B_{p\omega} = \begin{bmatrix} -1 \\ 0 \\ 0 \end{bmatrix} \quad C_{p\omega} = \begin{bmatrix} 0 & -m & 0 \end{bmatrix} \quad \text{and} \quad C_{pu} = \begin{bmatrix} 0 & 0 & -n \\ 0 & 0 & 0 \end{bmatrix}
$$

7.4.1.2 Voltage controller

For the voltage controller state equations and algebraic equations of (7.8) and (7.9), respectively, are linearized and the small-signal model can be written in a state-space form as:

$$
[\Delta\dot\varphi_{dq}] = [0][\Delta\varphi_{dq}] + B_{u1}[\Delta u^*_{odq}] + B_{u2} \begin{bmatrix} \Delta i_{ldq} \\ \Delta u_{odq} \\ \Delta i_{odq} \end{bmatrix} \tag{7.29}
$$

$$[\Delta i^*_{Ldq}] = C_u[\Delta\varphi_{dq}] + D_{u1}[\Delta u^*_{odq}] + D_{u2}\begin{bmatrix}\Delta i_{Ldq}\\\Delta u_{odq}\\\Delta i_{odq}\end{bmatrix}\qquad(7.30)$$

where the matrices of the coefficients of state variables are

$$B_{u1} = \begin{bmatrix}1 & 0\\0 & 1\end{bmatrix}; B_{u2} = \begin{bmatrix}0 & 0 & -1 & 0 & 0 & 0\\0 & 0 & 0 & -1 & 0 & 0\end{bmatrix}$$

$$C_u = \begin{bmatrix}k_{ui} & 0\\0 & k_{ui}\end{bmatrix}; D_{u1} = \begin{bmatrix}k_{up} & 0\\0 & k_{up}\end{bmatrix}; D_{u2} = \begin{bmatrix}0 & 0 & -k_{up} & -\omega_0 C_f & F & 0\\0 & 0 & \omega_0 C_f & -k_{up} & 0 & F\end{bmatrix}$$

7.4.1.3 Current controller

The current controller small-signal equations are derived by linearizing the state equations of (7.10) at steady-state operating point and equating with Eq. (7.11).

$$[\Delta\dot\gamma_{dq}] = [0][\Delta\gamma_{dq}] + B_{i1}[\Delta i^*_{Ldq}] + B_{i2}\begin{bmatrix}\Delta i_{Ldq}\\\Delta u_{odq}\\\Delta i_{odq}\end{bmatrix}\qquad(7.31)$$

$$[\Delta u^*_{dq}] = C_i[\Delta\gamma_{dq}] + D_{i1}[\Delta i^*_{Ldq}] + D_{i2}\begin{bmatrix}\Delta i_{Ldq}\\\Delta u_{odq}\\\Delta i_{odq}\end{bmatrix}\qquad(7.32)$$

where the matrices of the coefficients of state variables are

$$B_{i1} = \begin{bmatrix}1 & 0\\0 & 1\end{bmatrix}; B_{i2} = \begin{bmatrix}-1 & 0 & 0 & 0 & 0 & 0\\0 & -1 & 0 & 0 & 0 & 0\end{bmatrix}; C_i = \begin{bmatrix}k_{ii} & 0\\0 & k_{ii}\end{bmatrix};$$

$$D_{i1} = \begin{bmatrix}k_{ip} & 0\\0 & k_{ip}\end{bmatrix}; D_{i2} = \begin{bmatrix}-k_{ip} & -\omega_0 L_f & 0 & 0 & 0 & 0\\\omega_0 L_f & -k_{ip} & 0 & 0 & 0 & 0\end{bmatrix}$$

7.4.1.4 Low-pass filter

The LPF small-signal model is achieved by linearizing equations of (7.12)–(7.14) at the steady-state operating point, and thus the following small-signal model is attained.

$$\begin{bmatrix}\Delta i_{Ldq}\\\Delta u_{odq}\\\Delta i_{odq}\end{bmatrix} = A_{LPF}\begin{bmatrix}\Delta i_{Ldq}\\\Delta u_{odq}\\\Delta i_{odq}\end{bmatrix} + B_{LPF1}[\Delta u_{dq}] + B_{LPF2}[\Delta u_{gdq}] + B_{LPF3}[\Delta\omega]$$

$$(7.33)$$

where the matrices of the coefficients of state variables are

$$A_{LPF} = \begin{bmatrix} -r_f/L_f & \omega & -1/L_f & 0 & 0 & 0 \\ -\omega & -r_f/L_f & 0 & -1/L_f & 0 & 0 \\ 1/C_f & 0 & 0 & \omega & -1/C_f & 0 \\ 0 & 1/C_f & -\omega & 0 & 0 & -1/C_f \\ 0 & 0 & 1/L_t & 0 & -r_t/L_t & \omega \\ 0 & 0 & 0 & 1/L_t & -\omega & -r_t/L_t \end{bmatrix}; B_{LPF1} = \begin{bmatrix} 1/L_f & 0 \\ 0 & 1/L_f \\ 0 & 0 \\ 0 & 0 \\ 0 & 0 \\ 0 & 0 \end{bmatrix}$$

$$B_{LPF2} = \begin{bmatrix} 0 & 0 \\ 0 & 0 \\ 0 & 0 \\ 0 & 0 \\ -1/L_t & 0 \\ 0 & -1/L_t \end{bmatrix}; B_{LPF3} = \begin{bmatrix} i_{Lq} \\ -i_{Ld} \\ u_{oq} \\ -u_{od} \\ i_{oq} \\ -i_{od} \end{bmatrix}$$

$[\Delta u_{dq}]$ can be replaced by $[\Delta u_{dq}^*]$ assuming that the inverter produces the demanded voltage ($u_{dq} = u_{dq}^*$) by ignoring the power losses in the converter.

To connect an inverter to the whole system, the output variables (i_{odq}) need to be converted to the common reference frame. Using the transformation technique of Fig. 7.10, the small-signal output current on the common reference frame can be obtained. The small-signal equivalent of the reference transformation is

$$[\Delta i_{oDQ}] = [T_S][\Delta i_{odq}] + [T_C][\Delta \delta] \tag{7.34}$$

where

$$T_S = \begin{bmatrix} \cos(\delta_0) & -\sin(\delta_0) \\ \sin(\delta_0) & \cos(\delta_0) \end{bmatrix}$$

$$T_C = \begin{bmatrix} -i_{od}\sin(\delta_0) - i_{oq}\cos(\delta_0) \\ i_{od}\cos(\delta_0) - i_{oq}\sin(\delta_0) \end{bmatrix}$$

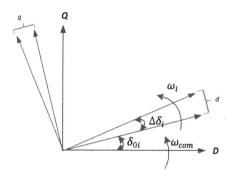

FIGURE 7.10 Reference frame transformation.

Similarly, the input signal to the inverter model (u_{gDQ}) is expressed on the common reference frame. The bus voltage can be converted to the individual inverter reference frame using reverse transformation, given by

$$[\Delta u_{gdq}] = [T_S^{-1}][\Delta u_{gDQ}] + [T_U][\Delta \delta] \tag{7.35}$$

where

$$T_S^{-1} = \begin{bmatrix} \cos(\delta_0) & \sin(\delta_0) \\ -\sin(\delta_0) & \cos(\delta_0) \end{bmatrix}$$

$$T_U = \begin{bmatrix} -u_{gD}\sin(\delta_0) + u_{gQ}\cos(\delta_0) \\ -u_{gD}\cos(\delta_0) - u_{gQ}\sin(\delta_0) \end{bmatrix}$$

Here, the axis set ($D-Q$) is the common reference frame rotating at a frequency ω_{com}, whereas axes ($d-q$)i and ($d-q$)j are the reference frame of ith and jth inverters rotating at ω_i and ω_j, respectively.

A complete state-space small-signal model of the inverter can be obtained by combining the state-space models of the power controller, voltage controller, current controller, and the output of LPF. Taking state-space equations of (7.27)–(7.35), the generalized state-space model of the inverter is derived as:

$$\Delta \dot{x}_{inv} = A_{inv}[\Delta x_{inv}] + B_{inv}[\Delta u_{gDQ}] + B_\omega[\Delta \omega_{com}] \tag{7.36}$$

$$\begin{bmatrix} \Delta \omega \\ \Delta i_{oDQ} \end{bmatrix} = \begin{bmatrix} C_{inv\omega} \\ C_{invI} \end{bmatrix}[\Delta x_{inv}] \tag{7.37}$$

The matrices of A_{inv}, B_{inv}, B_ω, $C_{inv\omega}$, and C_{invI} are determined with the matrices of Eqs. (7.27)–(7.35). The variations of $\Delta \omega_{com}$ and $\Delta \omega$ are designed to be the same.

7.4.1.5 Distribution network model
The state-space model for the distribution network is derived by linearizing the state equations of (7.15) at steady-state operating point, thus we can have the following state-space model.

$$\Delta\left(\frac{d(i_{lineDQ})}{dt}\right) = A_{NW}[\Delta i_{lineDQ}] + B_{NW1}[\Delta u_{gDQ}] + B_{NW2}[\Delta \omega] \tag{7.38}$$

7.4.1.6 Load model
The state-space model for the load is also deduced by linearizing the equations of (7.16):

$$\Delta\left(\frac{d(i_{LoDQ})}{dt}\right) = A_{Lo}[\Delta i_{LoDQ}] + B_{Lo1}[\Delta u_{gDQ}] + B_{Lo2}[\Delta \omega] \tag{7.39}$$

7.4.1.7 Virtual resistor model

When bus voltages were used as an input to the system, effects of load perturbation could not be accurately predicted. In practice, the only perturbation that occurs in the system comes from the step change in load. A method is needed to include the terms relating to the bus voltages in the system "A" matrix. To do this, a virtual resistor can be assumed connected at the inverter bus.

Using Kirchhoff's voltage law in Fig. 7.11, the equations describing the bus voltage in terms of the inverter, load, and line currents can be generally expressed as in Eq. (7.40).

$$u_{gD} = R_{vir}(i_{oD} + i_{lineD} - i_{LoD})$$
$$u_{gQ} = R_{vir}(i_{oQ} + i_{lineQ} - i_{LoQ})$$
(7.40)

The algebra equations of (7.40) can be described with the Lyapunov linearity method and are expressed in (7.41).

$$\Delta u_{gDQ} = R_{vir}\left(M_{inv}[\Delta i_{oDQ}] + M_{Lo}[\Delta i_{LoDQ}] + M_{NW}[\Delta i_{lineDQ}]\right)$$
(7.41)

The dimensions and the elements of matrices in (7.41) should be defined and stated carefully.

- Matrix R_{vir} depends on the number of buses m and hence its size is set by $2m \times 2m$, whose diagonal elements are equal to R_{vir}.
- Matrix M_{inv} depends on m and number of inverters s, then the matrix M_{inv} is with the size of $2m \times 2s$, which maps the inverter connection points onto distribution network buses. For example, if ith inverter is connected at jth bus, the element $M_{inv}(j, i)$ will be 1 and all the other elements in that row will be 0.
- Matrix M_{Lo} depends on m and number of loads connecting points p, then whose size is stated by $2m \times 2p$ which maps load connecting points onto the network nodes with 1.
- The size of matrix M_{NW} is devised by $2m \times 2n$; where n is the number of connecting lines in the distribution network, then the matrix maps the

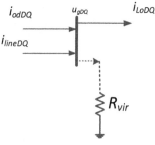

FIGURE 7.11 Virtual resistor model.

connecting lines onto the network buses. Here care should be taken to put either 1 or -1 based on whether the given line current is entering or leaving the bus.

As stated earlier, this section considers two inverters, one connecting line, and one load in the microgrid system for investigating the characteristics of the dynamic stability of the proposed control system. Substituting (7.37) for Δi_{oDQ} in (7.39) and (7.41); and substituting (7.41) for Δu_{gDQ} in (7.36), (7.38), and (7.39), we will obtain the complete microgrid small-signal state-space model and hence the system state matrix can be given by A_{sys}.

$$\begin{bmatrix} \Delta \dot{x}_{inv} \\ \Delta i_{lineDQ} \\ \Delta i_{loadDQ} \end{bmatrix} = A_{sys} \begin{bmatrix} \Delta x_{inv} \\ \Delta i_{lineDQ} \\ \Delta i_{loadDQ} \end{bmatrix} \tag{7.42}$$

The complete system state matrix A_{sys} is given in the following equation.

$$A_{sys} = \begin{bmatrix} A_{inv} + B_{inv}R_{vir}M_{inv}C_{invI} & B_{inv}R_{vir}M_{NW} & B_{inv}R_{vir}M_{Lo} \\ B_{NW1}R_{vir}M_{inv}C_{invI} + B_{NW2}C_{inv\omega} & A_{net} + B_{NW1}R_{vir}M_{NW} & B_{NW1}R_{vir}M_{Lo} \\ B_{Lo1}R_{vir}M_{inv}C_{invI} + B_{Lo2}C_{inv\omega} & B_{Lo1}R_{vir}M_{NW} & A_{Lo} + B_{Lo1}R_{vir}M_{Lo} \end{bmatrix}$$
$$\tag{7.43}$$

Considering Fig. 7.8 with two DERs, one connecting line and a load, a total of 30 state variables are contained in the system. The state variables of the system are given by

$$\Delta x = [\Delta x_{inv1}, \Delta x_{inv2}, \Delta i_{lineD}, \Delta i_{lineQ}, \Delta i_{loadD}, \Delta i_{loadQ}]^T \tag{7.44}$$

where

$$\Delta x_{inv1} = [\Delta \delta_1, \Delta P_1, \Delta Q_1, \Delta \varphi_{d1}, \Delta \varphi_{q1}, \Delta \gamma_{d1}, \Delta \gamma_{q1},$$
$$\Delta i_{Ld1}, \Delta i_{Lq1}, \Delta u_{od1}, \Delta u_{oq1}, \Delta i_{od1}, \Delta i_{oq1}]$$

$$\Delta x_{inv2} = [\Delta \delta_2, \Delta P_2, \Delta Q_2, \Delta \varphi_{d2}, \Delta \varphi_{q2}, \Delta \gamma_{d2}, \Delta \gamma_{q2},$$
$$\Delta i_{Ld2}, \Delta i_{Lq2}, \Delta u_{od2}, \Delta u_{oq2}, \Delta i_{od2}, \Delta i_{oq2}]$$

In this application, the inductance of the transformer is considered as a coupling inductance which can provide reasonable coupling impedance between the inverter output and the connection bus with a good bus voltage regulation.

7.4.2 Eigenvalues and analysis of state-space model of dynamic control

A complete model was devised by considering the initial conditions of the system. These initial conditions were obtained at the steady-state operating point by setting a Power Systems Computer Aided Design (PSCAD) software simulation time-step. It is also possible to obtain these conditions from a more general load-flow solution as is often done in a conventional power system. The system parameters and initial conditions are listed, respectively, in Tables 7.1 and 7.2.

TABLE 7.1 System parameters.

Parameter	Value	Parameter	Value	Parameter	Value
f	50 Hz	m	0.00628	k_{up}	100
L_f	200 µH	n	25.7 µH	u_{gD1}	247 V
r_f	0.1 Ω	L_{Lo}	0.2 mH	u_{gQ1}	−12 V
C_f	266 µF	R_{Lo}	1 Ω	u_{gD2}	247 V
R_d	0.4 Ω	k_{ii}	0.35	u_{gQ2}	−12 V
L_t	65 µH	k_{ip}	1.0^{-4}	ω_c	628 rad/s
r_t	0.026 Ω	k_{ui}	2		

The objective of this section is to investigate the stability of the storage units which provide for the chosen values of droop gains and designed parameters of the controllers along with system parameters. Accordingly, the eigenvalues for the matrix "A_{sys}" are determined and listed in Table 7.3.

7.4.2.1 Eigenvalue sensitivity to filter inductance L_f

This subsection presents the sensitivity of eigenvalues to the inductance L_f. Fig. 7.12 shows the locus of low-frequency modes of eigenvalues subsequent to a change in L_f in the range of 0.1 mH ≤ L_f ≤ 0.8 mH. Some higher-frequency mode of eigenvalues travel toward the left half plane, whereas most of the rest of the eigenvalues move toward the right half planes as L_f increases. The lower-frequency modes of eigenvalues quickly approach to the right half plane. As shown in Fig. 7.12, those of the eigenvalues close to the origin which are very-low-frequency modes move speedily and enter the unstable region.

While these inductance values are rather large, they represent the stability limit of the system. Hence, a robust method should be devised to significantly manage the parameter variations and uncertainties.

7.4.2.2 Eigenvalue sensitivity to the transformer inductance L_t

In this subsection, the sensitivity of eigenvalues to the inductance L_t is studied. Fig. 7.13 shows the locus of all eigenvalues of the system subsequent to a change in L_t in the range of 0.01 mH ≤ L_t ≤ 0.07 mH. Some eigenvalues travel toward left half plane (the stability region). However, the most move toward the right half plane as L_t increases. Eigenvalues with low-frequency modes travel toward the right half plane as L_t increases. Some sensitive eigenvalues quickly approach and enter the right half plane and become unstable. Some eigenvalues with low-frequency modes go further to the left half plane as L_t increases, and some eigenvalues also approach and enter speedily the right half plane.

TABLE 7.2 Initial conditions of the state variables.

Parameter	Value	Parameter	Value	Parameter	Value	Parameter	Value
δ_1	0.00	i_{Lq1}	−100 A	φ_{d2}	0.13152 Vs	i_{od2}	370 A
P_1	281.5 kW	u_{od1}	251 V	φ_{q2}	0.00342 Vs	i_{oq2}	−80 A
Q_1	33.5 kVAR	u_{oq1}	−18 V	γ_{d2}	0.86569 As	i_{LoD}	11.1 kA
φ_{d1}	0.13152 Vs	i_{od1}	730 A	γ_{q2}	0.00147 As	i_{LoQ}	−170 A
φ_{q1}	0.00342 Vs	i_{oq1}	−140 A	i_{Ld2}	730 A		
γ_{d1}	0.86569 As	δ_2	−0.55805	i_{Lq2}	−20 A		
γ_{q1}	0.00147 As	P_2	141 kW	u_{od2}	250 V		
i_{Ld1}	730 A	Q_2	19 kVAR	u_{oq2}	−25 V		

TABLE 7.3 Eigenvalues of the system matrix A_{sys} R_{vir1} = 0.024 Ω and R_{vir2} = 0.0075 Ω.

Index	Real (1/s)	Imaginary (rad/s)	ω_0^2 (Hz)	Major participants
1, 2	−5.28E + 03	± 1.30E + 04	14,052	i_{lineDQ}
3, 4	−4.54E + 03	± 1.29E + 04	13,707	i_{odq1}, i_{odq2}
5, 6	−1.40E + 03	± 1.18E + 04	11,909	$i_{Ldq1}, i_{Ldq2}, u_{odq1}, u_{odq2}$
7, 8	−1.35E + 03	± 1.12E + 04	11,305	u_{odq1}, u_{odq2},
9, 10	−5.26E + 03	± 2.95E + 02	5272	$u_{odq1}, u_{odq2}, i_{Ldq2}$
11, 12	−3.00E + 03	± 2.97E + 02	3013	u_{odq1}, u_{odq2}
13	−1.70E + 03	0.00E + 00	1699	
14, 15	−6.56E + 02	± 7.80E + 02	1020	$P_1, Q_1, i_{Ld1}, i_{oq1}$
16, 17	−8.94E + 02	± 4.59E + 02	1005	$P_2, Q_2, i_{Ldq2}, i_{odq2}$
18	−6.60E + 02	0.00E + 00	660	
19	−6.25E + 02	0.00E + 00	625	
20	−3.19E + 02	0.00E + 00	319	
21, 22	−4.98E + 00	± 5.21E + 00	7	φ_{dq2}
23, 24	−3.91E + 00	± 1.92E + 00	4	φ_{dq1}
25, 26	−2.86E-04	± 7.79E − 09	0	γ_{dq1}
27, 28	−2.86E-04	± 6.36E − 09	0	γ_{dq2}
29	0.00E + 00	0.00E + 00	0	δ_1
30	0.00E + 00	0.00E + 00	0	δ_2

7.4.2.3 Eigenvalue sensitivity to real power droop gain m

Fig. 7.14 shows the locus of the eigenvalues as a function of the real power droop gain. It is traced with the ranges of the droop gains $m_1 = 0.116 \times 10^{-3}$ to 12.116×10^{-3} and $m_2 = 0.133 \times 10^{-3}$ to 12.133×10^{-3}. The incremental value of the droop gains is along the arrow direction. The eigenvalues marked with an arrow are largely sensitive to the state variables of real power part of the power controller.

As m is increased, some of the low-frequency modes of eigenvalues move toward the right half plane. As shown in Fig. 7.14, some modes of eigenvalues which are close to the origin enter into the instability region earlier than the others. Of course, some of them stay in the stability region. It is to be noted that low droop gain is necessary to improve the transient response of DGs, whereas the LPF with low cutoff frequency is needed to achieve good attenuation of high frequency

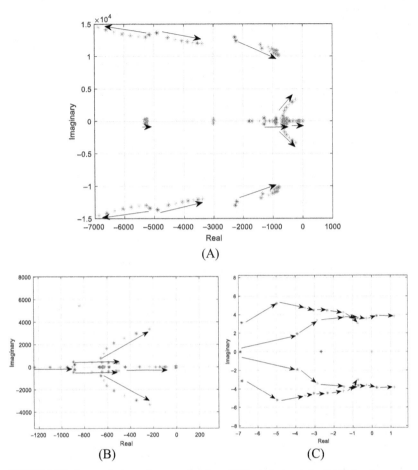

FIGURE 7.12 Locus of eigenvalues as a function of filter inductance L_f: (A) higher-frequency modes of eigenvalues, (B) lower-frequency, and (C) very-low-frequency modes of eigenvalues.

distortion components in the measured power and to avoid any interaction with inner current controllers. So, the appropriate droop gains corresponding to the LPFs should be determined so as to achieve a stable system along with good attenuation and to avoid any interaction with inner loop controllers. Generally, the inverter is the most critical element from the point of view of system stability.

7.4.2.4 Eigenvalue sensitivity to reactive power droop gain n

In this subsection, the sensitivity of eigenvalues to the reactive power droop gains n_1 and n_2 is analyzed. Fig. 7.15 shows the locus of eigenvalues to a change in reactive power droop gains for the values in the range of $n_1 = n_2 = 0.57 \times 10^{-5}$ to 6.57×10^{-5}. The eigenvalues located far away from the real axis exhibit slightly larger imaginary parts. On the other hand, the eigenvalues located near

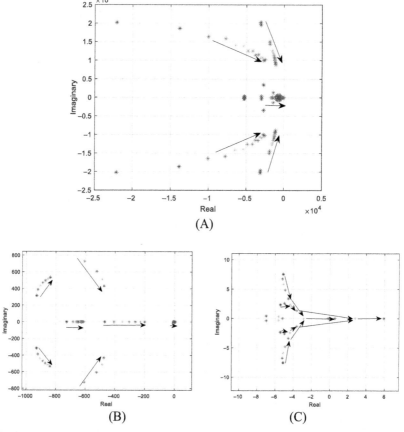

FIGURE 7.13 Locus of eigenvalues as a function of the inductance L_t: (A) higher-frequency modes of eigenvalues, (B) lower-frequency, (C) very-low-frequency modes of eigenvalues.

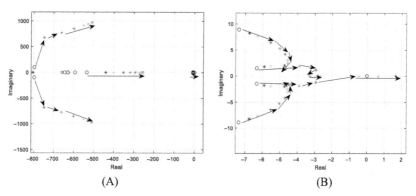

FIGURE 7.14 Locus of low-frequency modes of eigenvalues as a function of real power droop gain (m): (A) for all-range frequency modes and (B) lower-frequency modes of eigenvalues.

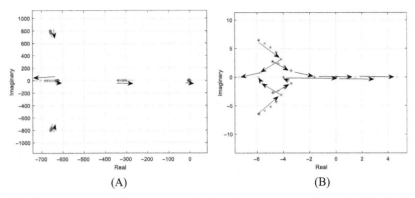

FIGURE 7.15   Locus of eigenvalues as a function of reactive power droop gain "n": (A) for all-range frequency modes and (B) lower-frequency modes of eigenvalues.

the origin show slightly reduced real and imaginary parts as shown in Fig. 7.15. Some eigenvalues stay in the stability region while n gets increased. However, one pair of very-low-frequency modes of eigenvalues travels to right half plane as n is increased and enter the instability region when n gets more increased. The droop gain values to let eigenvalues enter into the instability region are about $n = 4.5 \times 10^{-5}$. Generally, the low-frequency dominant modes are less sensitive to the reactive power droop gain compared to the active power droop gain.

7.4.2.5   Eigenvalue sensitivity to virtual resistance R_{vir}

Some eigenvalues' real parts are more negative compared to the rest of the eigenvalues associated with virtual resistance. The two high-frequency modes of eigenvalues with higher imaginary parts primarily correspond to the virtual resistor which become much more negative as the virtual resistor is increased. Some high-frequency modes of eigenvalues go to a convergence point while the virtual resistance is increased (shown in Fig. 7.16). Some low-frequency modes of eigenvalues fast travel to more negative values but decrease even though the virtual resistance is increased. One eigenvalue which is a low-frequency mode moves toward the positive direction but gets exhausted at some point in the stability region (shown in Fig. 7.16).

In general, as the virtual resistance is increased, the stability margin of the system is increased. The stability margin for other parameters also moves to the right side to have a wider stability region. Large values of virtual resistance are not without their disadvantages either, which will be elaborated in the later sections. Hence, the appropriate value of virtual resistance should be chosen.

Generally, the small-signal modeling and analysis based on state-space model is very helpful to investigate the effect of the LPF parameters and droop gains as well as the virtual resistance on the stability of microgrid system. It is also used to design the microgrid system which can contain many components. Accordingly, a

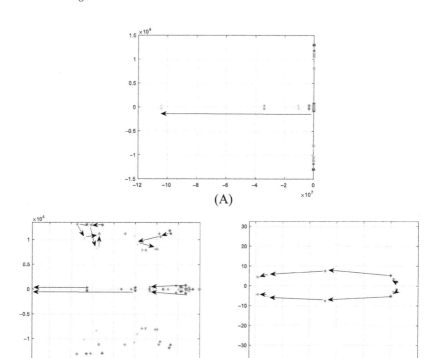

FIGURE 7.16 Locus of the eigenvalues as virtual resistance varies ($R_{vir1} = 0.024 - 240 \, \Omega$ and $R_{vir2} = 0.0075 - 750 \, \Omega$): (A) higher-frequency modes of the eigenvalues, (B) low-frequency modes of the eigenvalues, and (C) very-low-frequency modes of the eigenvalues.

better possible control strategy can be devised so as to achieve better performance along with reasonable values of system parameters. An advanced control strategy modeling is presented in the following section.

7.5 Active damping and impedance reconstruction for improving dynamic stability

In the previous section, the dynamic disturbance stability and its control system has been designed, discussed, and analyzed with simulated results. The filter parameters sensitivity analyses have been presented and discussed with the eigenvalue analysis approach. In this section, a control method to attain better performance is discussed.

In the literature, the technical limitations on filter parameters have been stated as:

- The value of the capacitance is limited by the decrease of the power factor, that has to be less than 5% at the rated power;

- The total value of the filter inductance has to be less than 0.1 p.u. for low power filters. However, for high power levels, the main aim is to avoid the saturation of the inductors; and
- In the design of the inverter output filter, the attenuation of ripple in the output voltage is possibly achieved in such a way that a common rule of thumb for the resonant frequency, which is 10 times more than the system frequency and less than half of the switching frequency, is likely required to avoid resonance problems in the lower and upper parts of the harmonic spectrum.

The LCL filter aims to reduce high-order harmonics on the grid side, but a poor filter design can cause a distortion increase because of oscillation effects. In fact, the converter current harmonics may cause saturation of the inductors or filter resonance. Therefore the inductors should be correctly designed considering current ripple, and the filter should be damped to avoid resonances. However, the damping level is limited by cost, the value of the inductors, losses, and degradation of the filter performance [6].

Generally, the value of the filter inductor is chosen to have low ripple content in the inductor current but, when combined with the capacitor value, it creates a resonant frequency since such a higher order LCL filter has inherently resonant poles. These poles are capable of degrading the control performance unless a proper damping method is introduced. A large coupling inductance also results in a poor bus voltage regulation. The resonant frequency which has an impact to increase the THD can be defined by

$$f_{res} \approx \frac{1}{2\pi} \sqrt{\frac{L_f + L_t}{L_f L_t C_f}} \tag{7.45}$$

The simulated results of microgrid bus voltages, power flows, and system frequency are shown in Fig. 7.17. The results are made when the damping

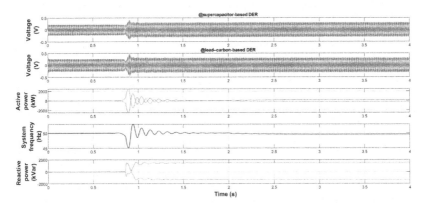

FIGURE 7.17   The simulated results without damping method.

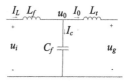

FIGURE 7.18 The LPF without damping element.

element is not used. When the power starts flowing in the network, the instability of voltage and frequency occur due to the effect of resonant frequency of the LPF.

Let us consider Fig. 7.18 with the absence of damping resistance R_d. The transfer function between output current versus the input voltage for the filter is given by Eq. (7.46). Substituting the filter parameters from Table 7.1 into (7.45), the resonant frequency f_{res} is computed to be as 1.757 kHz. This resonant frequency is greater than 10 times of the system frequency but greater than half of the switching frequency (which was chosen as 3000 Hz in our simulation and experimental cases), which violates the common rule of thumb and will be the cause of instability.

Taking Fig. 7.18, we can obtain the transfer function of filter output current versus input voltage as stated in (7.46). To compute the transfer function of the filter, the grid voltage is assumed to be an ideal voltage source and it represents a short circuit for harmonic frequencies, and for the filter analysis it is set to zero ($u_g = 0$).

$$H_1(s) = \frac{I_0}{U_i} = \frac{1}{s^3 L_f L_t C + s(L_f + L_t)} \qquad (7.46)$$

The bode plot diagram shown in Fig. 7.19 verifies the above explanation. This implies that when the DERs come together for sharing power, instability occurs. As the simulated results show in Fig. 7.17, when the power sharing between two DERs starts at about 0.8 seconds, the voltage stability is no longer ensured.

Thus the damping technique needs to be designed so as to suppress the resonance. There have been different damping techniques used, such as passive damping [damping resistor (R_d) connected in series with the filter capacitor, as shown in Fig. 7.20], and active damping techniques.

Passive damping can damp and suppress resonance in all conditions. The value of R_d can be determined by using Eq. (7.47) [7].

$$R_d = \frac{1}{3\omega_{res} C_f} \approx 0.1 \ \Omega \qquad (7.47)$$

This damping resistance is supposed to avoid resonance problems in the lower and upper parts of the harmonic spectrum.

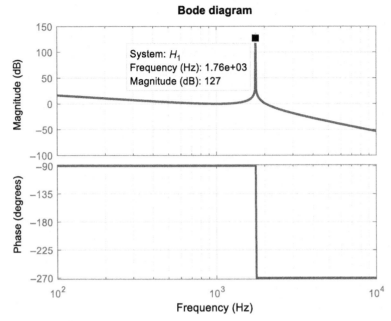

FIGURE 7.19   The bode diagram of Eq. (7.46).

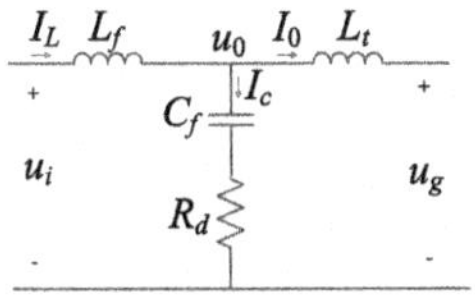

FIGURE 7.20   Series resistor passive damping circuitry.

Based on this passive damping topology, the transfer functions of the filter are to be defined. Thus the transfer functions for the output current versus inverter output voltage are given by:

$$H_2(s) = \frac{I_0}{U_i} = \frac{1 + CR_d s}{s^3 L_f L_t C + s^2 (L_f + L_t)CR_d + s(L_f + L_t)} \qquad (7.48)$$

In the real case, as also considered in the state-space analysis above, the inductances have their own resistances, and the transfer functions can be expressed by:

$$H_3(s) = \frac{I_0}{U_i} = \frac{1 + sCR_d}{(sL_f + r_f)(1 + s(Cr_t + CR_d) + s^2 CL_t) + (sL_t + r_t)(1 + sCR_d)} \qquad (7.49)$$

Nevertheless, since those resistances are small, using Eq. (7.48) is essentially valid for the sake of ease of analysis.

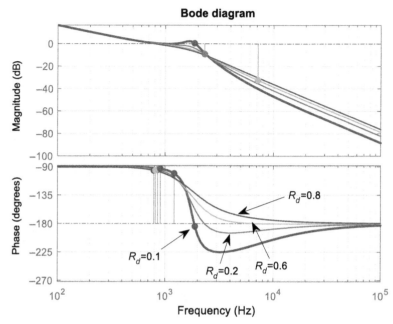

FIGURE 7.21 The bode diagram of Eq. (7.48) with the different damping resistance values.

Thus based on the transfer functions of (7.48), the bode plot diagrams are drawn as shown in Figs. 7.21 and 7.23. The bode diagram shows the effects of damping resistance variation. According to the computed value of the damping resistance, the bode diagram has negative gain margin and phase margin, which will cause the system to be still unstable. Basically, Eq. (7.47) is an approximated formula. When we increase the damping resistance, we will get larger phase and gain margins and will have a stable system. Thus the passive damping resistor, giving larger series resistance, can give better damping or lower "Q" as made clear by the transfer function after damping.

However, the issues of power level, damping requirement, and efficiency (or losses) are typically faced when such passive damping methods are applied. To solve such problems, active damping methods have been preferably used.

Active damping can be provided using the same concept as passive damping where the resonance oscillations in a network are damped by connecting a resistor in series to the filter capacitor while the resistor is a virtual one this time. By modifying the control algorithm, a similar behavior can be achieved without using a real resistor. Thus no additional power losses are generated. As illustrated in Fig. 7.22, the approximate circuit diagram of an active damping-based system is drawn to represent the equivalent passive damping circuit of Fig. 7.20.

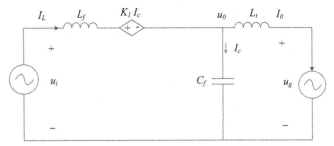

FIGURE 7.22 The circuit diagram representation for active damping.

Accordingly, the transfer functions are derived and expressed as

$$H_4(s) = \frac{I_0}{U_i} = \frac{1}{s^3 C L_f L_t + s^2 K_1 C L_t + s(L_f + L_t)} \qquad (7.50)$$

Active damping can effectively increase the lowering of gain margin while power flow varies in the system. Active damping is designed with the mathematical equation given by

$$u_{dqc}(s) = K_r * i_{dqc} \qquad (7.51)$$

where K_r is the damping coefficient and is determined based on how much gain margin needs to be increased (Fig. 7.23).

Even though the amplitude margin is well regulated by the active damping method, the phase margin is forcefully decreased as resonant frequencies exist while the power flow becomes larger. The simulated and experimental results both validate the concepts hypothesized and discussed in this chapter. As shown in Fig. 7.24, the phase margin (mainly) and the gain margin decrease when the power increases more than some limits.

To eliminate such oscillating frequencies and to increase the phase margin and gain margin importantly, the high-pass voltage-controlled voltage source can be designed as shown in Fig. 7.25, since the high resonant frequencies have a big impact in the operation, especially when high power is demanded.

Thus the transfer function of the virtually damped LPF circuit of Fig. 7.25 is defined by

$$H_5(s) = \frac{I_0}{U_i} = \frac{1}{s^3 C L_f L_t + s^2 K_1 C L_t + s(L_f + (H_{pass} + 1)L_t)} \qquad (7.52)$$

But H_{pass} is defined by $H_{pass}(s) = K_i \frac{s}{s + \omega_{hc}} u_0$.

Hence, by varying the value of K_i in $H_{pass}(s)$, the bode plot diagram for (7.52) is drawn as shown in Fig. 7.26. According to this figure, larger phase margin is found due to the high-pass element being added to the input of LPF.

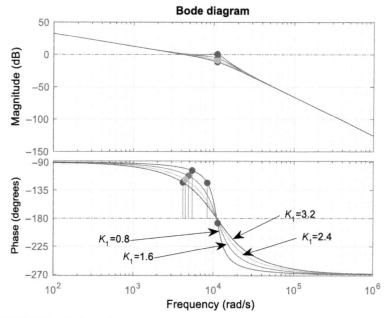

FIGURE 7.23 The bode diagram of Eq. (7.50) with different damping values.

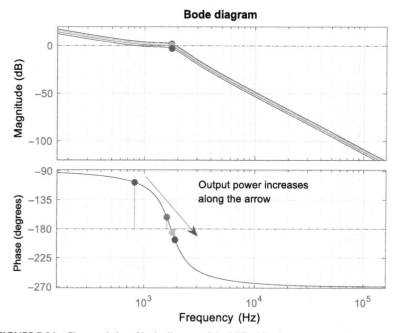

FIGURE 7.24 Characteristics of bode diagram of the LPF while the output power increases.

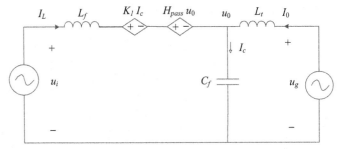

FIGURE 7.25 Circuit representation of impedance reconstruction along with active damping.

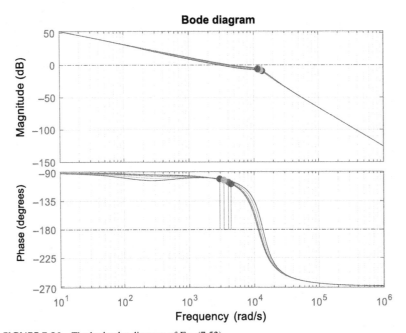

FIGURE 7.26 The bode plot diagram of Eq. (7.52).

Thus the concept raised above is verified by simulation result (shown in Fig. 7.27) with the response to increased power flow in the network. Fig. 7.27 demonstrates the phase margin characteristics when the output power increases. According to the figure, there is no significant phase margin reduction to make the system unstable while power increases in the network. The power is intentionally increased by 25% in each testing case, but the phase margin reduction is very minimal.

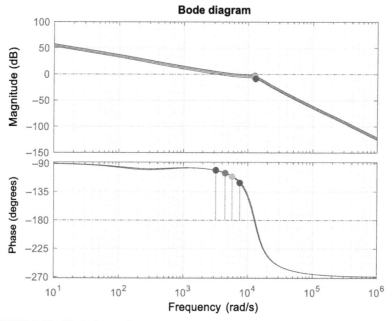

FIGURE 7.27 The bode plot diagram of Eq. (7.52) if the power is varying.

7.5.1 Realization of control strategy

The realization of active damping and impedance reconstruction control methods should also be carried out through the transfer functions of the overall system including circuit of the LPF and the control system. This procedure is presented in this section. We have already hypothesized and verified that active damping through the application of virtual resistance can be an effective approach. As the name implies, "virtual" resistances do not basically exist for real and they are realized through the control structure. Let us start with the state equations of the circuit of LPF system as in (7.53).

$$L_f \frac{di_L}{dt} = U_i - U_c$$

$$I_L = I_c + I_0 \tag{7.53}$$

$$C_f \frac{dU_c}{dt} = I_c$$

Due to the existence of inductance and capacitance voltage in (7.53), the equivalent model in the dq-coordinate system is characterized by the coupling of the dq-axis. This coupling determines the energy transformation in the system, which is different from the single-phase system model. The equivalent transformation of the model in the complex variable domain is carried out in this work. Laplace transformation for a complex variable is given by:

$$\mathcal{LG}\left(\frac{d^k U}{dt} = (s+j\omega)^k(U_d + jU_q)\right) \tag{7.54}$$

Thus applying the concept of (7.54) for (7.53), Eq. (7.55) is obtained.

$$L_f(s+j\omega)(I_{dL} + jI_{qL}) = U_{di} + jU_{qi} - (U_{dc} + jU_{qc})$$
$$I_{dL} + jI_{qL} = I_{dc} + jI_{qc} + I_{d0} + jI_{q0} \tag{7.55}$$
$$C_f(s+j\omega)(U_{dc} + jU_{qc}) = I_{dc} + jI_{qc}$$

Rearranging Eq. (7.55) and equating the real parts and imaginary parts separately, we can obtain the transfer functions of the circuit of LPF as in (7.56) and (7.57).

$$(1 + C_fL_fs^2 - \omega^2 C_fL_f)U_{dc} - 2s\omega C_fL_f U_{qc} = U_{di} - sL_f I_{d0} + \omega L_f I_{q0}$$
$$(1 + C_fL_fs^2 - \omega^2 C_fL_f)U_{qc} + 2s\omega C_fL_f U_{dc} = U_{qi} - sL_f I_{q0} - \omega L_f I_{d0} \tag{7.56}$$

$$I_{dc} = sCU_{dc} - \omega C U_{qc}$$
$$I_{qc} = sCU_{qc} + \omega C U_{dc} \tag{7.57}$$

The conventional PI control structure applied for regulating the system whose circuit transfer functions are expressed in (7.56) and (7.57) can be drawn as shown in Fig. 7.28. In this figure, the voltage U_{dqi} is the regulated output voltage of the controller and fed to the LPF as input voltage.

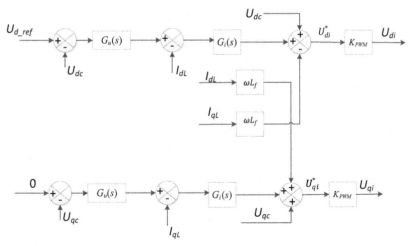

FIGURE 7.28   Conventional PI controller.

Fig. 7.28 is described by the transfer functions of (7.58).

$$U_{di} = \{[(U_{d\_ref} - U_{dc})G_u - I_{dc}]G_i + U_{dc} - \omega L_f I_{qL}\}K_{pwm}$$
$$U_{qi} = \{[(0 - U_{qc})G_u - I_{qc}]G_i + U_{qc} + \omega L_f I_{dL}\}K_{pwm} \tag{7.58}$$

Substituting (7.58) in (7.56) and computing with (7.57), we can have the following equations.

$$U_{dc} = HU_{d-ref} - Z_1 I_{d0} - Z_2 I_{q0}$$
$$U_{qc} = -Z_1 I_{q0} + Z_2 I_{d0} \tag{7.59}$$

The transfer function H indicates the voltage droop in the filter, and Z_1 and Z_2 are the dq-axis impedance components of the system impedance. But since I_{q0} is very small compared with I_{d0}, considering only Z_1 is essentially valid. Thus H and Z_1 are very important to determine the performance of the control structure, and hence are presented and analyzed as follows.

Based on Eq. (7.59), the transfer functions H and Z_1 are expressed by

$$H = \frac{G_u G_i K_{pwm}}{1 + C_f L_f s^2 - \omega^2 C_f L_f + G_i K_{pwm}(G_u + SC_f) + \omega^2 C_f L_f K_{pwm} - K_{pwm}}$$

and

$$Z_1 = \frac{G_i K_{pwm} + sL_f}{1 + C_f L_f s^2 - \omega^2 C_f L_f + G_i K_{pwm}(G_u + SC_f) + \omega^2 C_f L_f K_{pwm} - K_{pwm}}$$

7.5.1.1 Active damping

The active damping control structure, discussed above, is built as shown in Fig. 7.29. To emulate a real resistor R_d in series with C_f, an additional damping term $(K_r I_{dc})$ has to be added to the converter voltage reference.

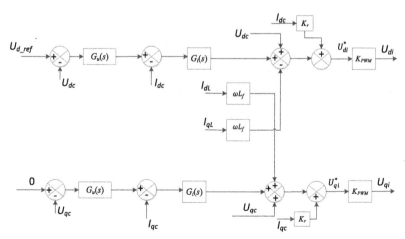

FIGURE 7.29 The PI controller with active damping.

Based on Fig. 7.29, the transfer functions are derived as:

$$U_{di} = \{[(U_{d-\text{ref}} - U_{dc})G_u - I_{dc}]G_i + U_{dc} - \omega L_f I_{qL} + K_r I_{dc}\}K_{pwm}$$
$$U_{qi} = \{[(0 - U_{qc})G_u - I_{qc}]G_i + U_{qc} + \omega L_f I_{dL} + K_r I_{dc}\}K_{pwm} \qquad (7.60)$$

Substituting these transfer functions in the filter circuit transfer functions of (7.56) and computing with (7.57), the transfer functions H and Z_1 are denoted by

$$H = \frac{G_u G_i K_{pwm}}{1 + C_f L_f s^2 - \omega^2 C_f L_f + G_i K_{pwm}(G_u + SC_f) + \omega^2 C_f L_f K_{pwm} - K_{pwm} - SC_f K_r K_{pwm}}$$

and

$$Z_1 = \frac{G_i K_{pwm} + s L_f}{1 + C_f L_f s^2 - \omega^2 C_f L_f + G_i K_{pwm}(G_u + SC_f) + \omega^2 C_f L_f K_{pwm} - K_{pwm} - SC_f K_r K_{pwm}}$$

The transfer functions H and Z_1 can be analyzed by comparing characteristics of the circuit of LPF, the PI controller with the addition of active damping. Accordingly, the models are developed and the simulated results are shown in Figs. 7.30 and 7.31. As depicted in the figures, the voltage drop in the LPF circuit is compensated by both controller methods. Specially, for some ranges of frequency which may cause resonance, the voltage drop significantly decreases while the active damping method is applied. That is because the inductive impedance is decreased for the active damping control method, as shown in Fig. 7.31.

By implementing this active damping technique and selecting an appropriate damping coefficient value of K_r, the simulated results are shown in Fig. 7.32A. According to this figure, the systems are stable while the power flows in the network, and the problems shown in Fig. 7.17 are solved. The two DERs are sharing power stably. Fig. 7.32B also shows the experimental test results. Both results verify the applicability of the concept of active damping.

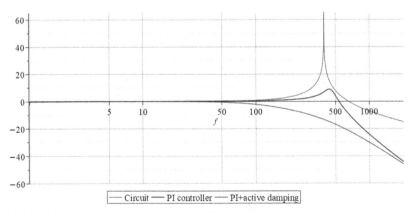

FIGURE 7.30   The bode plot characteristics of H transfer function.

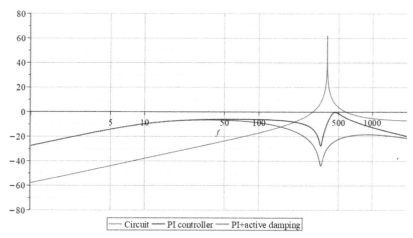

FIGURE 7.31 The bode plot characteristics of Z_1 transfer function.

According to Fig. 7.32, the active damping method has good performance for regulation of the microgrid dynamic disturbances. However, it becomes less effective when the power flow becomes larger (more than about 300 kW), which causes a large dynamic disturbance in our simulation and experimental testing cases.

As discussed above, even though the amplitude margin is well regulated by the active damping method, the phase margin is forcefully decreased as the resonant frequencies exist while the power flow becomes larger. This issue can be addressed by applying the impedance reconstruction method added with feedforward term in the current controller as shown in Fig. 7.33. The method features a high-pass function.

7.5.2 High-pass function damping

With the implementation of a proper method, the resonance poles in the LCL filter can be compensated and should remain effective under grid impedance, and filter parameters uncertainties without control tuning. Additionally, a proper method can achieve higher stability margins, highly decoupled current control, and better dynamic performance with a simple algorithm and lower level of reliance on the system modeling.

The impedance reconstruction method designed to perform as a high-pass filter is one such technique. It enables an increase in the phase margin. The high-pass filter equivalence of the impedance reconstruction can be defined by:

$$H_{pass}(s) = K_i \frac{s}{s + \omega_{hc}} u_{dq0} \qquad (7.61)$$

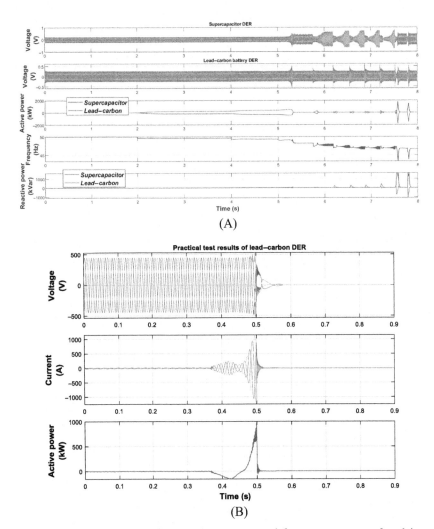

FIGURE 7.32 The responses for the voltage, power, and frequency responses of applying active damping method while the power flow increases in the network: (A) simulated results and (B) experimental test results.

Based on Fig. 7.33, the transfer functions are derived as:

$$U_{di} = \{[(U_{d-ref} - U_{dc})G_u - I_{dc}]G_i + U_{dc} - \omega L_f I_{qL} + K_r I_{dc} + H_{pass} U_{dc}\}K_{pwm}$$
$$U_{qi} = \{[(0 - U_{qc})G_u - I_{qc}]G_i + U_{qc} + \omega L_f I_{dL} + K_r I_{dc} + H_{pass} U_{dc}\}K_{pwm}$$

$$(7.62)$$

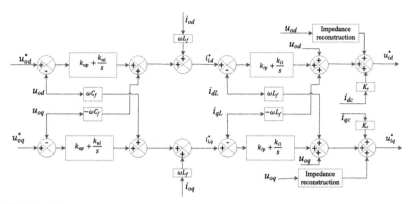

FIGURE 7.33 The control mechanism implemented for all methods.

Substituting these transfer functions in the filter circuit transfer functions of (7.56) and computing with (7.57), the transfer functions H and Z_1 are defined by

$$H = \frac{G_u G_i K_{pwm}}{1 + C_f L_f s^2 - \omega^2 C_f L_f + G_i K_{pwm}(G_u + SC_f) + \omega^2 C_f L_f K_{pwm} - K_{pwm} - SC_f K_r K_{pwm} - H_{pass} K_{pwm}}$$

and

$$Z_1 = \frac{G_i K_{pwm} + sL_f}{1 + C_f L_f s^2 - \omega^2 C_f L_f + G_i K_{pwm}(G_u + SC_f) + \omega^2 C_f L_f K_{pwm} - K_{pwm} - SC_f K_r K_{pwm} - H_{pass} K_{pwm}}$$

The comparison among the H and Z_1 transfer functions, for controller with and without high-pass damping element, is made though simulation tests and the results are shown in Figs. 7.34 and 7.35, respectively. According to these figures, the voltage drop in the LPF circuit is more compensated by applying an impedance reconstruction method with a high-pass function. In particular, for lower and higher frequencies in which the active damping method was less effective, this method performs better, as the inductive impedance is compensated by the method.

The method is verified by simulation and experimental tests. The simulation has been carried out using a PSCAD simulation model of Fig. 7.33 as a control mechanism. The control parameters used in the simulation model are also used to undertake the experimental tests whose setup is the same. The dynamic disturbance stability and performance of the methods are compared with the PI control without/with impedance reconstruction method under the DERs capacity variations via simulation and experimental results.

The results of simulation and experimental test are shown in Figs. 7.34 and 7.35, respectively. The incidence for both cases was power flow change in the lead–carbon battery by setting its reference value. The power flow was increased

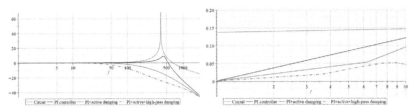

FIGURE 7.34 The bode plot characteristics of H transfer function to analyze the impedance reconstruction method.

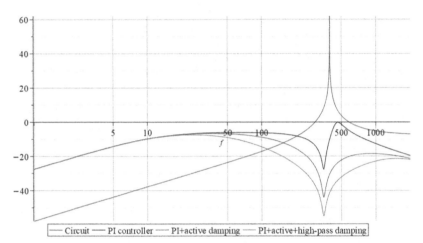

FIGURE 7.35 The bode plot characteristics of Z_1 transfer function to analyze the impedance reconstruction method.

beyond 300 kW which was unstable while only an active damping method along with PI controller was implemented. The performance of frequency and voltage dynamic stability was guaranteed even for the case of the power flow reaching up to 500 kW. Hence, this method can ensure the inverter works stably over a wide range of the typical inductive-resistive grid impedance and has good adaptability to the filter parameter perturbation (Figs. 7.36 and 7.37).

The addition of a virtual resistor loop to the voltage loop as a feedback block can also help synchronization between multiple DERs. In this case, the current used as input of the virtual resistor should be the LCL filter output so as to compensate for deviations that may occur during the control of inverter and LCL filter voltages. This is done as shown in Fig. 7.38, where i_{dqx} is the feedback current computed by LPF current along with the current related to the power variation which causes the frequency mismatch between the DER and the microgrid when they are being synchronized. Fig. 7.39 shows the simulated results when the virtual resistor loop is ignored in the control system of supercapacitor-based DER. Hence, a virtual resistor loop plays a very important role in the synchronization of

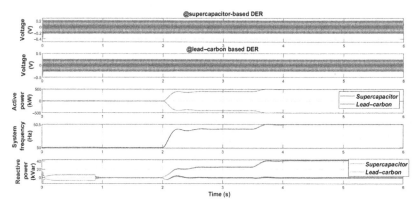

FIGURE 7.36 Simulation results for applying PI controller with active damping and high-pass filter methods while the power flow increases in the network.

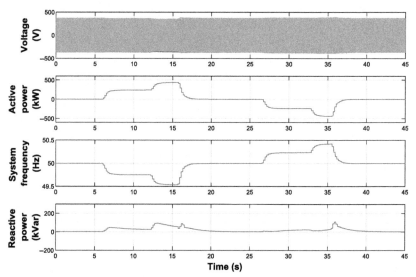

FIGURE 7.37 Experimental test results for applying PI controller with active damping and high-pass filter methods while the power flow increases in the network.

inverter-based DER with the microgrid. All of the above simulated and experimental tests have been done after the virtual resistor loop was included. For the sake of safety of the equipment, the experimental test was not carried out without this virtual resistor loop. The verification was made only using a simulation.

Other active damping techniques implemented in different applications include notch filter, filtered capacitor voltage feedforward, and active

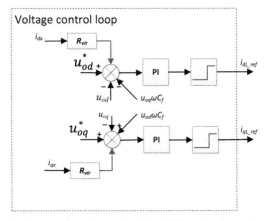

FIGURE 7.38 Virtual resistor loop for synchronizing DER to the microgrid.

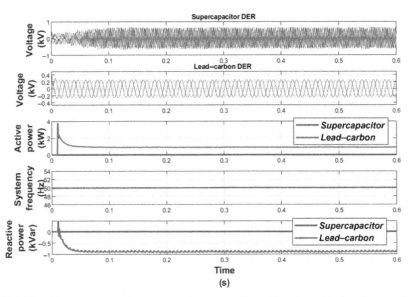

FIGURE 7.39 The voltage, frequency, and power profiles of the microgrid when the virtual resistor loop is ignored in the control system.

disturbance rejection control [8]. A notch filter, even though it is less expensive to implement, requires prior information about grid impedance and is sensitive to parameter variations. On the other hand, capacitor voltage feedforward is less sensitive to parameter variation, but it may not achieve higher dynamic stability margins on the basis of filter parameters [9,10].

References

[1] Y. Huang, D. Wang, L. Shang, G. Zhu, H. Tang, Y. Li, Modeling and stability analysis of DC-link voltage control in multi-VSCs with integrated to weak grid, IEEE Trans. Energy Convers. (2017). Available from: https://doi.org/10.1109/TEC.2017.2700949.

[2] C. Sun, G. Joos, F. Bouffard, Identification of low-frequency oscillation mode and improved damping design for virtual synchronous machines in microgrid, IET Gener. Transm. Distrib. (2019). Available from: https://doi.org/10.1049/iet-gtd.2018.7066.

[3] IEC, IEC TS 62898-3-1:2020, Microgrids—Part 3–1: Technical requirements—protection and dynamic control, 2020.

[4] S. Leitner, M. Yazdanian, A. Mehrizi-Sani, A. Muetze, Small-signal stability analysis of an inverter-based microgrid with internal model-based controllers, IEEE Trans. Smart Grid (2018). Available from: https://doi.org/10.1109/TSG.2017.2688481.

[5] M.N. Marwali, A. Keyhani, Control of distributed generation systems—Part I: Voltages and currents control, IEEE Trans. Power Electron. (2004). Available from: https://doi.org/10.1109/TPEL.2004.836685.

[6] M. Liserre, F. Blaabjerg, S. Hansen, Design and control of an LCL-filter-based three-phase active rectifier, IEEE Trans. Ind. Appl. (2005). Available from: https://doi.org/10.1109/TIA.2005.853373.

[7] H. Cha, T.K. Vu, Comparative analysis of low-pass output filter for single-phase grid-connected photovoltaic inverter, in: Proceedings of the 25th Annual IEEE Applied Power Electronics Conference and Exposition (APEC), 2010. Available from: https://doi.org/10.1109/APEC.2010.5433454.

[8] E. Rodriguez-Diaz, F.D. Freijedo, J.C. Vasquez, J.M. Guerrero, Analysis and comparison of notch filter and capacitor voltage feedforward active damping techniques for LCL grid-connected converters, IEEE Trans. Power Electron. (2019). Available from: https://doi.org/10.1109/TPEL.2018.2856634.

[9] W. Ma, Y. Guan, B. Zhang, L. Wu, Active disturbance rejection control based single current feedback resonance damping strategy for LCL-type grid-connected inverter, IEEE Trans. Energy Convers. vol. PP, no. 99 (2020). Available from: https://doi.org/10.1109/TEC.2020.3006151.

[10] A. Benrabah, D. Xu, Z. Gao, Active disturbance rejection control of LCL-filtered grid-connected inverter using padé approximation, IEEE Trans. Ind. Appl. (2018). Available from: https://doi.org/10.1109/TIA.2018.2855128.

Chapter 8

Transient control of microgrids

8.1 Introduction

One of the factors for the growing interest in microgrids is the improved power supply reliability due to the luxury of the possibility that they can disconnect themselves in the event of faults in the grid side and continue operation. This feature is especially important in the case of microgrids serving critical loads such as data centers, health facilities, and the like. However, the expected higher level of supply reliability in microgrids is influenced by the fact that they are prone to stability issues owing to their low system inertia coinciding with other factors [1−3].

The natures of transient performance and stability problems of microgrids are different from those of conventional power systems. Some of the factors that make transient control of microgrids special from the conventional grids are as follows:

- microgrids have low system inertia especially in case of island mode of operation;
- microgrids are of small size and usually constructed of a combination of small units;
- renewable energy resources (RES) with intermittent and fluctuating power output are very common and make the major share of generation in microgrids;
- most of the distributed generation (DG) units involve interfacing converters [4]; and
- microgrids usually involve low-voltage (LV) and on a few occasions medium-voltage distribution lines which have low reactance to resistance ratios [5]. This therefore results in strong coupling between active and reactive power and consequently the same for voltage and frequency. Thus all system parameters are influenced by any disturbance in microgrids.

Such and other factors indicate the need for the special consideration of transient stability and control of microgrids.

The switching process between grid-connected operation and island operation of the microgrid, and the startup and shutdown of high-power loads will cause serious transient disturbance to the system. These serious transient processes in the microgrid make the voltage and frequency of the system

Microgrid Protection and Control. DOI: https://doi.org/10.1016/B978-0-12-821189-2.00004-8
213

unstable, which may cause system load loss or system collapse. Hence, transient disturbances and their control appear to be critical aspects in the operation of microgrids. The development of a transient control system will also play a vital role in the actual construction and commercialization of microgrid projects.

Transient disturbances are different from dynamic disturbances, addressed in Chapter 7, Dynamic Control of Microgrids, in terms of the time period involved as well as the magnitude of transient parameters. Transient disturbance involves a smaller time period and higher magnitude changes, and, as a result, it requires a control system that can sense the disturbance and act fast.

The requirements and characteristics of transient control are different for the grid-connected and island modes of operations of microgrids. The stability issue is more pronounced and more critical in island mode than that of grid-connected mode. Voltage and frequency stabilizations are handled by the grid side during operation in grid-connected mode. The main phenomenon to be considered in this mode of operation is the "grid synchronization." This synchronization shall be done by satisfying the standard interconnection requirements to maintain the stability and reliability of the system from the customer's point of view. When these standards are not followed, problems related to imbalance in load, improper load sharing, transient instability, and so on, could possibly appear.

When the microgrid switches to island mode, many peculiar problems related to stability and power quality emanate. To alleviate these issues, control strategies are utilized, which in turn complicate the control of the microgrid. The task of maintaining the voltage and frequency stability and acceptable power factor should be taken care of by the microgrid itself without getting support from the main grid. Accordingly, many control strategies are used to mitigate the risk of complete system blackouts while islanding and after different types of transient disturbances during island mode. Another distinctive problem is the penetration of RES into the microgrid, as their outputs are intermittent [3,6].

There is a variation in the way transient disturbance is defined in different literature. Some opt to consider the time period involved as the sole factor while others go beyond that and place the magnitude of the disturbance into consideration as well. A term that is closely linked to transient disturbance is "large signal disturbance." These types of disturbances are characterized by large frequency and voltage excursions and power swings among multiple distributed energy resources (DERs). The events causing these disturbances happen for short time periods (usually in the order of milliseconds).

8.2 Transient characteristics of microgrids

Transients are sudden and severe voltage and current changes characterized by a large magnitude and a phase change in a time duration of 0−50 ms.

They are caused by the injection of energy due to DG switching, load switching, interruption of an inductive load, or energizing and deenergizing of a capacitor bank. The disturbances may be either oscillatory (damped or undamped) or impulsive. Transients arising from the switching of heavy loads or power factor correction capacitors may considerably affect normal system operation. Moreover, capacitor banks in combination with load and line inductances may even create resonant circuits, leading to the magnification of voltages due to harmonic sources.

One of the issues investigated and addressed in previous studies related to transient characteristics of the microgrid is the island mode microgrid consisting of doubly fed induction generator wind turbine and diesel generator, which is the heuristic nature of the hybrid system. In such a system, the microgrid fails to provide the required reactive power supply for the load. This is because the reactive power provided by diesel synchronous generator is absorbed not only by the load but also by the induction generator. This creates a gap between the supply and demand of reactive power. This kind of reactive power compensation was realized by a capacitor bank, flexible alternating current transmission system (FACTS), and proportional-integral (PI) controller. However, the capacitor cannot provide reactive power when the load changes in a large amount. The selection of gain parameters of the PI controller is also a cumbersome task in the capacitor-based system. The reference [7] proposed a method to solve this problem by combining FACTS based on an adaptive neuro-fuzzy inference system (ANFIS) with a PI controller. Unified power flow controller (UPFC) also acts as a FACTS controller. The gain constant and neural network parameters of PI controller can be optimized by using a fuzzy inference system. In the above study, the microgrid network without UPFC, with UPFC, and with UPFC combined with ANFIS are compared. Considering a 2% increase in reactive power demand and change in wind supply for each case, the results show that the combination of UPFC and ANFIS-based microgrid network provides better performance in terms of settling time, peak overshoot against all the load changes, and provides better reactive power compensation with good transient performances.

Transient characteristics of the microgrid can be investigated through the fault occurrence and the transitional operation of a microgrid between the isolated and grid-connected modes.

8.2.1   Causes for the transient disturbances in microgrids

The most common causes of transient disturbances in microgrids are:

1. switching between grid-connected and island modes;
2. heavy load on/off switching;
3. major DG on/off switching;
4. fault clearing.

8.2.1.1 Switching between grid-connected and island modes

When the microgrid is connected to the grid, it responds to the demands of the large power grid and the local load demand. The amount of power to be supplied or absorbed is decided according to the dispatching instructions. The microgrid acts as a power supply or as a load to the large power grid. The power delivered to or absorbed from the large power grid depends on the load demand within the microgrid. When the microgrid runs in an isolated mode, the microgrid does not have the voltage and frequency regulation support from the power grid. In this case, there shall be one or more DG within the microgrid that is dedicated to supporting the voltage and frequency regulation of the microgrid to maintain the power supply and stability of the system.

When the microgrid is switched from grid-connected to island mode, there will be a transient phenomenon causing disturbances such as voltage and frequency drops. Without transient disturbance control strategy, the microgrid will not operate properly. Hence, a transient control system that ensures a smooth and seamless switching process is needed.

8.2.1.2 Heavy load on/off switching

Loads are unexpectedly varying over time in addition to planned switching. The load dynamics are a factor in transient stability of a system. Abnormal operation or even a system failure can happen subsequent to a large or fast load change. Particularly, microgrids are vulnerable to unbalanced and nonlinear loads being switched on/off. Hence, transient control strategy should be appropriately provided to tackle such disturbances.

8.2.1.3 Major DG on/off switching

Switching on/off the major DG leads to an imbalance of active/reactive power flow in the microgrid network. Such situations can cause reverse power flows and lead to complications in protection coordination, undesirable power flow patterns, fault current distribution, and voltage control.

8.2.1.4 Fault clearing

When a microgrid is running in parallel with the large utility grid, if faults occur in the microgrid, the protection device will try to detect the fault and disconnect the faulty section, isolating some parts of the microgrid in the process. The remaining part of the microgrid can continue to run in parallel with the large power grid. From the time when the fault occurs to the time when the fault is cleared and normal operation is restored, there is a time delay due to the protection relay device. There will, as a result, be a significant change in the states of the voltage, frequency, and power flow of the system. This may bring a negative impact on the stability of the whole system unless it is properly controlled.

A fault current limiter (FCL) is a zero or minimal impedance device having little or no influence in normal operation of a utility grid but can be used for fault clearance. When a short-circuit fault occurs in the utility grid, it quickly becomes a high impedance to achieve fault current limiting [8]. According to the type of current limiting impedance it is in series with when the system is short-circuited, the FCL can be divided into superconducting (inductance type and resistance type) and solid-state type. Among them, the inductance type has better economic and technical performance, which can not only limit the short-circuit current but also effectively improve the transient stability of the system [9].

Though the FCL has been used for a while, it has limitations such as energy loss, large iron requirement, difficulty in switching, slow response time, so on, when applied in microgrids [8]. Hence, for transient conditions due to internal faults in the microgrid, some effective control methods should be devised to suppress the disturbances and reduce the impact of faults on the whole system.

8.2.2 System parameters during transient disturbances

When the microgrid is subject to either of the events discussed in Section 8.2.1, the electrical parameters of the microgrid will be affected. Some of the ways the transient disturbances are manifested in the system parameters are discussed in this section.

8.2.2.1 Voltage sag and frequency drop of microgrid

When the microgrid is subjected to major disturbances such as large load startup, the voltage waveform may show a significant sag and the frequency may also drop from the standard value [10]. One such scenario is shown in Figs. 8.2 and 8.3 where a microgrid with the arrangement shown in Fig. 8.1 is subject to starting of a large load. The magnitude of drop in magnitude and frequency of the voltage as well as the amount of time to recover to the normal values depend on the strength of the grid in grid-connected mode and the size and droop characteristics of the voltage and frequency (U/f) source in the island mode. It is however evident that the drop is more apparent in island mode of operation. The microgrid shall have the distributed power supply in the microgrid needed to adjust its own output to support the normal operation of the microgrid and restore nominal value of voltage and frequency.

As shown in Fig. 8.2, the voltage drop may be observed in the grid-connected mode as well, especially when the microgrid is weakly connected. The temporary deficiency of reactive power will require the large grid to react correspondingly to cover the gap and bring back the stability of the system. However, the system may collapse if the final overstepping reaches the

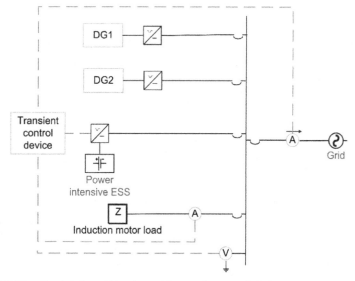

FIGURE 8.1 Simple microgrid topology with a transient control device.

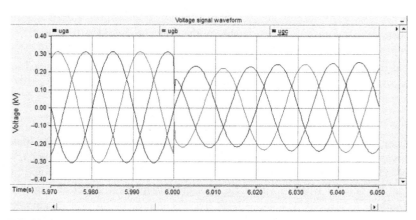

FIGURE 8.2 Voltage sag as a result of large load starting.

instability margin. Therefore to provide strong support for the power grid, and meanwhile to ensure that the local loads are supplied with the required quality power, the microgrid needs to be equipped with the transient control which makes use of a power-intensive energy storage that, according to the power deficit, adjusts its active and reactive power output.

In general, transient disturbances of the microgrid bring a considerable impact in its different operation modes including grid-connected mode, transit mode, and island mode, and need to be suppressed and regulated by devising and designing the appropriate control systems.

8.3 Design of transient disturbance control system

A microgrid transient disturbance control system is composed of system coordination control and a local measurement and control system. Among them, the system coordination control system mainly realizes the functions of real-time monitoring of the operation of power supply and load lines in the microgrid system and real-time monitoring of the frequency and voltage on both sides of the grid-connection point (POC) of the microgrid system. In this control system, based on real-time monitoring and power control of the microgrid lines, economic dispatch of grid-connected and off-grid modes and power balance adjustment, frequency and voltage emergency control should be established. The seamless switching of the microgrid from grid-connected mode to island mode and from island mode to grid-connected mode are to be realized.

The local measurement and control system can realize real-time monitoring of local line operation, local action, and other functions as well. Local information is uploaded to the main control system based on the monitoring of local lines, and action information is uploaded to the main control station based on the judgment of local faults and the action of protective devices.

In view of the relationship between islands and grid connection, the transient disturbance control system of the microgrid can realize the above functions from the perspectives of economic dispatch, system stability, and protection control.

Regarding economic dispatch being taken into account under island mode, it needs to consider the support capability of the backup power supply to the system and the stability of the system impacted by the ratio of the backup power supply to the power supply. In grid-connected mode, the large power grid should be used as the main power supply to provide frequency and voltage support for the system. DG needs to work in the form of power supply without considering the impact of renewable energy sources such as wind and solar.

In the aspect of system stability problem caused by load fluctuation in islanding mode, it is solved by the microgrid itself, which needs the microgrid transient stability control device. After the fault is removed, the master unit devices should be coordinated to consider the impact on the system. The system stability problem caused by load fluctuation in grid-connected mode is handled by the grid, and no transient disturbance control device of the microgrid is needed. The influence on the system is not considered following local fault clearance. These perspectives, in general, are illustrated in Fig. 8.3.

There are a variety of approaches suggested in the literature and some are already practically applied for transient disturbance control of microgrids. To give an idea about what the different approaches share in common, irrespective of the difference in the internal routine, the basic objective that the

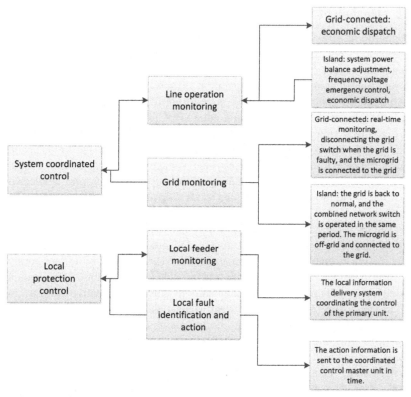

FIGURE 8.3 Transient disturbance control block diagram.

microgrid transient control system is expected to achieve is discussed first, and some of the common techniques are also presented in subsections of 8.3.2.1, 8.3.2.2 and 8.3.2.3.

8.3.1 Objectives of transient disturbance control system of a microgrid

As explained above, there are many causes for the transient conditions in the microgrid operation. However, the control objective of the microgrid under various transient conditions is similar and can be summarized into the following broad and common aspects:

1. *Power balance*: The power output of the microgrid should be coordinated with the dispatching order of the large power grid when connected to the grid. The power supply should meet the load demand of the microgrid system when operating in island mode. In the process of switching between the two operation modes, the microgrid changes from one equilibrium point to

another equilibrium point which causes a transient disturbance. Therefore power balance should be maintained within the system in both cases.

2. *Voltage and frequency control*: In grid-connected operation, the voltage and frequency requirements of the microgrid are consistent with and dictated by those of the larger power grid. In island mode, the task of keeping the voltage and frequency of microgrid within the normal range lies within the microgrid itself. When voltage sags or swells or frequency fluctuations in the system are caused by power shortage or other factors, the microgrid system should be equipped with a control system that drives the power flow within the microgrid so as to maintain the voltage and frequency within the acceptable ranges [11−13].

3. *Power quality*: There are two main implications of power quality in the microgrid. The first implication is the reactive power or power factor and the other is the harmonic content or harmonic distortion [12,14]. The transient control system can be tasked at providing power quality measures in both grid-connected and island modes. In the case of grid-connected mode, the microgrid may play a role in supporting the power quality of the large grid through the reactive power and harmonic compensation at the POC. The task of the control system in island mode will be limited to keeping the power factor and harmonic level at the main bus or selected point of the microgrid within the required ranges.

8.3.2   Control strategies for transient disturbances in microgrids

As shown in Fig. 8.4 with a typical example of a control topology of microgrid, the transient control strategy can be guaranteed by the primary control of the microgrid in a hierarchical control level platform. The primary control level featuring the fastest response is exclusively based on local measurements with no communication requirements. Considering their speed requirements and dependence on local measurements, detection of islanding, power-sharing control, power balance control, and output

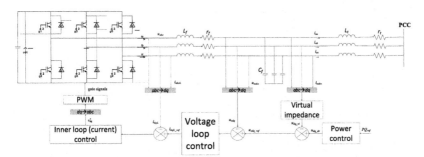

FIGURE 8.4   The overall control structure of VSI-based model of a DER applied for transient disturbance control.

control are all included in this category. In synchronous generators, output control and power sharing can be performed by the inertia of the machine itself, governors, and voltage regulators. To simulate the inertia characteristics of the synchronous generator and provide appropriate frequency regulation, the voltage source inverter (VSI), whether taken as a part of back-to-back converter or as the interface of the direct current (DC) power supply, needs a specially designed controller. With this aim, VSI controllers may comprise two stages of inverter output controller and DG power-sharing controller. Inverter output controllers should be responsible for the output voltages and currents control, whereas power-sharing controllers should be dedicated to the adequate share of active and reactive power mismatches in the microgrid network. The inverter output control usually consists of the inner loop for current control and outer loop for voltage control. By using the droop controllers of active power—frequency and reactive power—voltage which emulate the droop characteristics of synchronous generators, power sharing can be realized without a communication system. The overall control structure applied for the transient disturbance control system is illustrated in Fig. 8.4.

To implement this control structure, there could be a variety of control architectures and techniques that can be used for transient disturbance control in microgrids. The basic ones include frequency/voltage droop control, supplementary control, energy storage system control, and virtual synchronous control. These are discussed in the following sections.

8.3.2.1 Frequency/voltage droop control

The active power and reactive power sharing of DG units of the microgrid can be quietly and quickly regulated by this control strategy. In this control technique, the reference frequency f and voltage magnitude V are used to determine the control signals from the measured active power P and reactive power Q values implementing the two droop control Eq. (8.1) with unique and designed characteristics and ratings of DG units. Thus it is likely to eliminate the stress on any single unit.

$$f_i = f_i^* - m_i\left(P_i - P_i^*\right)$$
$$V_i = V_i^* - n_i\left(Q - Q_i^*\right)$$
(8.1)

where P_i^* and Q_i^* are the rated value of real and reactive power; P_i, and Q_i are the measured real and reactive powers; and f_i^* and V_i^* are the rated frequency and voltage magnitude, respectively. Here, the frequency droop coefficient m and the voltage magnitude droop coefficient n are derived from the maximum and rated value of the load [15,16]. Fig. 8.5 represents the droop characteristics of frequency and voltage versus with active power and reactive power, respectively.

The details of the design for droop control are presented and discussed in Chapter 7, Dynamic Control of Microgrids. In this chapter, the droop

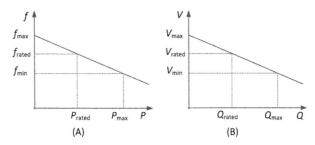

FIGURE 8.5 Frequency/active power (A) and voltage/reactive power (B) droop characteristics.

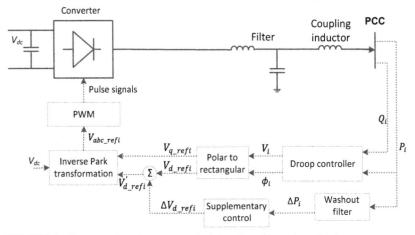

FIGURE 8.6 Structure of supplementary control for microgrid transient disturbances.

control is used for comparison with the virtual synchronous generator (VSG) control algorithm.

8.3.2.1.1 Supplementary control to the droop control strategy

Lead-lag compensator is applied in this control strategy to keep stability while ensuring better power sharing. During power sharing, it is desired to have high gain at low output power and low gain at high output power. Hence, for better load sharing, this control strategy allows the implementation of higher droop gains. It will damp out the small-signal oscillation to a considerable extent.

As shown in Fig. 8.6 the washout filter block detects the active power oscillations (ΔP). The supplementary control block generates the d component of the input signal (ΔE_{d_refi}) to the converter. The reference voltage generated for the converter is defined as in (8.2).

$$V_i^* \angle \phi_i^* = V'_{d\_refi} + j V_{q\_refi} \qquad (8.2)$$

8.3.2.2 Energy storage system

For the case of loss of DGs or rapid increase of unscheduled loads, an energy storage system control strategy can be implemented in the microgrid network. Such a control strategy will provide a spinning reserve for energy sources which can very quickly respond to the transient disturbances by adjusting the imbalance of the power in the microgrid network.

Batteries, pumped hydro, compressed air energy storage, flywheel, and supercapacitor are some of the energy storage systems featuring in the microgrids. Energy storage systems are a necessity for the stable operation of isolated microgrids or island mode of nonisolated microgrids. The electrical energy storage units are the most commonly utilized strategies in the microgrids. The electrical storage systems (ESSs) may be suited to either of the energy intensive or power-intensive applications based on their response rate and storage capacity. These ESSs can serve as controllable AC voltage sources to ensure voltage and frequency stability in the microgrids.

Power-intensive ESS shall be used to smooth the disturbances. If the disturbance is caused by a fault, the total harmonic distortion (THD) values of the branch currents and bus voltages are changed rapidly. The fault point should be isolated and the right breaker should be tripped. If there is no fault, the current and voltage THD values should return to the normal value, and hence no breakers are tripped. Transient controls against transient disturbances play a key role in the microgrid operational reliability. Since the fly wheel and supercapacitor have higher power density than other ESSs, they are very suitable for transient disturbance control of the microgrids. Testing and verification of the application of a supercapacitor for the transient disturbance control system in the microgrid is included in this chapter. Such a transient disturbance control system based on a single energy storage system with no communication network is proven to be an economic and reliable solution for voltage and frequency control in the transient disturbances.

Large changes in the current and voltage waveforms due to the switching of large-generating units, unplanned mode switching, and fault clearance are how severe transient disturbances are manifested in microgrids. The complexity and diversity of the internal load of the network increase the degree of fluctuations in the network due to load-side changes. The transient disturbance control of the microgrid can be applied widely in large-scale distributed energy systems. Such a control system discussed in this book can realize smooth mode transfer during unplanned switching from grid-connected mode to island mode and vice versa. The switching between the two modes should be performed with the grid-connection requirements fully obeyed. As discussed earlier, power-intensive type ESS, such as supercapacitor and flywheel, have high-power density and very fast response. Thus they can quickly supplement the energy gap in the microgrid, which in turn suppresses the transient disturbance of the microgrid, and ensures voltage and frequency stability.

In case of the grid-connected mode of operation of the microgrid, while transient disturbances occur, the active/reactive power control (PQ control) strategy can be suitably applied. In this mode of operation, the frequency and voltage are possibly regulated by the main grid control system, and the microgrid system is supposed to appropriately manage the power flow into or from the grid and within its network as well. Fig. 8.7 shows the PQ control structure of the microgrid likely used to regulate the power balance control during disturbances in the grid-connected mode of operation.

8.3.2.3 Virtual synchronous generator in transient disturbance control

From the conventional synchronous generator application, the swing equation is defined by

$$\frac{d\left(\frac{1}{2}J_m\omega^2\right)}{dt} = P_m - P_e \tag{8.3}$$

$$J_m\frac{d\omega}{dt} = P_m - P_e \tag{8.4}$$

where J_m, P_m, and P_e are, respectively, moment of inertia of the combined components in the rotating machines, mechanical power of the prime mover, and electrical output power. The general control structure of the synchronous generator supported by droop regulation is depicted in Fig. 8.8.

In synchronous generators, the relation between deviations of frequency Δf, generated power ΔP_m, and load ΔP_L is expressed from the above swing equation as

$$\Delta P_m - \Delta P_L = J_m s \Delta f + D \Delta f \tag{8.5}$$

where D is load-damping coefficient. Once J_m and/or D decrease (increase), the rate of frequency deviation is increased (decreased). In microgrids,

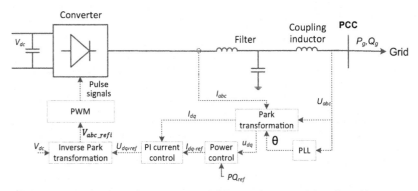

FIGURE 8.7 The active/reactive power control (PQ control) system of the microgrid.

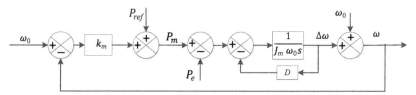

FIGURE 8.8   The basic control structure of a synchronous generator supported by droop regulation.

however, the damping factor D is a small value while the total moment of inertia J_m changes mainly depending on the number of DGs in the system. In such systems, the relation between DG's inertia constant (H_{DGi}) and system's inertia constant (H_{sys}) is expressed as in (8.6) [17].

$$H_{sys} = \sum_{i} H_{DGi} \times \frac{S_{DGi}}{S_{sys}} \qquad (8.6)$$

where S_{DGi} and S_{sys} are nominal system's power and nominal DG's power, respectively. Therefore by increasing the penetration and proportion of none/less-inertia DGs in the system, the H_{sys} is decreased.

Thus a VSG can be developed from the properties of a synchronous generator. The inverter mimics the inertia and damping properties which are the main features of a synchronous generator to regulate the frequency and voltage. Basically, VSG consists of energy storage, inverter, and a VSG control loop. On one side of the VSG there can be a DC bus, source, or distributed generator, and on the other side, a utility grid or a microgrid. The synchronous machine interface power supply has large enough inertia, but slow response time, whereas the power electronics power supply has small inertia and fast response time. Thus the VSG control system adopts the advantages of synchronous generator operation and the characteristics of fast time response of micropower supply [18]. Hence, the electrical energy storage provides the virtual inertia and will be able to absorb short-term power transients. A VSG's ultimate role is to increase the inertia constant of a given power system [19]. As the synchronous generator can adjust its output power according to the frequency modulation device with active power versus frequency characteristics and the excitation device with reactive power versus voltage characteristics, the distributed generators in the microgrid can do so. Thus the VSG control algorithm is adopted to maintain power balance, and voltage and frequency stability of the microgrid [20,21].

Setting $P_m - P_e = P_{VSG}$ and expanding (8.3), a basis is set for how a VSG should handle active power, as shown by Eq. (8.7), where P_{VSG} is the power exchange between the VSG and grid.

$$P_{VSG} = J_i \omega_0 \frac{d\omega}{dt} \qquad (8.7)$$

Eq. (8.7) describes that power can be absorbed or generated by the VSG based on the rate of change of frequency ($d\omega/dt$), and J_i defines the inertial property of the VSG. As seen in Eq. (8.7), J_i acts as an amplifying constant and should be set such that the VSG exchanges its maximum active power when the maximum defined rate of change of frequency (1 Hz/s) occurs. J_i should also be negative to counter the direction of the rate of change and that in effect is how virtual inertia is emulated by the inverter.

However, since $d\omega/dt$ is by nature an error signal because the balance point is at zero, power will be exchanged only during the transient without guaranteeing that the frequency of the power system will return to the normal state, that is, 50 Hz. To observe the above and have the VSG act accordingly, a frequency droop part should be added to (8.7) which becomes:

$$P_{VSG} = J_i \frac{d\Delta\omega}{dt} + k_\omega \Delta\omega \qquad (8.8)$$

where k_ω is the proportional gain that dictates the droop characteristics of the inverter. In (8.8) the reference value of frequency is always subtracted from the measured frequency. This creates an error signal whenever there is a frequency deviation within the power system. k_ω is chosen so that the P_{VSG} will equal the nominal power of the VSG when the frequency deviation is at its maximum (2.5 Hz).

Examining (8.8) indicates that power should flow in both directions, from the DER device to power system or from the power system to the DER device. So, in effect a VSG can either act as a load or as a generator dictated by the frequency disturbance and (8.8).

In the synchronous generator as shown in Fig. 8.7, the damping of the load is taken into consideration and here it is emulated for the VSG system too. Thus Eq. (8.8) is redefined, as in (8.9).

$$P_{VSG} = J_i \omega_0 \frac{d\Delta\omega}{dt} + D_\omega \Delta\omega + k_\omega \Delta\omega \qquad (8.9)$$

where D_ω is damping coefficient of the load in the network.

$D_\omega \Delta\omega$ in (8.9) is the load change due to frequency sensitive loads. D_ω can be determined as change in load divided by change in frequency.

As explained above, with the principle of VSG which imitates the transient behavior of the synchronous generator, Fig. 8.8 can be developed. Eq. (8.9) explains this figure and describes the frequency-active power characteristics of the DER with the concept of VSG along with droop characteristics of the microgrid and damping effect of the load in the network system [22].

Fig. 8.9A and B show the generic structure of VSG control with Fig. 8.9B showing simplified version of Fig. 8.9A.

The small-signal analysis of this control algorithm, is discussed next. The state-space models designed and presented in Chapter 7, Dynamic Control of

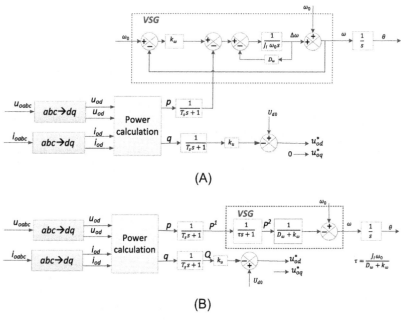

FIGURE 8.9 Block diagram of the VSG control system; (A) detailed structure, (B) simplified structure.

Microgrids, are applied for this chapter but the state parameters used for power controller are replaced by the VSG control algorithm parameters.

$$P_1 = \frac{\omega_c}{s + \omega_c} p \Rightarrow \frac{d(p_2)}{dt} = -\omega_c P_1 + 1.5\omega_c \left(u_{od} i_{od} + u_{oq} i_{oq} \right)$$

$$P_2 = \frac{1/\tau}{s + 1/\tau} P_1 \Rightarrow \frac{d(P_2)}{dt} = (1/\tau)P_1 - (1/\tau)P_2 \qquad (8.10)$$

$$Q = \frac{\omega_c}{s + \omega_c} q \Rightarrow \frac{d(Q)}{dt} = -Q\omega_c + 1.5\omega_c \left(u_{od} i_{oq} - u_{oq} i_{od} \right)$$

The droop characteristics is expressed according to Fig. 8.9B and defined as

$$\omega = \omega_0 - \frac{1}{D_\omega + k_\omega} P_2$$

$$u^*_{oq} = u_{oq\_0} - nQ = 0 \qquad (8.11)$$

In the same fashion as undertaken in Chapter 7, Dynamic Control of Microgrids, the angle δ, which represents the angle between an individual inverter reference frame and the common reference frame, is defined by

$$\delta = \int (\omega - \omega_{com}) dt = \int \left(\omega_0 - \frac{1}{D_\omega + k_\omega} P_2 - \omega_{com} \right) dt \qquad (8.12)$$

Differentiating (8.12) with respect to time gives us

$$\frac{d(\delta)}{dt} = \omega_0 - \frac{1}{D_\omega + k_\omega} P_2 - \omega_{com} \qquad (8.13)$$

Based on the state equations of (8.10)–(8.13), and by linearizing and rearranging the equations, the small-signal power controller model can be written in a state-space form as:

$$\begin{bmatrix} \Delta\dot{\delta} \\ \Delta\dot{P}_{12} \\ \Delta\dot{Q} \end{bmatrix} = A_p \begin{bmatrix} \Delta\delta \\ \Delta P_{12} \\ \Delta Q \end{bmatrix} + B_p \begin{bmatrix} \Delta i_{Ldq} \\ \Delta u_{odq} \\ \Delta i_{odq} \end{bmatrix} + B_{p\omega com}[\Delta\omega_{com}] \qquad (8.14)$$

$$\begin{bmatrix} \Delta\omega \\ \Delta u_{odq}^* \end{bmatrix} = \begin{bmatrix} C_{p\omega} \\ C_{pu} \end{bmatrix} \begin{bmatrix} \Delta\delta \\ \Delta P \\ \Delta Q \end{bmatrix} \qquad (8.15)$$

where the matrices of the coefficients of state variables are

$$A_p = \begin{bmatrix} 0 & 0 & -1/D_\omega + k_\omega & 0 \\ 0 & -\omega_c & 0 & 0 \\ 0 & 1/\tau & -1/\tau & 0 \\ 0 & 0 & 0 & -\omega_c \end{bmatrix}$$

$$B_p = \begin{bmatrix} 0 & 0 & 0 & 0 & 0 & 0 \\ 0 & 0 & 1.5\omega_c i_{od} & 1.5\omega_c i_{oq} & 1.5\omega_c u_{od} & 1.5\omega_c u_{oq} \\ 0 & 0 & 0 & 0 & 0 & 0 \\ 0 & 0 & 1.5\omega_c i_{oq} & -1.5\omega_c i_{od} & -1.5\omega_c u_{oq} & 1.5\omega_c u_{od} \end{bmatrix}$$

$$B_{p\omega} = \begin{bmatrix} -1 \\ 0 \\ 0 \\ 0 \end{bmatrix}; C_{p\omega} = \begin{bmatrix} 0 & 0 & -\frac{1}{D_\omega + k_\omega} & 0 \end{bmatrix} \text{ and } C_{pu} = \begin{bmatrix} 0 & 0 & 0 & -n \\ 0 & 0 & 0 & 0 \end{bmatrix}$$

The state variables in one inverter is increased to 14, whereas it was 13 in Chapter 7, Dynamic Control of Microgrids. Thus the matrix configurations in each inverter and the whole system should be carefully designed. Subsequently, the two DERs and one load system connecting with one line will have the size of 32 state variables, and the system matrix A_{sys} becomes 32×32. Considering the system and initial parameters of Chapter 7, Dynamic Control of Microgrids, and the additional parameters of moment of inertia J_ω and damping coefficient D_ω, the eigenvalues are obtained and discussed as follows.

8.3.2.3.1 Eigenvalues analysis for the damping coefficient D_ω

The eigenvalues of the system from varying the damping coefficient D_ω are illustrated in Fig. 8.10. The figure is provided to show the characteristics of eigenvalues which are close to the critical axis and susceptible to parameter

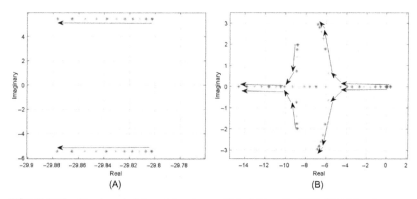

FIGURE 8.10 The characteristics of damping coefficient ($D_\omega = 0-1250$) (A) higher-frequency modes of eigenvalues, (B) lower-frequency modes of eigenvalues.

variations. As D_ω increases these eigenvalues move away from the critical axis. The value of D_ω varies from 0 to 1250 in the direction of the arrow. This implies that as D_ω increases, the stability of the system is ensured while some perturbation is imposed to the microgrid system.

Table 8.1 shows the eigenvalues with the modes of frequency and the state parameters which are major participants in the status of the eigenvalues. The moment of inertia J_ω is set to be 0.35 and the damping constant D_ω is set as 1250.

8.3.2.3.2 Filter inductance sensitivity analysis with eigenvalues by comparing the VSG and droop control systems

A complete model was devised by considering the initial conditions of the system used in Chapter 7, Dynamic Control of Microgrids. The additional parameters used in this chapter are virtual inertia J_i and virtual damping coefficient D_ω. The proportional gain that dictates the droop characteristics of the inverter k_ω is the reciprocal of droop constant m which is used in Chapter 7, Dynamic Control of Microgrids. In Figs. 8.11 and 8.12, the star marks and circle marks represent the eigenvalues of VSG and droop control systems, respectively. The arrows on the lines show the direction of filter inductance increase. The figure demonstrates the filter inductance sensitivity for two methods of control systems (VSG and droop control). As shown in Fig. 8.11, for the eigenvalues of low modes of frequency which are close to the critical axis, the eigenvalues of the VSG control method move behind those of the corresponding ones in the droop control method. Even some eigenvalues of the VSG control method retain in fixed position in the stability region while those of the droop control method move to the right-half plane.

Hence, the VSG control method can be vitally implemented in the transient disturbance control application for the inverter-based units. As traditional power electronic sources are rigid sources by their virtue of low

TABLE 8.1 Eigenvalues of the system matrix A_{sys} at $J\omega = 0.35$ and $D\omega = 12500$.

Index	Real (1/s)	Imaginary (rad/s)	ω_0^2 (Hz)	Damping ratio (ζ)	Major participants
1, 2	−2.04E + 08	± 3.14E + 02	9.86E + 04	−1.00	i_{lineDQ}
3, 4	−2.99E + 07	± 3.14E + 02	9.86E + 04	−1.00	i_{odq1}, i_{odq2}
5, 6	−4.67E + 03	± 1.12E + 04	1.25E + 08	−0.38	i_{Ldq2}, u_{odq2}
7, 8	−3.49E + 03	± 1.12E + 04	1.25E + 08	−0.30	i_{Ldq2}, u_{odq2}
9, 10	−1.99E + 03	± 8.09E + 03	6.54E + 07	−0.24	i_{LtoDQ}
11, 12	−2.47E + 03	± 7.92E + 03	6.27E + 07	−0.30	u_{odq1}, i_{Ldq2}
13, 14	−2.53E + 03	± 1.43E + 02	2.04E + 04	−1.00	i_{Ldq1}, u_{odq1}
15, 16	−9.06E + 02	± 1.37E + 02	1.88E + 04	−0.99	i_{Ldq1}
17	−9.71E + 01	0.00	0.00	−1.00	P_1^1, Q_1, P_2^1, Q_2
18	−9.91E + 01	0.00	0.00	−1.00	P_1^1, P_1^2, Q_2
19	−1.00E + 02	0.00	0.00	−1.00	P_1^1
20	−1.00E + 02	0.00	0.00	−1.00	P_1^1, Q_1, P_2^1, Q_2
21, 22	−2.99E + 01	± 5.45E + 00	29.70	−0.98	$\varphi_{dq1}, \varphi_{dq2}$
23	−1.46E + 01	0.00	0.00	−1.00	$P_1^2, \varphi_{q1}, P_2^2$
24	−1.16E + 01	0.00	0.00	−1.00	P_1^2, γ_{d1}
25, 26	−6.83E + 00	± 2.95E + 00	8.70	−0.92	φ_{dq1}

(Continued)

TABLE 8.1 (Continued)

Index	Real (1/s)	Imaginary (rad/s)	ω_0^2 (Hz)	Damping ratio (ζ)	Major participants
27, 28	−2.86E−04	±2.97E−09	8.82E−18	−1.00	$\gamma_{dq1}, \gamma_{dq2}$
29, 30	−2.86E−04	±6.26E−10	3.92E−19	−1.00	$\gamma_{dq1}, \gamma_{dq2}$
31	0.00	0.00	0		δ_1
32	0.00	0.00	0		δ_2

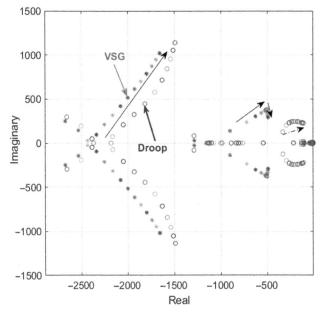

FIGURE 8.11 Comparison of VSG and droop control systems with the eigenvalues characteristics as filter inductance increases.

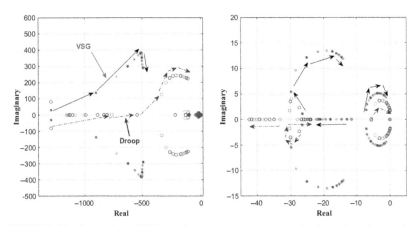

FIGURE 8.12 Comparison of VSG and droop control systems with the eigenvalues characteristics as filter inductance increases.

inertia, introducing the VSG algorithm can be considered as equivalent to the flexible source of rotary machine to emulate the inertial increment of the microgrid system. On the other hand, the design of control system needs to remember that the synchronous machine interface power supply has large enough inertia, but slow response time, while the power electronics power supply has small inertia and fast response time, so, the control algorithm using power electronic interface is inherently defective, which is not suitable for the control strategy of a microgrid. Hence, a microgrid transient disturbance control system needs to improve the control system by taking the advantages of VSG algorithm and getting an improved strategy comprising an inverter, a control mechanism, and energy storage [20].

A direct torque control (DTC) method can be effectively applied in the VSG control strategy to maintain the power balance of the microgrid and to suppress the power imbalance, oscillation, and circulation phenomena in the transient state. This control method can ensure seamless switching between grid-connected and island modes of the microgrid. It can also be used for regulating the transient disturbances due to its fast response characteristics. When the prime mover of DG operates in transient state, it is suggested that applying DTC on the VSG is quite important.

The six-pulse converter controlled is adjusted by DC voltage (E_d) via the pulse width modulation (PWM), so that the DC voltage could generate rotating magnetic field φ_{Gs}, which is controlled by the gate of insulated gate bipolar transistor (IGBT). With the condition satisfying the control variables $\varphi_{min} \leq \varphi_t \leq \varphi_{max}$ and $T_{min} \leq T_t \leq T_{max}$, and the limit variables $\theta_{min} \leq \theta_t \leq \theta_{max}$, the synchronous speed lags the rotor speed, which influences the VSG generating the corresponding active power.

Note:

φ_t is time-dependent variable of the φ_{Gs};

T_t is time-dependent variable of the rotor electromagnetic torque T_m; and

θ_t is time-dependent variable of phase angle of VSG.

Thus the phase angle and the active power become

$$\varphi_{Gs} = \frac{kE_d}{\omega_{Gr}}, P_r = E_{2N}I_1 \cos\theta_s, T_m = \frac{9.55P_r}{n_s} \times 3 \qquad (8.16)$$

$$p_e = \frac{E_{GL}E_{SL}}{X'_T}\sin\theta_{GS}, \theta_{GS} = \sin^{-1}\left[\frac{P_e}{E_{GL}E_{SL}/X'_T}\right] \qquad (8.17)$$

where k is the constant parameter, E_d is the DC power supply of VSG's converters on two sides, ω_{Gr} is the rotor speed, P_r is the rotor power of single phase, E_{2N} is the equal stator voltage, I_1 is the stator current, θ_s is the angle between E_{2N} and I_1, n_s is the synchronous speed, $P_e = 3 \times (P_r - P_{jr})$ is the three-phase active power generated by VSG, P_{jr} is the single-phase active power, E_{GL} is the line voltage of VSG, E_{SL} is the line voltage of grid, X'_T is the equivalent reactance between E_{GL} and E_{SL}, and θ_{GS} is the angle between E_{GL} and E_{SL}.

The variables of DTC are vectors and vary in a certain range over a particular time period, which constitutes the fuzzy set. The DTC is thus applied to analyze all of the fuzzy set through controlling the IGBT of six-pulse converters of the grid, so that φ_{Gs} would be controlled accurately. Below the key points are listed while DTC is in process:

1. keep comparing the fuzzy set of strategy control $W_{(+)}$ with the fuzzy set of rotating magnetic field $\varphi_{(GS)}$;
2. obtain the difference between $W_{(+)}$ and $\varphi_{(GS)}$; and
3. reduce the difference to the value approaching zero.

The $W_{(+)}$ is the intersection of $A_{(+)}$ and $B_{(-)}$, thus

$$W_{(+)} = \mu_{W(+)}(x), \varphi_{Gs} = \mu_{\varphi_{Gs}}(x), A_{(+)} = \mu_{A(+)}(x), B_{(-)} = \mu_{B(-)}(x) \qquad (8.18)$$

$$W_{(+)} = \mu_{W(+)}(x) = \mu_{A(+) \cap B(-)}(x) = min[\mu_{A(+)}(x), \mu_{B(-)}(x)] = A_{(+)} \cap B_{(-)}$$
$$(8.19)$$

The $\mu_{A(+)}(x)$ is the fuzzy set of vectors $A_{(+)}$, $\mu_{B(-)}(x)$ is the fuzzy set of vectors $B_{(-)}$, and x is the variable set, that is, $x = \{x_1, x_2, x_3, \ldots\ldots, x_n\}$.

The $x_1 = \varphi_{Gs}$, $x_2 = T_m$, $x_3 = \theta_{Gs}, \ldots\ldots$ are the variables of φ_{Gs}, T_m, and θ_{Gs}, collected in real time over a particular time period.

DTC would play a dominant role in solving the unstable operating problem of distributed generator due to its capability in changing of prime power. Fig. 8.13 shows the DTC topology can be applied in the transient control compensation. This concept will be verified with the case study of a field test and will be included in the next edition.

Accordingly, in the application subsection of 8.6, the two control strategies including droop control and VSG algorithm are presented and discussed with their simulation and field testing results.

8.3.3 Hardware requirements of transient control systems

The tasks of microgrid transient disturbance control system design include hardware and software design. The required hardware device is to function the configurable components of sampling control, open-in, open-out, data management, and managing multiple functional boards. In the aspect of analog and digital circuit design, signal integrity, and electromagnetic compatibility, standard interfaces in high-speed circuit boards for protection requirements are taken into consideration, and the 802.3 media access control (MAC) layer optical fiber communication for customized data transfer design is implemented.

Software design includes device software and central computer software. The installed software is divided into platform layered software design, that is, each layer is maintained separately according to the functional interface. The strategic allocation is based on the engineering application to implement the functional perfection of the hardware drive layer, power data processing layer, power function logic layer, and configurable strategy processing layer. The central computer software is mainly used in the local debugging work including production, field engineering, and testing.

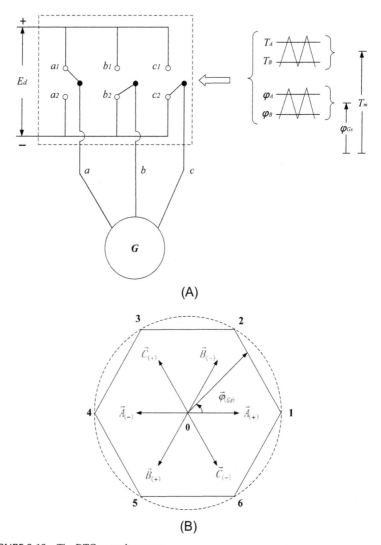

FIGURE 8.13 The DTC control strategy.

8.4 Identifying different kinds of faults from transient disturbances

Faults and transient disturbances are different based on their characteristics and severity of their impacts on the whole power network or part of it, even though the fault clearance is categorized in the transient disturbance. Identifying faults from the transient disturbances will be carried out by applying a method of fault detection applications. Fault detection is possible to identify the type and location of the faults. Such identification has been performed with the help of Fourier transform characterizing the transient waveforms occurring in the time and frequency domain.

The fault current of DER with synchronous generators is characterized typically with 4–6 p.u. of its positive and negative sequence currents; and the current magnitude and phase angle dependency on the design of the synchronous generator, but very little on controllers. Whereas if the DER is configured with inverters, the positive sequence fault current is typically less than 1.2 p.u. The current magnitude and phase angle depend on inverter controllers and control mode [23].

The beginning characteristics of fault occurrences and clearances are closely related to that of the transient disturbances. So, unless the differences between faults and transient disturbances are identified, the microgrids may be collapsed.

In the microgrids, during the fault clearances, the system may be collapsed because of the transient disturbance. A transient disturbance control is required to prevent the system from collapsing and make sure the right breakers will trip off.

In the microgrids, when transient disturbances happen, even if the initial characteristics of the transient are very similar to that of the faulty ones, the breakers should not be tripped by transient control. Especially in an isolated mode of configuration, when there is a switching of heavy loads, the transient characteristics of bus bar voltage and frequency are almost the same as the fault characteristics. Transient disturbance control is also needed to deal with this situation.

The requirements of transient disturbance control are to:

1. prevent the system from collapsing during fault clearing, unintentional switching processes of DG and heavy loads, and mode transfers (from grid-connected to island or vice versa);
2. avoid serious consequences such as blackout and system collapse from the undesired operation of the microgrid;
3. restore the power balance of microgrid and damp system oscillation during fault clearing or the switching of DG and heavy loads;
4. prevent the tripping of DGs, and compensate system inertia through ESS during fault clearing in isolated microgrids and weakly connected nonisolated microgrids; and
5. prevent abnormal conditions in the microgrid due to power ramping by a synchronized group of converter operations or dynamic setpoint changes.

According to IEEE 1547 standard stipulating that DER units shall stop supplying power to the local power system for faults in their connected local circuits, DER, or even synchronous generator DER have a very limited contribution to the fault current provided by the power grid. In the case of inverters, the contribution of fault current is actually nil. As a practical problem, DER is often impossible to detect feeder faults through overcurrent or distance relays, and feeders are still powered by substations. After substation circuit breakers (or line reclosures or transverse fuses) are tripped or cleared, DER is more likely to detect faults through overcurrent or distance relays if it has sufficient fault current contribution. This is known as "sequential tripping" and is generally accepted as conforming to the IEEE 1547. For inverters, current is usually not a satisfactory method to detect faults. The fault detection of inverter DERs is usually designed based on undervoltage protection means. This often detects faults, even though the substation power supply still drives the fault current. However, the fault current may not be detectable in some cases until the feeder is powered off.

Applying the undervoltage fault detection method for the DER, inherently a very nonselective protection method, the undervoltage caused by the faults on the DER feeder is usually indistinguishable from the faults on the transmission system. Thus the minimum time is placed to detect the undervoltage fault by imposing low-voltage ride-through (LVRT) requirements due to its inherent characteristics. The main requirement of DER fault detection is that DER does not cause long-term contribution of current to power grid fault, thereby increasing the damage that may be caused by the fault, and the possible maloperation of fuses or other switchgear.

Inherently, the types of DER which are dependent on undervoltage fault detection method have a very little current contribution. Thus if the requirement of detecting all faults is not overinterpreted to necessitate the DER fast tripping from the network that does not have a significant contribution to fault current magnitude, LVRT requirements are possibly accommodated and are unlikely to cause maloperations of fuse or relay.

8.5 Frequency and voltage ride-through

The underlying goal for ride-through capability is the avoidance of significant system security deterioration as a result of generation units tripping during critical events that result in voltage or frequency deviations. Ride-through is not a "setting," it is a minimum capability of the DERs to restore the normal operations following disturbances. To clear out the ambiguity in understanding different terminologies related to frequency ride-through (FRT) and voltage ride-through (VRT), it is better to define some of the common terms first.

Fault ride-through is the ability of a generating unit to remain connected during a fault that stays for a short period and resumes operation directly after clearance of the fault without any mechanical switching operation.

FRT time is the required fault duration that a generating unit shall ride-through for a particular fault disturbance

VRT is the ability of a generating unit to remain connected during a voltage disturbance condition that stays for a short period and resumes operation directly after clearance of that condition without any mechanical switching operation

FRT is the ability of a generating unit to remain connected during frequency disturbance condition that stays for a short period and resumes operation directly after clearance of that condition without any mechanical switching operation [24].

The sudden disconnection of DGs could affect the stability and reliability of the microgrid. The ride-through strategy of the microgrid should be considered to improve the system stability and reliability. It is not reasonable to require all sources to ride-through all possible grid events, but rather ride-through requirements should be designed to be generally consistent with the severity of events. By limiting the ride-through requirements to realistic grid needs, equipment limitations, and conflicts with other distributed resources (DERs), integration requirements can be greatly alleviated.

8.5.1 Frequency ride-through

The system frequency should be kept constant throughout the power grid network, or a part of the power grid when it is disconnected from the grid following serious contingencies. Because of the transient swing of the angle, there may be short-time and low-amplitude frequency changes at the specific location. The instantaneous power balance between generation and load determines the system frequency, hence any disturbance can perturbate the grid frequency. Frequency response which is defined as the automatic corrective response provided depends partly on the system inertia for balancing load and generation. An underfrequency record caused by power loss in the system can be further escalated when generation loss is made for responding to underfrequency incident; for the reliability of large-scale power systems, this is called cascading [25]. The rules of reliability require power companies to implement automatic load shedding schemes to respond to underfrequency events, while large-capacity generators need to have underfrequency ride-through capability, which has to exceed these load shedding frequency-duration curves.

Generation loss caused by overfrequency seems to be developing toward correcting generation load mismatch, but excessive generation loss can quickly transform the overfrequency events into underfrequency events [25]. Therefore large-capacity generators also need to define the overfrequency ride-through capability. Distributed resources are part of the overall load balance of power generation by connecting with the grid. With the increasing penetration of distributed resources, it is undoubtedly very important that distributed resources have the FRT capabilities equivalent to those required for large-scale generation.

The FRT requirements are set with different standards in different countries and power grids. The microgrid can withstand certain abnormal system frequencies and can operate under the grid frequency deviation. IEEE 1547, for example, states that the frequency range for FRT within which a DER can operate without tripping is between 59.8 and 60.5 Hz for a 60 Hz system for a duration not larger than 0.16 seconds [26]. The lower limit can be as low as 57 Hz in the case of larger DERs (>30 kW) [26]. The national standard of China, GB/T 34930-2017, specifies the operation and control requirements for microgrids connected to a distribution network. The standard specifies that a microgrid may be considered operating normally as long as the frequency is within the range of 49.5−50.2 Hz [27]. The microgrid should also stop exchanging power with the grid if the microgrid frequency stays out of the normal range for a period of 0.2 seconds with a minimum limit of 48 Hz and a maximum limit of 50.5 Hz though it may stay physically connected [27]. However, in the extreme events of the frequency dropping below 48 Hz or increasing above 50.5 Hz, it shall be switched to island mode immediately or after 0.2 seconds, respectively [27]. The transient control system shall thus be able to assist the microgrid to operate according to such standards in the respective countries.

8.5.2 Voltage ride-through

Transient behavior of large-capacity transmission systems during and post faults may lead to large-scale voltage sags. In the postfault stage, due to the dynamic "backswings" of the power grid, voltage swells may occur in a wide range of the power grid. These voltage variations tend to propagate into distribution systems connected to microgrids. If there is no requirement of VRT, a large number of DER capacities may trip instantaneously due to large power grid events, which will have a stress impact on the power grid, and may aggravate the interference of the large power grid, violating the basic principles of large system planning. Within the planning criteria, power generation should not be lost due to grid failures. The study of power grid planning is based on unexpected events at selected locations. It is impractical to conduct specific transmission studies on DER interconnections. Therefore it is necessary to define unified DER−VRT performance requirements based on generic criteria of voltage and frequency versus time. These standards reflect the voltage and frequency conditions that can be observed at the interconnected DER points during key large-scale grid events within the range of reliability planning criteria.

The distribution system affected by three-phase faults may have near-zero voltage amplitude at the moment of fault. However, the impacted distributed resource capacity has little impact on the security of large-capacity power grids. With the increase of the distance between fault point and transmission system, the voltage sag of the distribution system decreases gradually, but the potential impact of distributed resource transfer increases, which may lead to LV. The lowest part of the LVRT demand curve should be based on the loss of distributed resources. It is not meaningful for a large power grid to experience more severe voltage sags.

Currently, the registration standard of the North American Electric Reliability Corporation does not consider a single generator unit with a total capacity of less than 20 MW or a power generation facility with a total capacity of less than 75 MW. Therefore a reasonable criterion is that the potential loss of DER capacity less than 20–75 MW is lower than the significant threshold of a large power grid. LVRT requirements can be defined so that for typical fault voltage versus radius gradients and typical DER capacity density, the potential DER loss is below this threshold. However, both the fault voltage gradient and the potential DER capacity density are location dependent. For the more load-intensive part of the grid, the potential DER capacity density will also be greater. However, the geographic range of voltage sags during faults is often more limited for the more intensive network transmission systems needed to support this dense load area. Therefore these two factors tend to track each other, making the LVRT level and potential losses more widely applicable.

In addition to the severity of voltage deviation, the requirement of VRT duration should be coordinated with the reliability standard of large power grid. The duration of the initial minimum voltage portion required by LVRT performance should be compared with the duration of the fault of the large-capacity transmission system design. The maximum clearance time of single-phase faults in large power grids is often longer than that of three-phase faults. However, at the distribution level, the severity of voltage sags in single-phase transmission systems is reduced considerably by the wye-delta transformer connection between transmission and distribution systems. Therefore the minimum voltage part of the LVRT curve should have the same duration as the three-phase fault clearance time of the large power grid, and only low LVRT performance is required for single-phase fault clearance time. If part of the LVRT performance requirement is to cover transient voltage recovery after the fault, this part of the curve should be based on the actual voltage recovery curve of the well-designed transmission system, rather than the threshold voltage instability of the marginal system with very large voltage recovery in the worst case. Large-capacity systems with such LV recovery rates need to be adjusted (e.g., with FACTS devices, revised operating restrictions, reinforcements, etc.) and DER should not be required for adjustment.

Very low distribution voltages for longer periods can be caused by faults occurring on subtransmission or local transmission systems rather than large-capacity transmission faults. However, these subtransmission faults are not important to the large-scale power grid, and these faults should not be the basis of VRT performance requirements. The main purpose of LVRT requirements is to maintain the security of the large-scale power grid by avoiding the aggravation of disturbances in large-scale power systems.

The voltage sag caused by the fault incidents has an impact on the transient recovering performance of DGs and the reliability of the power supply. DGs might be enforced to trip off quickly during fault for the safe operation, but result in many adverse effects on system stability and reliability. Any sudden disconnection of DG or microgrid from utility grid should be avoided. Thus the design and development of a microgrid system should consider LVRT capability to improve its stability and reliability. To meet the requirements of LVRT and to improve the grid voltage during voltage sag, the reactive power should be adequately injected by DGs.

The distribution static synchronous compensator, the dynamic voltage regulator, the superconducting FCL, and the superconducting magnetic energy storage system are the methods being used for compensating the voltage drop caused by transient disturbances. So as to improve the LVRT capability of the microgrid system, a distributed control based on dynamic consensus algorithm in the system can be used. This method enables the microgrid to maintain accurate reactive power sharing and to restore the nominal voltage of the system. Comparing this control method with the centralized control one, it shows better performance as verified through several case studies.

Thus to maintain the voltage within the defined limits in the microgrid for both conditions of high-voltage ride-through (HVRT) and LVRT, additional reactive current support must be provided by the microgrid system. The compensation of the voltage is effectively realized by injecting inductive reactive current in HVRT or capacitive reactive current in LVRT. In spite of any reactive power, the load is required in the microgrid network, and reactive current injection must meet the requirements.

Voltage response requirements that define the VRT requirements for microgrids are stated in the standards and grid codes of different countries. The Chinese GB/T33589-2017 considers microgrids operating with POC voltages in the range of 90%−110% of the nominal POC voltage as operating normally [28]. However, it states that microgrids with POC voltage out of such range may stay connected up to 2 seconds when the POC voltage is not less than 50% and not higher than 120% of the nominal value, and shall be disconnected from the grid within 0.2 seconds otherwise [28]. The voltage response requirement for microgrids in the United States is similar to that of the Chinese except trip time required for the 110%−120% is 1 second and the trip time for the extreme cases is 0.16, according to Ref. [24]. The other standard for Europe, International Electrotechnical Commission (IEC) 61727, is different from the Chinese one as it relaxes the lower limit of 120%−135% though it reduces the trip time under the extreme cases of the POC voltage to being less than 50% and higher than 135% to 0.1 and 0.05 seconds, respectively [29].

In general, for both VRT and FRT, the following process is deployed. When the system is subjected to disturbances, it may get into the state of ride-through or must trip, depending on the severity of the disturbances or the capability of DER to withstand the disturbances. Following the ride-through performance, the system passes momentarily through a permissive operation and goes to the operating level in which the DER may either continue operation or may cease to energize. To continue the operation, that is a mandatory operation, a proper delivery of active and reactive current is required, otherwise it ceases to energize following momentarily cessation. In the postdisturbance, in either of the paths the operation passes, the restored output, which is a DER recovery to normal output following a disturbance that does not cause a trip, is performed. However, if the system is imposed on severe disturbances or fails to ride-through capability, a must-trip state is required and reentry of DER to service is made following the trip.

According to the IEEE 1547 standard, the restored output of the system with ride-through performance should satisfy the following:

1. Time begins when applicable voltage returns to mandatory operation or continuous operation ranges.
2. If DER provides dynamic reactive power support (not mandatory), it must continue for 5 seconds before returning to predisturbance reactive control mode.
3. DER must restore output to 80% of predisturbance active current within 0.4 seconds.
4. If positively damped, oscillatory power output is acceptable.

8.6   Application examples: practical experiment and simulation of transient disturbance control system

This section provides an application example of transient disturbance control in microgrids. A set of experiments was carried out on a typical demonstrational microgrid in China. One type of load variation is presented and analyzed here using both simulated and experimental methods for two control methods including droop control and VSG control methods. The general layout of the section of the microgrid in which these experiments were made is shown in Fig. 8.14, and the simulation and experimental results of the designed models discussed above are also analyzed. As asserted in the above, fluctuations of large loads and high penetration of intermittent renewable energy sources cause abnormal conditions in a specially isolated microgrid or nonisolated microgrid which is weakly connected to the grid or operating in the island mode. Such situations may result in transient disturbances.

8.6.1   Transient control

As discussed above, heavy load variation is one of the events which cause transient disturbance and is used to showcase the role of a transient control system for system stability. The results of simulation and experimental tests are presented below.

8.6.1.1 Simulation results for transient control systems

A simple microgrid consisting of two DERs connected with a 400 V voltage source (considered as a grid) and a resistive load, as shown in Fig. 8.15, was considered for the simulation of the VSG control method. The simulating

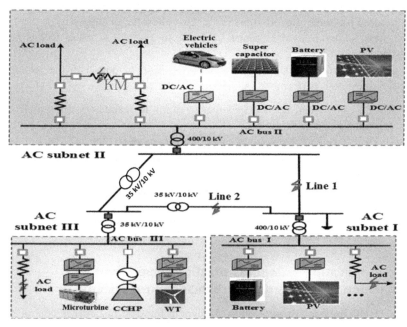

FIGURE 8.14 The platform of goldwind microgrid for testing transient control.

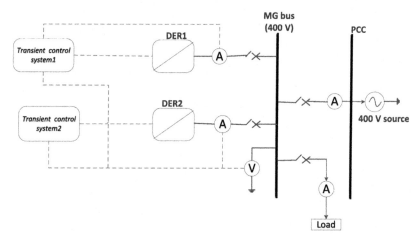

FIGURE 8.15 Simple microgrid platform operating in island mode for testing transient disturbance due to heavy load variation.

application software platform is PSCAD. The control parameters used for the simulation and field tests are listed in Table 8.2.

The procedure followed in the simulation testing is that the microgrid with the two DERs and light load of 30 kW was running in island mode. At 1.8 seconds, a heavy resistive load of 170 kW was added while the microgrid was still running in island mode. The resulting waveforms of power flow in each element and the frequency profiles for both droop control and VSG control methods are shown, respectively, in Figs. 8.16 and 8.17.

According to Fig. 8.17, the profiles of the frequency between droop control and VSG control method show that the responses to heavy load variation are different; the VSG control method performs better than the droop control with regard to magnitude deviation and recovery time. The deviation of the frequency in the VSG control method is within 49.7 and 50.1 Hz, whereas in droop control, the deviation reaches 48.2 and 50.7 Hz. Even during steady-state operation, the deviation is insignificant in the VSG control method compared with the droop control method. As

TABLE 8.2 Parameters used for droop control and VSG control systems.

Lithium battery			Lead–carbon battery		
Parameters	Symbol	Value	Parameters	Symbol	Value
Droop gain	m_{Li}	$1e^{-3}$	Droop gain	m_{PbC}	$2.5e^{-3}$
Moment of inertia	J_{Li}	0.1	Moment of inertia	J_{PbC}	0.04
Damping coefficient	D_{Li}	1250	Damping coefficient	D_{PbC}	500

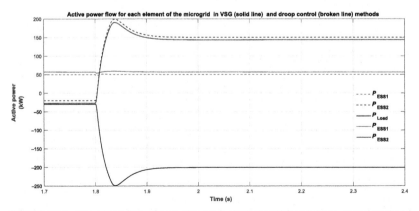

FIGURE 8.16 The power flow of each element for both the VSG and droop control methods' implementation in the microgrid network.

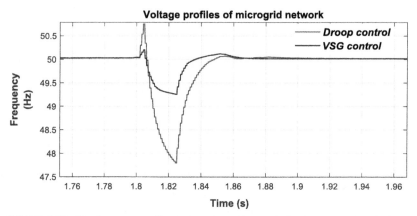

FIGURE 8.17 The frequency profiles in the microgrid network for droop control method and VSG control method implementation.

explained above, the power difference between supply and consumption causes the frequency deviation, and hence needs to be damped quickly. The results shown in Fig. 8.16 confirm that there is more surge power in droop control compared with the VSG control method around the times of heavy load change. This verifies that the virtual moment of inertia and damping coefficient help to damp the transient disturbances well and instantly.

To address the voltage transient stability issue caused by the islanding mode and heavy load change, the VSG transient control system has the same topology with that of the droop control method. The control strategy employed has a reactive current compensating control. It involves voltage and current dual closed-loop control to stabilize the voltage at common coupling point (PCC). The gate signals generated by this controller are supplied to the converter of the DERs.

Significant voltage drop at the instant of the heavy load insertion was expected when there is no transient control device. This may lead to the circuit breaker tripping due to LV. However, through the application of the trainset control device, reactive power can be compensated. The simulation result shown in Fig. 8.18 verifies this fact. As can be seen from Fig. 8.18, during heavy step load change, the voltage drop is too small to trip the breaker. Hence, the VSG transient control method was effective at achieving voltage stability within a short period (within one cycle) as shown in Fig. 8.18. But, basically, the voltage control was managed by the feature of droop control in the VSG strategy. Hence, as asserted in the conventional power system, the VSG is more pronounced in the control of frequency instability.

To further validate the effectiveness of these control strategies, tests were conducted in field experiments and are described in the following section.

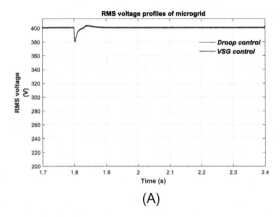

(A)

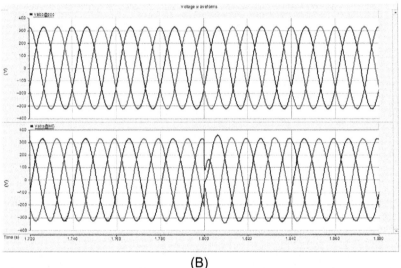

(B)

FIGURE 8.18 The voltage profiles of the microgrid for the transient control system response while disturbance occurs: (A) root mean square (RMS) voltage profiles and (B) instantaneous voltage waveform showing the response of heavy load step change.

8.6.1.2 Field testing results for the transient control device

The transient control device that applies the aforementioned strategy of the VSG control method was tested in an actual microgrid system. The test was made with two different cases with different values of damping coefficient D_ω. The voltage, current, and the frequency waveforms in a microgrid are presented in Figs. 8.20 and 8.21.

Case 1: The lithium battery is set with $J_\omega = 0.35$, $D_\omega = 0$, and resistive load change of 200 kW.

Case 2: The lithium battery is set with $J_\omega = 0.35$, $D_\omega = 1250$, and resistive load change of 200 kW.

Fig. 8.19 and 8.20 are the three-phase voltage, current, and frequency waveforms during step change of resistive load in the experiment. Comparing the above two cases of experiments, it can be concluded that the setting of damping coefficient D_ω will affect the frequency response of the system. Only setting J_i reflects the characteristics of VSG. After adding D_ω, it will return to the droop control characteristic. It verifies that in the design model D_ω describes the droop coefficient characteristics of the power control system. Following these testing cases, comparative experiments between droop control and VSG control method were carried out.

During these comparative experimental tests, the main components involved were lithium battery-based DER with the capacity of 500 kVA, lead−carbon

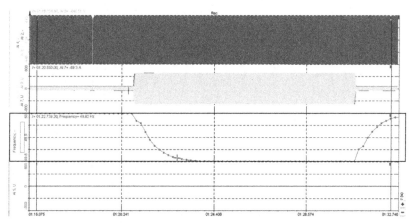

FIGURE 8.19 The profiles of output voltage and current of the lithium battery and frequency of the system in the experimental setup for case 1.

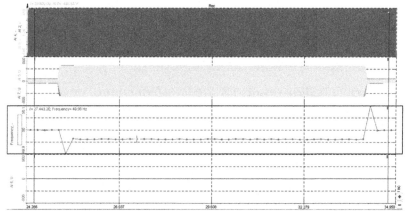

FIGURE 8.20 The profiles of output voltage and current of the lithium battery and frequency of the system in the experimental setup for case 2.

battery-based DER with the capacity of 200 kVA and load. The parameters of the control system were taken as optimal values according to the above testing results and listed in Table 8.2. The damping coefficients of the VSG system are set to characterize and show the impact on the frequency and power sharing. At T1, 200 kW load is added in the system and removed at times T2 while droop control is applied. When the VSG control system was used, the 200 kW load is added at T3 and T5 and switched off at times T4 and T6. During the VSG control method implementation, different damping coefficients were used; $D_{Li} = 1250$ and $D_{PbC} = 500$, which are proportional with the capacity of the DERs during T3 to T4, and $D_{Li} = 1000$ and $D_{PbC} = 1000$ during T5 to T6. The test results of current, voltage, and frequency waveforms and profiles in a microgrid are illustrated in Figs. 8.21−8.23. These figures are plotted using MATLAB® after taking the data of test results from the storage of the control system devices.

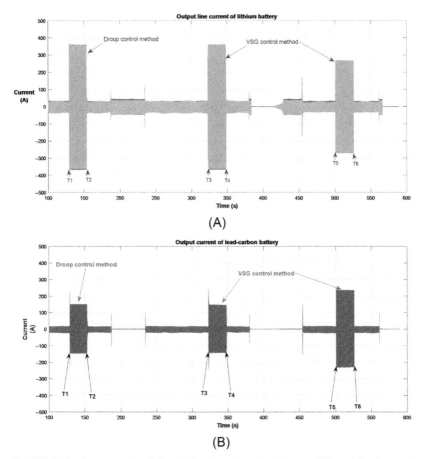

FIGURE 8.21   Current profiles during field testing of large load change while applying droop and VSG control methods in (A) lithium battery-based DER and (B) lead−carbon battery-based DER.

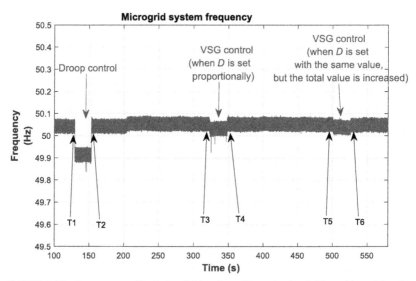

FIGURE 8.22   Frequency profile during field testing of heavy load switching with applying the droop and VSG control methods.

The power sharing between the DERs, according to Fig. 8.21A and B, depends on the control parameters in both control methods. The power is shared based on the capacity of each DER. It was supposed to operate each DER with the same capacity in different control methods, but the frequency response is different in that the VSG control method performs better than the droop control with regard to frequency deviation and features of dynamic response. This all verifies the analysis of mathematical model and simulation results discussed above of the VSG control method.

If the same parameters are set in the VSG control method, for example, during T5−T6 of Fig. 8.21A and B, the powers are shared equally in the two DERs, but a little difference of frequency deviation is demonstrated, since the overall damping coefficient D_ω increases, which verifies the effect of damping coefficient on the frequency dynamics and deviation. It shows that determining the appropriate values of the control parameters is significantly essential.

Fig. 8.23A and B show the three-phase voltage waveforms in the event of connecting a resistive load by applying the droop and VSG control methods, respectively. The experimental results are consistent with the simulation results presented in Fig. 8.20. It can be seen from Fig. 8.23 that the voltage falls are small to trip the circuit breaker and the recovery time is within one cycle. This is achieved through the reactive current compensation by the well-regulated transient disturbance control equipment. If such robust control strategies were not implemented, significant voltage fall, as shown in Fig. 8.2, would occur.

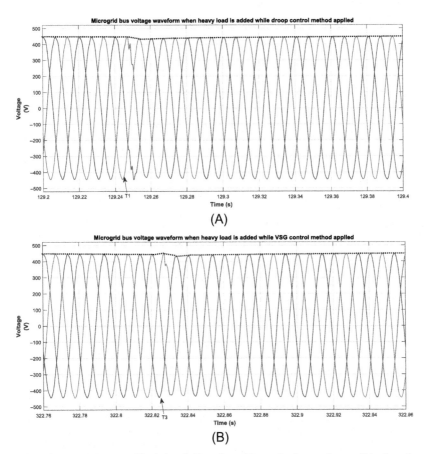

FIGURE 8.23 Voltage profile during field testing of heavy load step change: (A) when the droop control method is applied and (B) when the VSG control method is applied.

Compared with the droop control strategy, VSG control has a fast dynamic frequency response to the transient disturbances due to inertial features emulating a rotary synchronous generator. However, the setting of values such as moment of inertia J_ω and damping coefficient D_ω needs to be undertaken carefully and properly.

References

[1] D.E. Olivares, et al., Trends in microgrid control, IEEE Trans. Smart Grid (2014). Available from: https://doi.org/10.1109/TSG.2013.2295514.

[2] A. Bidram, A. Davoudi, F.L. Lewis, Z. Qu, Secondary control of microgrids based on distributed cooperative control of multi-agent systems, IET Gener. Transm. Dis. (2013). Available from: https://doi.org/10.1049/iet-gtd.2012.0576.pdf.

[3] J. Choi, S. Oh, R. Horowitz, Distributed learning and cooperative control for multi-agent systems, Automatica (2009). Available from: https://doi.org/10.1016/j.automatica.2009.09.025.
[4] T. Morstyn, B. Hredzak, V.G. Agelidis, Cooperative multi-agent control of heterogeneous storage devices distributed in a DC microgrid, IEEE Trans. Power Syst. (2016). Available from: https://doi.org/10.1109/TPWRS.2015.2469725.
[5] X. Wu, X.G. Yin, X. Song, J. Wang, Research on microgrid and its application in China, Gaoya Dianqi/High Volt. Appar. (2013). Available from: https://doi.org/10.4236/epe.2013.54b033.
[6] T. Morstyn, B. Hredzak, V.G. Agelidis, Distributed cooperative control of microgrid storage, IEEE Trans. Power Syst. (2015). Available from: https://doi.org/10.1109/TPWRS.2014.2363874.
[7] A. Mohanty, M. Viswavandya, P.K. Ray, S. Patra, Stability analysis and reactive power compensation issue in a microgrid with a DFIG based WECS, Int. J. Electr. Power Energy Syst. (2014). Available from: https://doi.org/10.1016/j.ijepes.2014.05.033.
[8] S. Parhizi, H. Lotfi, A. Khodaei, S. Bahramirad, State of the art in research on microgrids: a review, IEEE Access (2015). Available from: https://doi.org/10.1109/ACCESS.2015.2443119.
[9] V. Dabeer, Fault current limiters, Water Energy Int. (2009). Available from: https://doi.org/10.1201/9781420050271.chh6.
[10] O. Ipinnimo, S. Chowdhury, S.P. Chowdhury, J. Mitra, A review of voltage dip mitigation techniques with distributed generation in electricity networks, Electr. Power Syst. Res. (2013). Available from: https://doi.org/10.1016/j.epsr.2013.05.004.
[11] M. Ramezani, S. Li, Voltage and frequency control of islanded microgrid based on combined direct current vector control and droop control, in: IEEE Power and Energy Society General Meeting, 2016. Available from: https://doi.org/10.1109/PESGM.2016.7741786.
[12] Q. Liu, Y. Tao, X. Liu, Y. Deng, X. He, Voltage unbalance and harmonics compensation for islanded microgrid inverters, IET Power Electron. (2014). Available from: https://doi.org/10.1049/iet-pel.2013.0410.
[13] M. Farrokhabadi, C.A. Cañizares, K. Bhattacharya, Frequency control in isolated/islanded microgrids through voltage regulation, IEEE Trans. Smart Grid (2017). Available from: https://doi.org/10.1109/TSG.2015.2479576.
[14] F. Nejabatkhah, Y.W. Li, Overview of power management strategies of hybrid AC/DC microgrid, IEEE Trans. Power Electron. (2015). Available from: https://doi.org/10.1109/TPEL.2014.2384999.
[15] M. Ramezani, S. Li, Y. Sun, Combining droop and direct current vector control for control of parallel inverters in microgrid, IET Renew. Power Gener. (2017). Available from: https://doi.org/10.1049/iet-rpg.2016.0107.
[16] C.N. Papadimitriou, E.I. Zountouridou, N.D. Hatziargyriou, Review of hierarchical control in DC microgrids, Electr. Power Syst. Res. (2015). Available from: https://doi.org/10.1016/j.epsr.2015.01.006.
[17] A. Fathi, Q. Shafiee, H. Bevrani, Robust frequency control of microgrids using an extended virtual synchronous generator, IEEE Trans. Power Syst. (2018). Available from: https://doi.org/10.1109/TPWRS.2018.2850880.
[18] Q.C. Zhong, Virtual synchronous machines: a unified interface for grid integration, IEEE Power Electron. Mag. (2016). Available from: https://doi.org/10.1109/MPEL.2016.2614906.
[19] V. Karapanos, S.W. De Haan, K. Zwetsloot, Testing a virtual synchronous generator in a real time simulated power system, in: Proceedings of the International Conference on Power Systems Transients (IPST 2011), 2011.
[20] Y. Zheng, Virtual Inertia Emulation in Islanded Microgrids With Energy Storage System, Delft University of Technology, 2016.

[21] Q.C. Zhong, G. Weiss, Synchronverters: inverters that mimic synchronous generators, IEEE Trans. Ind. Electron. (2011). Available from: https://doi.org/10.1109/TIE.2010.2048839.

[22] H. Bevrani, T. Ise, Y. Miura, Virtual synchronous generators: a survey and new perspectives, Int. J. Electr. Power Energy Syst. (2014). Available from: https://doi.org/10.1016/j.ijepes.2013.07.009.

[23] J.C. Boemer, R. Walling, DER Ride-Through Performance Categories and Trip Setting, Presentation at PJM DER Ride-Through Workshop, Philadelphia, PA, October 1−2, 2018.

[24] North American Electric Reliability Corporation, Performance of Distributed Energy Resources During and After System Disturbance: Voltage and Frequency Ride-Through Requirements, 2013.

[25] S. Gonzalez, R. Walling, A. Ellis, Implementation of Voltage and Frequency Ride-Through Requirements in Distributed Energy Resources Interconnection Standards, 2014.

[26] IEEE, 1547-2018—IEEE Standard for Interconnection and Interoperability of Distributed Energy Resources With Associated Electric Power Systems Interfaces, IEEE, 2018.

[27] The Standardization Administration of the People's Republic of China, GB/T 34930-2017 Standard for Operation and Control Specification for Microgrids Connected to Distribution Network, National Standard Administration of China, 2017.

[28] The Standardization Administration of the People's Republic of China, GB/T 33589-2017 Standard for Technical Requirements for Connecting Microgrid to Power System, National Standard Administration of China, 2017.

[29] International Electrotechnical Commission, IEC 61727 Standard for Photovoltaic (PV) Systems—Characteristics of the Utility Interface, International Electrotechnical Commission (IEC), 2004.

Chapter 9

Tertiary control of microgrid

9.1 Introduction

A microgrid is a local, smaller grid comprising a number of distributed generators, energy storage systems (ESSs), and loads, with proper coordinated controllers and capability of operating in both grid-connected and island modes.

As part of the effort into standardizing the operation and functionalities of microgrids, hierarchical control has been proposed recently with a slightly varying nomenclature and classification. However, based on a wide review of available literature and the naming trend in the industry, the following three main hierarchical control levels are used according to the functions and timescales of the control:

1. *Primary control* is usually considered to refer to local and basic controls for frequency and voltage stability. This level of control is achieved exclusively based on local measurements, calculations, and implementation without the need for a communications network. Functions such as output control, load sharing, islanding detection, and plug-and-play of distributed generations (DGs) may lie under the jurisdiction of primary control. More discussion on the primary control designs and functions are available in Chapter 7, Dynamic Control of Microgrids, and Chapter 8, Transient Control of Microgrids, of this book.
2. *Secondary control* is usually a centralized control designed with a dynamic response relatively slower than primary control. Secondary control complements the impacts of primary control functions so as to improve the power quality and system performance of the microgrid by acting on steady-state errors caused by actions of the primary control. Secondary control may also provide ancillary services such as load regulation, spinning reserve control, voltage regulation, and reactive power supply.
3. *Tertiary control* is a central control involving intensive communication among the DGs and loads of the microgrid and the distribution system operator. It is mainly implemented through a central calculation and global command that decides the import or export of power from/to the microgrid, scheduling and energy dispatching of DGs, load curtailment

Microgrid Protection and Control. DOI: https://doi.org/10.1016/B978-0-12-821189-2.00001-2
255

or load shifting, etc. The power exchange with grid and optimization of local production or consumption are performed based on optimization of features such as cost of energy, emission level, efficiency, reliability, security, reserve capacity, renewable energy utilization rate, or others, as per the specified objective functions (OFs).

9.2 Optimal energy dispatching control in microgrids

9.2.1 Introduction

In older microgrids, the microgrid operator (MGO), which is responsible for the energy management of the microgrid, is considered to be a central entity while the end-users act as passive entities. The MGO is dedicated to minimizing the total cost of the purchased energy from the utility grid and the cost of generation from distributed energy resources (DERs). In recent times, the operation and management of microgrids have been shifted to a mechanism consisting of multiparties and have been paying attention to better involvement of end-users. The end-users may consist of DERs, prosumers, flexible loads, etc. These parties are presumed to be rational, autonomous, and independent entities.

Prosumer refers to an end-user who consumes energy at some time and produces energy at a certain time [1]. Greater involvement of end-users has made optimal energy dispatching in microgrids more complex with different prosumers having different objectives while consuming, selling, or purchasing electricity. It has become dependent on the specific interests and roles of individual end-users. Therefore the objective of energy management can no longer be articulated as minimizing the parties' operation cost, but to address different interests of the parties. Hence, customized optimal energy dispatching control plays a crucial role. This section is dedicated to the current issues related to optimal energy dispatching control in microgrids. The mathematical modeling and dispatching algorithms to develop optimal energy dispatching control in microgrid are discussed in this section.

9.2.2 Mathematical modeling

The core of the mathematical model in optimal energy dispatching is the OF, $f(x)$, which defines the parameter which is going to be maximized or minimized in the optimization process. The variables $x_1, x_2, x_3, \ldots, x_n$ are the inputs that we can control. The model is intended to determine the optimal commitment, active and reactive power exchange by the DERs including renewable energy resources (RESs), EESs, and loads in the microgrid. ESSs play a key role in facilitating optimal energy dispatching in microgrids. The ESS planning and operation involves consideration of factors including site, size, and costs.

In the mathematical modeling process, variables can be continuous or discrete; some problems are dynamic while some are static; some systems can be stochastic while some others can be deterministic; some equations of the system can be nonlinear or others can be linear. Hence, the modeling process needs a curious outlook to consider the mentioned characteristics and features revealed from the microgrids to come up with a reliable and robust model.

Many of the descriptive models can be represented by the mathematical expressions such as unit commitment and optimal power flow. In general, the optimization of the power system containing ESSs has the form as in (9.1) [2]:

$$
\begin{aligned}
&min: \quad \text{objective function (9.1a)} \\
&s.t. \quad \text{network constraints (9.1b)} \\
&\quad\quad \text{energy storage constraints (9.1c)}
\end{aligned}
\tag{9.1}
$$

The OF in (9.1) can be any of cost minimization of active power, voltage deviation minimization, active power loss minimization, reactive power loss minimization, power transfer maximization, emission minimization, and others. Normally, the OFs in (9.1a) are convex. However, due to the type and size of the constraints in (9.1b) and (9.1c), the OFs may be nonconvex and computational burden of the model could be larger. The constraints in (9.1b) have been presented in detail in Refs. [3,4].

For different power system optimizations on modeling of ESS for grid integration system, the literature is noted to include several indices. The ESS can be characterized by its rated power and energy, efficiencies of charging and/or discharging cycles, and others. The models remain part of reaching attainments where necessary statistics like the network's topology, admittance matrix, and others are to be included. Some typical categories of the models are listed in the following sections.

9.2.2.1 Linear models

1. *Mixed integer linear model*

To formulate the charge/discharge state of the ESS, the efficiencies of charge/discharge are taken into consideration, while binary variables are introduced to represent the status of the ESS. Eq. (9.2) describes the mixed integer linear model of the ESS in a microgrid.

$$
P_{ESS,n,t}^{net} = Y_{ch}P_{E,n,t}^{ch} - (1/Y_{disch})P_{E,n,t}^{disch}
$$

$$
E_{E,n,t}^{min} \leq E_{E,n,0} + \Delta t \sum_{i=1}^{t} P_{E,n,i}^{net} \leq E_{E,n,t}^{max}
$$

$$
0 \leq P_{E,n,t}^{ch} \leq \lambda_{n,t}^{ch} P_{E,n,t}^{max}
\tag{9.2}
$$

$$
0 \leq P_{E,n,t}^{disch} \leq \lambda_{n,t}^{disch} P_{E,n,t}^{max}
$$

$$
\lambda_{n,t}^{ch} + \lambda_{n,t}^{disch} \leq 1
$$

$$
\lambda_{n,t}^{disch}, \lambda_{n,t}^{disch} \in \{0,1\}, n \in B, t = 0,1,\dots,T-1
$$

where the net absorbed/injected power (P^{net}) to the ESS is described in the first expression of (9.2); P_E^{ch} and P_E^{disch} are charge and discharge variables, respectively, at each ESS connected to bus n at time t; λ^{ch} and λ^{disch} are binary variables to restrict the ESS from charging and discharging respectively at the same time. Υ_{ch} and Υ_{disch}, respectively, indicate the losses during the charging and discharging of the ESS. E_E^{min} and E_E^{max} are the lower and upper tolerable limits of the voltage at each bus n where each ESS connected. T is the planning time horizon. B is the number of all buses to which the ESSs are connected.

2. *Continuous linear model*

For the sake of ignoring the binary variables while remaining to meet the basic requirements, a simplified lossless continuous linear model is used by some researchers, as in (9.3).

$$-P_{E,n}^{max} \leq P_{E,n,t} \leq P_{E,n}^{max}$$
$$E_{E,n}^{min} \leq E_{E,n,0} + \Delta t \sum_{i=1}^{t} P_{E,n,i} \leq E_{E,n}^{max} \qquad (9.3)$$
$$n \in B, t = 0, 1, \dots, T-1$$

9.2.2.2 Nonlinear models

1. *Nonlinear model adapted from linear models*

These models are designed by adapting any of the linear models. They are used to prevent the coinciding action of charging and discharging during the same operating time period. The adaption is made based on the principle that the application of integer variables is superseded by the control variables related to ESS. Such a model is constructed by (9.4).

$$P_{E,n,t}^{ch} * P_{E,n,t}^{disch} = 0$$
$$0 \leq P_{E,n,t}^{ch} \leq P_{E,n,t}^{max} \qquad (9.4)$$
$$0 \leq P_{E,n,t}^{disch} \leq P_{E,n,t}^{max}$$

2. *Nonlinear model for ESS*

A model based on a circuit of a battery energy storage system (BESS) is developed in Ref. [2] as shown in Eq. (9.5).

$$P_{E,n,t}^{loss} V_{n,t} = r_n^{eq} (P_{E,n,t})^2 + r_n^{CVT} (Q_{E,n,t})^2$$
$$P_{E,n,t}^{loss} = P_{E,n,t} + P_{E,n,t}^{loss} \qquad (9.5)$$
$$(P_{E,n,t})^2 + (Q_{E,n,t})^2 \leq (S_{CVT,n}^{max})^2$$

where $r_n^{eq} = r_n^E + r_n^{CO}$ is the equivalent resistance of the battery storage and converter units. The Ohmic power losses due to battery storage system and converter units is related with the terms of active and reactive power outputs in the expression of Eq. (9.5).

3. *Convex nonlinear model for ESS*
The optimal scheduling problem for distributed ESSs designed in Ref. [2] is a convex nonlinear model, which is a novel convex relaxation equation.

$$P_{E,n,t}^{loss} V_{n,t} \geq r_n^{eq}(P_{E,n,t})^2 + r_n^{CO}(Q_{E,n,t})^2$$
$$r_n^{eq}\left(S_{CO,n}^{max}\right)^2 \geq P_{E,n,t}^{loss} V_n^{min} + r_n^E(Q_{E,n,t})^2 \qquad (9.6)$$
$$\left(S_{CO,n}^{max}\right)^2\left(V_{E,n}^{min} + V_{E,n}^{max}\right) \geq \left(S_{CO,n}^{max}\right)^2 V_{E,n,t} + P_{E,n,t}^{loss}\left(V_{E,n}^{min}*V_{E,n}^{max}\right)$$

The state of charge (SOC) constraints are applicable here. The convex hull relaxation is within system bounds ($V_{E,n}^{min} \leq V_{E,n,t} \leq + V_{E,n}^{max}$). The details of the model are presented in Ref. [5].

A similar work, multistage and stochastic mathematical modeling, is presented in Ref. [6]. The optimal sizing, timing, and placement of distributed energy technologies are determined by this model, which is constructed by the system operator's viewpoint, while the technologies are coordinated with the reactive power sources and ESSs. Maximizing the size of renewable power absorbed by the system is the ultimate goal of this model. The acceptable power quality and required system stability at a minimum possible cost are guaranteed in this optimization model. The merit of this model is verified with the numerical demonstration in Ref. [7].

9.2.2.3 Multiobjective optimization modeling
The objectives in multiobjective optimization problems mostly conflict each other. To solve this problem, a mechanism is needed which is transferring the multiobjective problem into a single-objective model. To do this transferring, a weighted aggregation technique can be implemented. This OF of the optimization model $G(\gamma, E, C)$ is defined in (9.7) [8]:

$$min: G = \gamma E + (1 - \gamma)\frac{1}{C} \qquad (9.7)$$

where γ is the weight between comfort level and economy level, E is the total economic OF in an optimization cycle, and C is the total comfort OF in the cycle.

9.2.2.4 Uncertainties modeling
According to Ref. [9], the absorbed power at any load bus in the grid is indeterminate. The probability density function is thus utilized to well explain this uncertainty. It is made based on a principle of normal distribution theory. It gets through relevant assumptions. Thus the function is given by (9.8).

$$f(P_L) = \frac{1}{\sqrt{2\pi}\sigma_{PL}}exp\left(-\frac{(P_L - \mu_{PL})^2}{2\sigma_{PL}^2}\right) \qquad (9.8)$$

where μ_{PL} is the mean value of the function, σ_{PL} is the value of the standard deviation, and P_L is the absorbed load power.

9.2.2.5 Costs modeling

The main objective of such modeling is to have a minimized cost function while meeting the requirement of providing appropriate services and power supplies and to satisfy the constraints that possibly exist. The cost incurred in microgrid generation and operation may assume a variety of forms. The topmost ones are presented in the following sections.

1. *Modeling of generated power's cost by units*

   The total generation fuel cost and the emission level reduction of every online unit in the generation system are determined by the main objective functional model which states the economic load dispatch problem. It will also satisfy a system constraint. The economic load dispatch formulation of the generators in a microgrid is presented in Ref. [10]. Thus the fuel cost function of generators is defined by (9.9).

$$F_c = \sum_{i=1}^{G} F_i(P_i) \qquad (9.9)$$

where F_c is the total generation fuel cost; $F_i(P_i)$ is the fuel cost of the ith generator; G is the number of generators; P_i is the active power generation of ith generator.

The fuel cost at each ith generator is defined by

$$F_i(P_i) = v_i P_i + u_i P_i^2 + w_i \qquad (9.10)$$

where, v_i, u_i, and w_i represent the ith generator cost coefficients in the indices of [\$/MWh], [\$/MW^{2}h], and [\$/h], respectively.

Thus the optimal function is described by

$$\text{Min: } F_c = \sum_{i=1}^{NG} u_i P_i^2 + v_i P_i + w_i \qquad (9.11)$$

To reduce various emissions, some criteria are included in a formulation of the optimal dispatch problem and the total emission value E_T is given by (9.12).

$$E_T = \sum_{i=1}^{n} \left(x_i P_i^2 + y_i P_i + z_i \right) \qquad (9.12)$$

where x_i, y_i, and z_i represent the ith generator emission coefficients in the indices of [kg/MW2h], [kg/MWh], and [kg/h] respectively.

According to Ref. [11], price penalty factor (PPF) h_i can be applied to convert the multiobjective model of integrated economic-emission deduction

(IEED) problem into a single-objective optimization model. But the total IEED cost is expressed by

$$F_T = \sum_{i=1}^{n} \left[(u_i P_i^2 + v_i P_i + w_i) + h_i (x_i P_i^2 + y_i P_i + z_i) \right] \quad (9.13)$$

Thus the optimal PPF in [$/kg] is formulated as:

$$h_i = \frac{u_i P_i^2 + v_i P_i + w_i}{x_i P_i^2 + y_i P_i + z_i} \quad (9.14)$$

The cost of generated power by units is usually determined by the cost of primary energy. However, the fuel cost of energy generated from renewables such as wind and solar sources is zero.

Fuel cells (FCs) are becoming common components of microgrids in recent years due to their high efficiency and other advantages. In modeling the cost function of FCs, taking a sample of s in the system, the efficiency of the FC ($\eta_{FC,s}$) depends on the power output. The efficiency of FC will be lower while increasing the power output. Thus the FC power generation cost can be determined by (9.15) [12].

$$Cost_{FC,s} = \frac{C_{nl}}{L} * \frac{P_{unit,s}}{\eta_{FC,s}} \quad (9.15)$$

Whereas the efficiency of microturbine (MT), unlike FC, increases while increasing the power output. Thus the MT power generation cost can be determined by (9.16).

$$Cost_{MT,s} = \frac{C_{nl}}{L} * \frac{P_{unit,s}}{\eta_{MT,s}} \quad (9.16)$$

The microgrid accommodates the combined heat and power (CHP) plants which may gear surge economics. The efficiency of the CHP system can possibly reach up to 80%−85%. Thus fuel cost of the system $(Cost_{CHP,s})$ with MT and CHP is calculated as in (9.17).

$$Cost_{CHP,s} = Cost_{MT,s} - B_{CHP,s}$$

$$B_{CHP,s} = Cost_{MT,s} \left(\frac{\varepsilon_{rec}(\eta_{T,CHP,s} - \eta_{e,CHP,s})}{\eta_b} \right)$$

$$Cost_{MT,s} = Cost_{MT,s} \left(1 - \frac{\varepsilon_{rec}(\eta_{T,CHP,s} - \eta_{e,CHP,s})}{\eta_b} \right)$$

$$(9.17)$$

where $B_{CHP,s}$ is cost reduction, due to exhaust gas heat usage, of MT power generation; η_b, $\eta_{e,CHP,s}$, and $\eta_{T,CHP,s}$ represent boiler efficiency,

MT electrical efficiency, and total CHP efficiency, respectively; and ε_{rec} represents a factor of heat recovery.

2. *Costs for operation and maintenance (O&M) and installation*

In a microgrid system with wind turbine (WT) and photovoltaics (PV) units, the cost function considers only O&M and installation costs. The cost of O&M and installation for units is as follows:

$$C_{OM, \ unit,s} = K_{OM,s}P_{unit,s}$$
$$C_{ins,unit} = K_{ins}P_{unit,r} \tag{9.18}$$

where $K_{OM,s}$ is coefficients of all units for the O&M cost function; K_{ins} is installation cost coefficient; and $P_{unit,r}$ represents each unit rated power.

3. *Modeling of costs for sold and purchased power*

   a. *Multiobjective model*

   The cost OF, which is a multiobjective model, includes all units having generated power cost, sold and purchased power costs, O&M, and installation cost. In a generic nonlinear problem of this model, thus the minimal power generation cost *OF* is defined as in (9.19).

$$Min:OF = \sum_s [F_{s,Pow} + F_{s,EM}] \tag{9.19}$$

where $F_{s, \ Pow}$ represents the units of cost function for power generation cost, power transaction cost, O&M cost, and installation cost in a sample s. $F_{s,EM}$ represents emission cost function in a sample s. The optimization problem can be minimized by the techniques such as particle swarm optimization (PSO) algorithm and imperialist competitive algorithm (ICA), which will be discussed later.

4. *Solar generation forecast and wind generation forecast*

   The cost functions of the generation forecast of solar source in (9.20) and wind source in (9.21) are defined.

$$F(P_{pv}) = aI^P P_{pv} + G^E P_{pv} \tag{9.20}$$

$$F(P_{wind}) = aI^P P_{wind} + G^E P_{wind} \tag{9.21}$$

$$a = \frac{r}{1 - [(1+r)^{-N}]} \tag{9.22}$$

where P_{pv} and P_{wind} stand, respectively, for solar generation and wind generation capacity in kilowatts; a, r, and N stand, respectively, for annuitization coefficient, interest rate, and investment duration; I^P is ratio of investment cost to unit established power; and G^E represents O&M costs.

9.2.2.6 Constraint functions

In a microgrid in island mode, there is no energy transaction with the main grid. However, many constrains inherently avail and need to be satisfied while the optimization process is carried out. The following are some of the major ones.

1. Power balance constraint: the generation of all units including RESs is balanced by the load consumption including storage and the system losses.

$$\sum_{i=1}^{N} P_i = \sum_{j=1}^{L} P_j + \sum_{k=1}^{NL} l_i \qquad (9.23)$$

2. Power generation constraint: each generator output is bounded by minimum and maximum active and reactive power boundaries.

$$\begin{aligned} P_i^{min} &\leq P_i \leq P_i^{max} \\ Q_i^{min} &\leq Q_i \leq Q_i^{max} \end{aligned} \qquad (9.24)$$

where PQ_i^{min} is the minimum output active/reactive power of the ith generator, and PQ_i^{max} is the maximum output active/reactive power of the ith generator.

3. Bus voltage constraint: each bus voltage is limited by the minimum and maximum tolerable boundaries.

$$V_i^{min} \leq V_i \leq V_i^{max} \qquad (9.25)$$

where V_i^{min} is the minimum voltage at ith bus, and V_i^{max} is the maximum voltage at ith bus.

4. Power flow limits: the apparent power flow ($P_j^2 + Q_j^2$) in a transmission or distribution line should be not more than the rated value as expressed as:

$$P_j^2 + Q_j^2 \leq \left(S_j^{max} \right)^2 \qquad (9.26)$$

5. Energy capacity limits: energy stored in ESS devices should be limited in the designed range denoted as:

$$E_k^{min} + E_k \leq E_k^{max} \qquad (9.27)$$

9.2.3   Optimal energy dispatching algorithms for microgrid

Optimal dispatching problem (ODP) is one of the fundamental energy management models in microgrid networks, which programs the input/output power of all DERs connected in the system to fulfill the load demand requirements with a possible minimum operating cost while satisfying the constraints of all units in the system. So far various algorithms haven been used to tackle ODP. Some common algorithms are presented in this section.

9.2.3.1   Jaya algorithm

Jaya is a Sanskrit word, meaning victory. Its algorithm is proposed in Ref. [13]. It strives to win by getting the best solution by trying to approach success and avoid failure. In Ref. [10], this algorithm is deployed to solve the optimal scheduling problem of microgrids.

Suppose $f_n(s)$ is the OF to be minimized or maximized, "n" is the number of candidate solutions, "m" is the number of design variables (i.e., $j = 1, 2, \ldots, m$), $f(s)_b$ is the best candidate value of $f(s)$ and $f(s)_w$ is the worst candidate value of $f(s)$. Thus during the ith iteration for kth candidate, if $S_{j,k,i}$ is the value of the jth variable, then it is determined by the iteration equation of (9.28).

$$S'_{j,k,i} = S_{j,k,i} + r_{1,j,i}(S_{j,b,i} - |S_{j,k,i}|) - r_{2,j,i}(S_{j,w,i} - |S_{j,k,i}|) \qquad (9.28)$$

where

$S'_{j,k,i}$ is the ith iteration updated value of $S_{j,k,i}$;
$S_{j,b,i}$ is the jth variable ith iteration value of for the *best* candidate;
$S_{j,w,i}$ is the jth variable ith iteration value of for the *worst* candidate;
$r_{1,j,i}$ and $r_{2,j,i}$ is the random numbers for the jth variable during the ith iteration in the range [0, 1];
$r_{1,j,i}(S_{j,b,i} - |S_{j,k,i}|)$ shows the solution converging to the best solution;
$r_{2,j,i}(S_{j,w,i} - |S_{j,k,i}|)$ shows the tendency of the solution to keep away from the worst solution;
$S'_{j,k,i}$: The updated value is accepted as far as it provides better functional value. Once the iteration is completed, all the accepted functional values are maintained and used as an input to the next iteration.

9.2.3.2 Whale optimization algorithm

Solving the dispatch problem of the system targeting economic and carbon emission is a basic problem in microgrids. Metaheuristic optimization techniques have been applied for optimizing the solution in such problem [11]. The algorithm WOA, which always tries to find accurate value in less iteration, is used in Ref. [14] for achieving a better solution in such problem. The algorithm WOA is motivated by the concept of the bubble-net hunting method, which explains the humpback whale special hunting behavior. During its hunting, the whale follows distinctive bubbles causing the creation of a "9-shaped" or circular path and encircling a prey in it. The expressions for the WOA algorithm are developed in two approaches [11], as follows.

1. *Encircling prey equation*

As the whale encircles the prey, it gets close to the prey and the new position is updated for reaching the optimal value. The iterating equation can be defined by (9.29).

$$\vec{P} = |B.\vec{X}^*(i) - X(i)|$$

$$\vec{X}(i + 1) = \vec{X}^*(i) - \vec{Q}.\vec{P} \qquad (9.29)$$

where $X(t)$ is position of vector; $\vec{X}^*(i)$ is optimum solution of the position of vector, i is current iteration, and $\vec{Q}.\vec{P}$ are coefficient vectors. The vectors $\vec{Q}.\vec{P}$ are computed in (9.30).

$$\vec{Q} = 2\,\vec{y}\,*s - \vec{y}$$

$$B = 2*s \qquad (9.30)$$

where s is a random number in the range of [0, 1], and $\vec{y}$ is a linearly decreasing variable from 2 to 0 in the iteration.

2. *Bubble-net attacking method*

There are mathematical equation mechanisms presented for bubble-net behavior of the whale hunting strategy.

a. *Shrinking encircling mechanism*

This mechanism is applied by reducing the value of $\vec{y}$ from 2 to 0 linearly. The random value for vector $\vec{Q}$ is in the range between [−1, 1].

b. *Spiral updating position mechanism*

For getting an updated and optimal position of the whale and prey, a mathematical spiral equation which describes the helix-shaped movement is given in (9.31):

$$\vec{X}\,(i+1) = \vec{P}'\,e^{kt}\cos(2\pi z) + \vec{X}\,(i) \qquad (9.31)$$

where k is a constant defining shape as logarithmic; z is a random number in the range of [−1, 1], $\vec{P}' = |\vec{X}(i) - X(i)|$ expressing the distance between ith whale movement to the prey mean best solution.

A 50/50 probability is assumed during optimization, in that the whale follows either a logarithmic path or the shrinking encircling. Its mathematical model is given by (9.32):

$$\vec{X}(i+1) = \begin{cases} \vec{X}(i) - \vec{Q}.\vec{P} & \text{if } z < 0.5 \\ \vec{P}'e^{kt}\cos(2\pi z) + \vec{X}(i) & \text{if } z \geq 0.5 \end{cases} \qquad (9.32)$$

3. *Search for prey*

To reach an optimum solution in this method (i.e., get close to prey fast), vector $\vec{Q}$ which has taken its value outside of the range [−1, 1], can be used for exploration. The algorithmic equation used in this method is expressed in (9.33).

$$\vec{P} = |\,\vec{B}\,X_{rnd}^{\rightarrow} - \vec{X}\,|$$

$$\vec{X}\,(i+1) = \vec{X}_{rand} - \vec{Q}.\vec{P} \qquad (9.33)$$

The exploration tracks two conditions.

- If $|\vec{Q}| > 1$, it enforces the exploration of the algorithm to discover global optimum;
- For $|\vec{Q}| < 1$, it updates the current search agent position.

9.2.3.3 Biogeography-based optimization algorithm

The BBO algorithm is a novel evolutionary algorithm proposed by Simon in 2008. The basic idea of this algorithm is to find the global optimal solution using *migration* and *mutation operations* to realize the information exchange among individuals and their information update. Each individual is considered as a habitat, then the suitable index vector (SIV) and the habitat suitably index (HSI) are used to reflect the living environment of the habitat and the extent to which the environment is suitable for the species to live within, respectively. Higher HIS represents that the environment of the habitat is more suitable for the species to live in (the corresponding SIV is superior), and the population here will be larger. SIV can be used to denote the power decision variable in the dispatch model and HIS is for denoting the corresponding objective fitness value of the decision variable. The interest in this algorithm by scholars has risen due to its simple implementation and good convergence properties [15].

The species migration model is defined by

$$\begin{cases} \lambda_i = \dfrac{I}{2\left[\cos(S_i \pi / S_m) + 1\right]} \\ \mu_i = \dfrac{E}{2\left[-\cos(S_i \pi / S_m) + 1\right]} \end{cases} \tag{9.34}$$

where E and I are the maximal immigration and emigration rates, respectively; S_i is the species number of ith habitat; and S_m is the maximum species population number in the habitat. To produce new intermediate individuals in the differential evolution algorithm, a mutation strategy based on the random deviation disturbance of multiobjective optimization problems is developed in Ref. [15]. In this system, the current global optimal is used to guide the individual mutation and ensure the individual diversity simultaneously. The mutation operator is expressed by

$$X_i^{(k+1)} = \begin{cases} X_i^{(k)} + M_i^{(k)} & \text{if } \text{rand}(0,1) > m_i \\ X_i^{(k)} & \text{otherwise} \end{cases}$$

$$M_i^{(k)} = F\left(\overline{X}_{go}^{(k)} - X_i^{(k)}\right) + (1 - F)\left(X_{r1}^{(k)} - X_{r2}^{(k)}\right) \tag{9.35}$$

where $\overline{X}_{go}^{(k)}$ is the global optimal solution reserved at the iteration k; F is the constant scale factor belonged to (0, 2); r_1 and $r_2(r_1 \neq r_2 \neq i)$ are two randomly selected individuals in the population.

9.2.3.4 Markov decision process algorithm

To handle the power exchange at the point of common coupling and smoothen the fluctuation of energy management in the microgrid, the MDP algorithm has been developed based on improved entropy [16]. The optimal entropy value and optimal policy, through policy iteration, have been acquired with the decision process. A fuzzy control of wavelet packet probability is advised to reduce the policy set. Such implementation should be a specific stable policy, which can be implemented as mapping from the energy storage SOC state-set to the hybrid energy storage charge/discharge action-set. The MDP algorithm is thus a method of getting the best policy to use for the optimal solution of the reward function. A basic MDP optimal algorithm of microgrid power management can be computed as in (9.9).

$$\langle P, O, R, G, T \rangle \tag{9.36}$$

where P designates the set of terminal node joint actions (output power of DGs and charge/discharge power of energy storage); O is a table of observation probabilities; R is a reward function (OF); G is a set of global states (SOC of energy storage); and T is transition function of a state.

9.2.3.5 Stackelberg game approach algorithm

A Stackelberg game theory investigates the multilevel decision-making processes in which some players (called the leaders) declare their strategies first and enforce them on the other players (called the followers). The leaders' decisions are made from their own perspective and purpose, but the followers execute the decisions as per the decision of the leader.

The problem of energy sharing management, considering Fig. 9.1 for this case, can be defined by Stackelberg game as in (9.37) [17]. Here, the leader considered is the MGO that establishes the internal prices for energy sharing to the consumers, and the followers are the prosumers that act in response to the prices by switching their energy consumption to either their schedule or demand level.

$$A = \left\{ (F \cup \{MGO\}), \{X_n\}_{n \in G}, \{P\}, \{O\}, \{U_n\}_{n \in G}, R \right\} \tag{9.37}$$

where F is the set of prosumers (followers) and MGO is the set of the leaders of the game; X_n is the prosumer n set of strategies, whose constraint is set by $x_n^{min} \le x_n \le x_n^{max}$; P and O are the vectors, which are the MGO sets of strategies constrained with the internal prices of $\lambda^g \le o < p \le \lambda_n^m$; U_n is prosumer n utility function that gains the benefit of the prosumer from sharing energy $(E_n^g - x_n)$ and consuming energy in the microgrid network. R stands for the profit function of the MGO gaining the total profit from transacting energy with the utility grid and sharing energy in the microgrid as well.

The aims of each prosumer and MGO are to maximize the benefit of OFs. Thus a viable solution for this purpose can be achieved in the Stackelberg equilibrium (SE) at which optimal prices are obtained by the

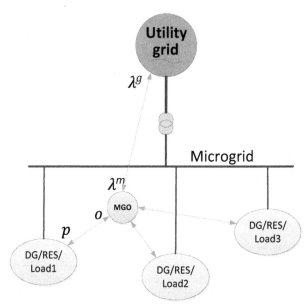

FIGURE 9.1 Energy sharing management.

leader taking the best responses of the followers, and the followers will take decision for their optimal energy consumptions. At the SE neither of them can benefit by making strategies unilaterally change.

Suppose the Stackelberg game A given in (9.37), thus a set of strategies (x^*, p^*, o^*) constitutes the SE of the game, if and only if the following set of inequalities are satisfied in it:

$$U_n(x^*, p^*, o^*) \geq U_n\left(x_n, \mathbf{X}^*_{-n}, o^*\right), \forall n \in G, \forall x_n \in \mathbf{X}_n \qquad (9.38)$$

and

$$R(x^*, p^*, o^*) \geq R(\mathbf{X}^*, p, o), \forall p_n \in \mathbf{P}, \forall o_n \in \mathbf{O} \qquad (9.39)$$

where $\mathbf{X}^* = [x^*_1, x^*_2, \ldots, x^*_N], \mathbf{X}^*_{-n} = [x^*_1, \ldots, x^*_{n-1}, x^*_{n+1}, \ldots, x^*_N]$.

Hence, when SE is achieved by both players in $(F \cup \{MGO\})$, the MGO profit cannot be increased by setting the prices from the SE prices p^* and o^*. Likewise, prosumer utilities cannot be improved by selecting the strategies other than x^*_n.

9.2.3.6 Consensus theory-based algorithms

Most of the existing algorithms of distributed system for multiagent coordination are based on consensus theory [18]. In such a system, each bus assigned as an agent has the properties of:

1. the collected data are processed, and the active power of generators is regulated;

2. a set of sensors is installed to acquire local bus measurements; and
3. a limited number of neighbor agents are set to communicate with.

These properties enable the consensus theory-based algorithms to implement an approach of a plug-and-play for connecting the components in the microgrid network. Moreover, the algorithms are vigorous to tackle failures and to make the microgrids stay operational after the unavoidable failures. They are also scalable in their computation and communication process. Utilizing multiagent coordination, these algorithms can also be used to obtain the feasible solution of the economic dispatch problem. This is possibly made by determining the mean value of each generator's additional cost and carrying out precise estimation of a proper weighted average value of variables.

By setting the imbalance between supply and demand as a feedback parameter to the leader iteration method and selecting a consensus variable on the incremental cost of each generator, a distributed algorithm with a quadratic cost function has been developed to solve the economic dispatch problem. But an agent of the leader is required to ensure the balance between supply and demand, and it shows that the convergence rate of the algorithm is affected by the choice of the leader. Therefore for the "leader-election," it needs to design some methods. Furthermore, as the algorithms show, the convergence rate is strongly impacted by the feedback gain. Along this way, the incremental cost of each generator is selected as a consensus variable and the imbalance of demand and supply, as an innovation parameter added to each agent consensus variable, is assigned.

Wang et al. [18] has proposed two approaches: a distributed algorithm with and without generation constraints.

For the case without generation constraints, letting $\lambda_i(k)$ be the optimal incremental cost estimations; $p_i(k)$, at the kth iteration, be an optimal power generation by the generator i; and $s_i(k)$ be the imbalance of total demand and power generation local estimation; the algorithm is established as in (9.40).

$$
\begin{cases}
\lambda_i(k+1) = \displaystyle\sum_{j\in N_i} a_{ij}\lambda_i(k) + \epsilon_i(k)s_i(k), \\[2mm]
p_i(k+1) = \dfrac{\lambda_i(k+1) - \beta_i}{2\alpha_i}, \\[2mm]
s_i(k+1) = \displaystyle\sum_{j\in N_i} a_{ij}s_j(k) - [p_i(k+1) - p_i(k)],
\end{cases}
\tag{9.40}
$$

where $\epsilon_i(.)$, for all $i = 1, 2, \ldots, n$, is a continuous nonnegative function having an impact on the algorithm convergence rate.

Note: In this algorithm:

- $\sum_{j\in N_i} a_{ij}\lambda_i(k)$ is the main consensus term that confirms converges of λ_i.
- $\epsilon_i(k)s_i(k)$ is the surplus term providing a feedback for the power demand and supply mismatch with the factor $\epsilon_i(k)$. This factor is the control

gain or feedback gain. $s_i(k)$ is the mismatch variable or feedback variable.

- The convergence of λ_i to the optimal value λ_i^* is ensured by the feedback variable.

Whereas for the case of having generation constraints, the algorithm is defined by the following equations.

$$
\begin{cases}
\lambda_i(k+1)=\sum_{j\in N_i} a_{ij}\lambda_i(k)+\epsilon_i(k)s_i(k), \\[2mm]
p_i(k+1)=\begin{cases} \underline{p}_i, & \lambda_i(k+1)<\underline{\lambda}_i \\[2mm] \frac{\lambda_i(k+1)-\beta_i}{2\alpha_i}, & \underline{\lambda}_i \leq \lambda_i(k+1)\leq \overline{\lambda}_i \\[2mm] \overline{\lambda}_i, & \lambda_i(k+1)>\overline{\lambda}_i \end{cases} \\[2mm]
s_i(k+1)=\sum_{j\in N_i} a_{ij}s_j(k)-\left[p_i(k+1)-p_i(k)\right],
\end{cases}
\qquad (9.41)
$$

where $\underline{\lambda}_i = 2\alpha_i\underline{p}_i + \beta_i$ and $\overline{\lambda}_i = 2\alpha_i\overline{p}_i + \beta_i$ for all $i = 1, 2, \ldots, n$. The initial values of the algorithm are made the same as the above one.

9.2.3.7 Particle swarm optimization algorithm

Having observed the social scavenging behavior of some animals, such as the schooling behavior of fish and flocking behavior of birds, the main idea of the PSO algorithm was coined by Kennedy and Eberhart in 1995 [19]. Particles in the swarm fly through a place following the stronger swarm members and basically biasing their movement toward historically good areas of their environment.

The aim of PSO in multidimensional hyperspace is to make all particles find the optimal solution. The realization of this achievement is that a random position is initially assigned to all particles in the initial random velocity and space. The algorithm is then performed like a simulation; each particle's position is advanced based on its speed. After each position update, the OF is sampled. Over time, particles gather around one or several best points through the exploration and exploitation of known good positions in search space.

The main steps of this algorithm:

- *Initialization*: initialization of random position and velocity,
- *Update velocity*: updating of the particles' velocity with the designed equation,
- *Update position*: updating of the position vector of the all particles,
- *Memory update*: updating of optimal value of power and position.

This algorithm is discussed in detail and verified with the case of a multimicrogrid system in Ref. [12].

9.2.3.8 *Imperialist competitive algorithm*

Based on the observation of imperialist competition, a new evolutionary optimization method ICA, is developed to solve the problems demanding optimal values [20]. ICA is a population-based metaheuristic approach in which every individual of the population represents a country and imperialists represent some of the best selected countries in the initialization. ICA, like other evolutionary algorithms, begins with an initial population, which is called a country. Each country could be expressed as a vector characterized by sociopolitical traits like culture, language, and religion. The algorithm is split into two parts as imperialists and colonies forming empires together. It is like the particles in the PSO algorithm and the chromosomes in the genetic algorithm.

The main steps of this algorithm are [21]:

Step 1: Randomly generate the initial population. Choose the least expensive imperialist solution and assign the remaining countries to imperialists (initialize empire).

Step 2: In every empire, assimilation is carried out in every colony, and some colonies are revolutionized (assimilation and revolution).

Step 3: If possible, exchange colonial and imperialist positions.

Step 4: Considering the empire total cost, the weakest colonies of the worst empire are redistributed to the winning empire, (imperialist competition).

Step 5: If there are no members in the empire, it will be directly eliminated.

Step 6: If the termination condition is true, the search ends; otherwise, go to step 2.

The application and comparative verification with PSO are presented in Ref. [12].

9.2.4 Role of soft computing tools in microgrid control

Soft computing is a method that is tolerant of uncertainty, imprecision, approximation, and partial truth. It opposes conventional "hard" computing. The concept of soft computing has come into eminence in the last decade. It still is a very recent approach in power system applications [22].

Some of the important types of soft computing are artificial neural networks (NNs), genetic computing, fuzzy logic, and probabilistic reasoning. They are well established conventional computing methods and attractive alternatives to the standard hard computing method. Evolutionary algorithms like genetic algorithm and others are vigorous tools used for solving the optimization problem in recent years. In any multiobjective functions, NNs and fuzzy logic are universal approximators since they can be utilized for developing nonlinear complex systems.

To evaluate the performance of the current power system and to ascertain the effectiveness of alternative plans for the system expansion, planning, design, and operation require comprehensive and continuous assessment and analysis. The optimal energy dispatching task in microgrids also involves a very complicated set of equations and process. The computational task of manual methods to determine power flows, frequency, and voltage levels is very bulky and insurmountable. The study of forecasting and estimation of power system parameters for any sort of time frame is difficult to envisage. Modeling power system security is equally complicated to perform based on manual computation. These and related challenges in power systems can be solved using soft computing methods which should have sophisticated and advanced techniques.

Lot of work has been done so far related to applications of soft computing techniques in the microgrid. The fuzzy control approach has, for example, been investigated remarkably. New nonconvex design algorithm for optimal polynomial fuzzy control in Ref. [23]; nonconvex stabilization criterion for polynomial fuzzy systems in Ref. [24]; a new sum-of-squares design framework for robust control of polynomial fuzzy systems with uncertainties in Ref. [25]; a new intelligent online fuzzy tuning approach for multiarea load frequency control—self-adaptive modified bat algorithm in Ref. [26]; and optimization of fuzzy controller design using a new bee colony algorithm with fuzzy dynamic parameter adaptation in Ref. [27] have been studied. However, there still is considerable room for the application of soft computing tools generally in the power system and microgrid applications and even more contributions are expected from soft computing for the better implementation of tertiary control in microgrids.

9.3 Demand side management and control of microgrids

9.3.1 Introduction

The rapid increase of electricity demand is well thought as one of the most significant challenges that the power grids are facing. To reach a higher level of reliability, stability, and robustness of the power grid system, it is designed in such a way that it should provide peak demands rather than the average load. This may lead to underutilization of the generation and distribution of power system. Thus utility companies are continuously tackling the variations and contingencies by adjusting the balance of power generation and the total loads. Undeniably, fast-responding generators such as fossil-fuel generators are pricey in operation and have a substantial climate impact. Hence, the planners of the power system are facing a persistent challenge to meet the surge demand of their customers while guaranteeing the integrity of power systems. Similar challenges exist in the operation of microgrids. To ease those problems, regulating the users' consumption profiles is considered to be an option and is implemented in many countries. Various methods have been proposed to achieve this. The general term demand side management

(DSM) refers to a group of such methods used to deploy the current generation capacity more efficiently without changing the existing grid infrastructure. In this section, the concept of DSM in microgrids and different methods and algorithms are briefly discussed.

9.3.2 Demand side management in microgrids

DSM is a robust tool that refers to initiatives, technologies, and a group of actions deployed to optimize and efficiently manage the consumers' energy usage and consumption. The development of DSM emerged in 1990 while significant energy crises occurred [28]. As such to solve the developed DSM problem efficiently and to enhance the smartness and robustness of the DSM system, the technologies such as the smart grid evolution, RES integration, smart meters of advanced metering infrastructure, electric vehicles (EVs), and dynamic pricing have all added momentum. Moreover, the existing power grid structure is being upgraded, due to widespread deployment of home energy management systems and communicating devices, and has been transformed into a more decentralized, robust, and intelligent system.

DSM plays a significant role in energy planning, power industry development, and environmental protection. With regard to microgrids, it can bring advantages such as alleviating effective investment pressure on energy production and power distribution, providing reasonable and transparent transactions, providing instant information exchange of supply and demand, restraining effective market power, promoting efficient market operation, speeding up and improving the price mechanism in energy transactions, facilitating to develop new scenarios in energy horizon, significantly reducing emissions, effectively relieving congestions during peak hours, and improving system reliability.

RESs, ESSs, and EVs are key components in microgrids. However, due to the intermittent nature of RES and randomness of EV's behavior, the integration of these components will bring supplementary aggravation. This phenomenon can lead to power imbalances and load uncertainties in the system. ESSs and EVs can, on the other hand, provide new openings for the DSM as they can be used to store energy during off-peak hours. Charging and discharging of EVs, energy demand for domestic usage, and the renewable energy production contrast at time of operation with the predicted profile. Thus using the stored energy through ESSs and EVs at peak hours is thus one mechanism for deploying the DSM method.

Another strategy for deploying DSM methodology can be made based on the conservation voltage reduction (CVR) technique. The principle of CVR is that when the load bus voltages in the network are forced to decrease within permissible ranges, the total peak demand of some loads can be lowered [29]. In conventional utility grids, it has been usually used to lower peak demand and energy consumption [30,31]. The same approach can be applied in isolated microgrids by letting the peak demand of some loads be lowered while the microgrids operate within the lower voltage bounds. This is because, DGs, according to the IEEE 1547.2 Standard, can safely operate within 88%−110% of the nominal value of voltage levels [30]. Various loads, based on their voltage sensitivity, exhibit different responses to CVR. For example, constant power type of loads cannot be affected by CVR, whereas it has a slight impact on the peak demand of constant impedance loads. CVR is also likely utilized for energy savings in unregulated motor loads and lighting loads. CVR does not have an impact on the energy consumption of furnaces and controlled motor loads. CVR can lower the transformer losses along with the energy savings [31]. CVR can be employed in long-term and short-term applications while off-peak and peak demands are variably imposed; in both cases, it will effectively decrease losses and conserve energy [32].

Nevertheless, having the slow response feature of voltage regulation devices, supercapacitors, load tap changers, and others performed in grid-connected microgrids affects the CVR efficiency [32]. Moreover, the distance of the voltage regulation devices from the microgrid is potentially enough to limit the CVR range for responding accordingly [33].

9.3.3   Demand response alternatives

Demand response (DR) in the microgrid network refers to creating a flexible energy usage in the operation of the network on the customer side. The DR resources include all intentional modifications to the electric consumption patterns of end-use customers that are intended to modify the quantity of customer demand on the microgrid in total or at specific time periods. It aims to adjust the demand instead of the power supply. DR is implemented by designing controllable loads either by curtailment with an associated cost or by price elasticity coefficients. The load behavior in response to various DR controls shall be captured. Such DR controls include peak demand limits or limits based on externalities such as weather conditions. The demand requests can be signaled and communicated to customers from the utilities through methods such as smart metering (at times of explicit requests or price changes) and simple off-peak hour metering (at certain times in the day the energy price is lower). Upon receiving such a request, customers will decide either to reschedule some tasks provided with adjusted power demand or to pay a higher price for their electricity demand. Some customers may shift their consumption to other alternate sources. DR, in many aspects of electric power supply, thus can be taken as a technology-enabled economic rationing system. A mechanism of price incentive in DR practice involves providing lower power pricing during peak load periods. The customer, in exchange, is expected to reduce its power consumption, or otherwise it will be subject to a high power price. This scheme can be called voluntary rationing. On the other hand, there could also be an involuntary rationing, though not commonly practiced, that leads to blackout during the peak load hours.

In DR programs, customers' baseline energy consumption level should be established and reduction below this level shall not be allowed. According to a certain load management strategy, the nonflexible customers' load request shall be the minimum amount of energy provided by the utility at a specific time, whereas the flexible load request can be reallocated over time. Different modalities of DR programs and methods are presented as follows.

9.3.3.1 Load management in the demand response

Load management programs seek to reduce peak demand during specific and limited periods by shifting the usage to other periods or temporarily curtailing electricity usage. Typically, in these programs, robust control and communication technologies are employed. Load management applies different strategic functions including the major ones which are load shedding and load shifting.

1. *Load shedding*

    The load shedding function establishes and monitors the balance of generated and consumed power by temporarily cutting off the consumer feeders with low priority. Following a large disturbance, automatic load shedding is the only way to prevent frequency collapse. Parameters to be considered for this program shall include shed priority, shed type, minimum time between sheds, minimum shed time, maximum shed time, and rated shed amount.

    a. *Fast power—based load shedding*: During critical events, to restore the balance between generated and consumed power in the grid system, consumer power with lower priority must be shed very fast. The balance of the active power shall be calculated periodically for each contingency to determine how much power to be shed. The shedding shall be made based on the available power, spinning reserve and network topology. Fast load shedding will occur within 30 ms. Fast load shedding may automatically result in smaller microgrids operating independently.

    b. *Dynamic power—based load shedding*: When the microgrid operates in island mode, the load balance shall be calculated periodically. If the spinning reserve falls below a defined threshold value, consumed power shall be shed until sufficient spinning reserve is regained. After the load balance is calculated it shall be mapped to the priority list of consumer feeders to be shed. If the typical power consumed by the loads is larger than the available reserve at the time, the loads shall be prevented from starting.

   c. *Frequency-based load shedding*: This approach of load sharing works independently of power-based load shedding and uses a distributed architecture. It is usually used as a backup shedding function. The design of frequency-based load shedding contains a frequency relay connected to each busbar which supervises the frequency based on some thresholds. When a threshold value is reached, fast shedding of predefined feeders is automatically initiated. Every stage of the frequency relays includes a time delay to prevent unwanted shedding. For having faster response, the rate of change of the parameters (*df/dt*) shall be monitored.

2. *Load shifting*

   Load can be shifted from peak consumption periods to nonpeak periods to take advantage of time-of-use (TOU) rates or real-time pricing (RTP). These two concepts will be elaborated upon later. Load shifting strategy aims to level a given load profile by shifting the load throughout a selected time horizon. During high power prices, the end-users can shift their energy consumption to another time period with lower price to capture the value benefit. Load shifting guarantees advantages and financial returns to the utilities as well.

   During load shifting for demand response deployment, a thorough analysis of the loads and processes are required. Activities such as switching off unnecessary operational devices, rescheduling of activities, and switching to onsite generation are carried out to achieve the goal of the analysis and the processes. The following two programs can be used in this method.

   a. *Power-based load shifting*: A typical example for this program is managing the EV charging in a way that the shape of the total load is changed.

   b. *Time-based load shifting*: With this program, the starting and ending times of the loading pattern are changed, for example, washing machine and dishwasher are utilized.

9.3.3.2 Price-based demand response

The goal of most of the price-based DR programs is to induce customers to shift their consumption between different time periods [34]. In this model, applicable programs are presented hereunder.

1. *TOU pricing*: In this category, prices at certain times are constant and set in advance. This program typically offers a high price for the peak hours and a cheap price for the off-peak hours. However, it does not adjust to reflect actual conditions.

2. *Critical peak pricing (CPP)*: In this DR program, different prices are put into effect for various hours on various days. When the utilities observe emergency conditions in the power system or high wholesale market prices in energy transaction, they may call critical events for a specified time period. Two types of CPP are available. One is that the time and duration of the price increase are predetermined while events are called to the end-users. The other one is made based on the grids demanding the loads to be lowered.

3. *RTP*: This DR program provides the best available signal about the marginal value of power at a location as it reflects the current conditions of the utility and energy transactions.

4. *Variable peak pricing*: This type of program is designed by integrating TOU and RTP programs. The principle is that the pricing for the different times is predefined, but utility and market conditions are allowed to have an effect in terms of varying the price established for the on-peak periods. An example for this DR program is setting on-peak prices during summer weekday afternoons and off-peak prices for all other hours in the summer months.

5. *Critical peak rebates*: This type of program can be used when utility companies anticipate emergencies in power systems or high prices in the wholesale market. By then, the utilities may invoke critical events for a prespecified time period. Electricity prices remain the same during this period, but any decrease in consumption is reimbursable at a single, predetermined value relative to what the utilities expect the customer to consume. For example, during summer weekday afternoons (3:00 p.m.−6:00 p.m.), this program may operate.

All the above pricing programs except TOU are usually described as dynamic DR pricing programs since prices are not known in advance for these programs. The TOU program is static and the rate of the schedule is predefined. Dynamic DR pricing programs therefore provide utilities and customers to yield better benefits from the variability of energy market transactions and the capabilities of smart grid customer systems.

9.3.4 Intelligent demand response algorithms

9.3.4.1 Decision-making auction algorithm

The auction algorithm, which was applied to maximize the net benefits in a bipartite graph as well as solve a variety of applications, such as finding the shortest path, was introduced by Bertsekas in 1979 [35]. Today many generic network optimization problems widely utilize this algorithm [36].

The principle of auction algorithm is derived based on how a mouse finds a cheese. Assume a smart mouse needs to get a cheese in a place of unfamiliar paths. A junction of each path is considered as a node. Whenever the mouse reaches a node, it faces two nodes, the next node and the previous node and must choose either of the two. If it returns to the previous node, it will note and compute the two distances: the distance between the current node and previous node and the distance between the current node and the next node. Once the mouse knows where the next node is by calculating and comparing the known distances, it can lead to the shortest cheese path, and the mouse moves forward.

In general, in the implementation of shortest path problem, this algorithm adopts two typical techniques commonly applied: label-correcting (Bellman–Ford-like) and label-setting (Dijkstra-like) algorithms [16,17]. For the case of label-correcting algorithm, since the distances determined based on the observations at different labeled nodes may fluctuate, the algorithm assumes the mouse to unceasingly update the optimal distance. Whereas in a label-setting algorithm, the mouse uses its featured to search the unlabeled node. The labeled nodes with the shortest path to the start point are connected by the unlabeled node. The searched node is made labeled, and the mouse remains on the same iteration process for the search until it gets the cheese.

This algorithm urges the end-users in the microgrid to independently satisfy the DR program. Consequently, the DR program operates in a distributed approach, where the information dissemination in neighboring consumers using the short-range ZigBee communication is quite limited. This significantly reduces the initial bulk investment for a centralized communication system. It also improves the data privacy compared to a centralized technique. A path with the a minimum loss can be selected as the preferred constraint for the decision-making process of the DR [36].

9.3.4.2 Heuristic-based evolutionary algorithm

This algorithm offers an efficient and cost-effective solution to a problem by easily adapting heuristics to the problem. This algorithm has a flexible trait to develop its structures and features. Such flexible development could not be presented by most of the other algorithms. The flexibility trait offers the option of implementing the features which can be used to design the patterns of the load demand based on the lifestyles of the end-users. As a result, the inconvenience that energy consumers could face is reduced to a minimum. For instance, suppose two controllable loads are a washing machine and coffee maker. The consumers begin using the coffee maker in the morning; but the washing machine has no significant time preference and so its operating time can be shifted to off-peak hours of the day. Hence, all factors such as customers' lifestyle can be taken into consideration in this algorithm which will easily adapt and schedule the operating time for these two devices as conveniently as possible in the day [37].

9.3.4.3 Greedy ratio algorithm

The greedy ratio algorithm can be deployed for event-based demand response management to tackle the challenging task of optimally deciding the loads to be curtailed in a microgrid during contingencies. In event-based management, it is an important algorithm to reduce loads as fast as possible to keep the microgrid operating with stability in a scalable approach during the island mode of microgrid operation. This algorithm can promptly compute a best optimal load reduction to maximize the aggregate customer benefits even for a large number of customers within a fraction of seconds. Some results demonstrate that this technique efficiently optimizes the island operation mode of the microgrid while maintaining the voltage levels in tolerable ranges and satisfying network constrains [38].

9.3.4.4 Distributed demand response algorithm

A distributed DR strategy is also another algorithm applied for shaving peak load [39,40], which is applied to make peak load shaving practical under the incentive of minimizing the customers that will be affected with the load interruption in a network of multiple types of end-users. This algorithm is effective in that it decreases the ratio of peak-to-average consumption and lowers the overall dissatisfaction level at the end-user's side [39]. It is adopted from the concept of congestion pricing in internet traffic control in which pricing information is very useful to regulate user demand. Thus such algorithm can be applied to balance network load in the microgrid system. The dynamics and convergence behavior of this algorithm are investigated and a strategy for plug-in hybrid-EVs was proposed and verified in which users, based on their preferences, can adapt their charging rates in Ref. [40].

9.4 Energy efficiency of microgrids

People across the globe are using energy for various purposes including cooking, lighting, heating, transportation, cooling, manufacturing, entertainment, and many others. The ways that people use energy will impact their lives and the environment and the ecosystem at large. Choosing to have fuel-efficient vehicles, energy-efficient appliances, or making machines/devices switch off when they are not being operated should be taken to avoid or reduce such impacts. The International Energy Agency (IEA) report (IEA-March 1, 2020) attaches a special importance to energy efficiency services development. This can help substantially to reduce carbon dioxide emissions. In this section, the concept of energy efficiency and its data analysis and optimization models in microgrid system is briefly presented.

9.4.1 Advanced energy efficiency services

The main concept of energy efficiency service is technological actions. Getting involved in taking measures for the design and manufacture of devices which consume less energy to give the same operational services while ensuring their long-lasting operation is taken to be an energy efficiency service. Energy efficiency measures can also, depending on the timing of the equipment used, produce significant reductions in peak demand.

Conservation of energy, though it has a different concept with energy efficiency services, helps to build an energy-efficient microgrid. It involves lowering the level of energy services to save energy and/or reduce demand.

For example, turning off corridor lights when people are not present, setting the level of the thermostat a bit higher in summer and lower in winter, taking slightly shorter showers, etc., aggregately can bring about the conservation of energy. It typically demands individual behavioral changes.

9.4.2 Classified energy consumption data analysis

For the sake of efficient energy management provided in the power grid and microgrids, the energy consumption data could be analyzed with standard, simplified, and interactive ways and categories. The very common classifications include residential, commercial, and industrial/manufacturing consumption data analysis.

1. *Residential data analysis*: involves household characteristics, home energy use and costs, and detailed household data processing. Housing characteristics data collected should include fuels used and end-uses in structural and geographic characteristics, such as electronics, lighting, heating and cooling, household unit, demographics, energy insecurity, etc. For primary housing units, energy consumption and expenditures can be calculated with end-uses and fuel consumption. In residential energy consumption investigation, energy-related random data are collected using methods like complex multistage and area-probability sampling models or a sample survey for primary housing units. This microdata will be used for the end-use modeling process and should contain consumption and climate-related variables.
2. *Commercial data analysis*: involves building characteristics, commercial energy use and costs, as well as detailed buildings' data processing. Detailed tabulated data may be applied and may contain characteristics of commercial buildings, such as buildings size, age, and their usage, energy sources used in the buildings, even geographic location, and other energy-consuming devices and units. The data can be complied based on a single respondent with a scope of sampled buildings from the sampling areas.
3. *Manufacturing data analysis*: involves the characteristics of consumption of energy for all purposes by an industry consisting of systems, buildings, devices, and other energy consumers. The process of collecting data is the same as in the commercial category but differs from the size of buildings and energy consumption perspectives. It enforces intensive data collection and minimizes the random data sampling.

9.4.3 Energy efficiency assessment and analysis model

Energy efficiency and its process can go along with update of digital technologies. Digitalization defines the usage of digital technologies and leads to greater connectivity amongst people, between people and machines, and between machines. The growing application of information communication technology across the economy and sociopolitics allows the greater increase of data capacity and fast growth in advanced analytics. Thus digitalization can enhance the capacity to escalate energy efficiency of the system through technologies as it enables the energy-consuming systems, units, and devices to be new sources of flexible load to the energy system. In the case of microgrids, such a trend may support the microgrid operator to minimize curtailment of renewable energy sources and improve their utilization. Digital technologies have the capacity to optimize energy consumption for numerous energy-consuming activities in all dimensions. Increasing system efficiency and end-use efficiency with the implementation of digital technologies remains substantial for the global energy transformation and eventually brings countless returns to the total energy system by avoiding huge investments in energy infrastructure. It also has significant impacts on improving the integration of renewables and power reliability, energy security, and flexibility of the demand side.

All energy consumer sectors have been implementing digital technologies. Sensors in gas and oil reservoirs are applied for how energy is produced and consumed. Residential and commercial households have been increasingly furnished with smart and intelligent appliances and equipment. The industrial parks have been applying advanced robotics and three-dimensional printing devices. The transportation sector has become electrified, connected, and automated. This all will remarkably transform the trends of future energy consumption and hopefully the enhancement of efficient energy services.

To investigate the possible way of utilizing digitalization for the sake of more improvement in energy efficiency and forwarding suggestions for policy makers, the IEA has introduced a cross-agency initiative [41]. In this regard, the IEA is pursuing emerging trends, such as active EMSs, new performance-based revenue models, and data-driven consumer engagement platforms, to reach a greater energy efficiency benefits. These not only deliver value in terms of improved energy performance within specific sectors and end-uses, but also in the wider energy system context by balancing supply and demand in modern electricity grids. Combining digital technologies with the right policy frameworks and innovative business and technical models enables greater control, optimization, and analytics of energy production, consumption, and trading for ultimately offering better end-use efficiency and system efficiency.

9.4.4 Energy efficiency diagnosis and optimization model

Energy consumption modeling seeks to determine energy requirements as a function of input parameters. Models may be used for determining the requirements of energy supply and the consumer consumption variations while an upgrade or addition of technology exist. Decisions can be carried out on various issues to support energy supply. Therefore so as to address the issues and maximize the energy efficiency, the following energy consumption optimization models can be applied in microgrids.

1. *Regression*: regression analysis is used to compute the coefficients of the model as a function of input parameters. Such a model determines the aggregate residence energy consumption onto parameters or combinations of parameters which are expected to affect energy consumption. The model involves calculation and analysis based on fitting characteristics. For the sake of simplicity, input variables which have an insignificant effect during the computation can be ignored.

2. *Conditional demand analysis (CDA)*: CDA accomplishes its purpose based on the existence of end-user equipment [42]. The coefficients computed indicate the level and rating of the end-uses. The major advantage of this model is the ease of determining input information, which may include energy billing data, simple appliance surveys, and others. However, it needs a dataset with a variety of appliances. This technique utilizes the variations in end-uses to compute every device of the entire energy consumption. To get trustworthy findings in this technique, big data are required regarding the end-uses.

3. *NN*: this technique utilizes information of all end-uses that involve one another through an interconnected parallel structure. Each component in the structure has an array of coefficients and a bias term, which are multiplied with the value of the previous layer's components. To account for nonlinearity, this model implements scaling and activation functions. The coefficients computed have no physical importance due to the parallel configurational model.

   Back-propagation neural network (BPNN) was proposed and presented in Ref. [43]. This method basically utilizes a forward NN model and chooses the second layer back-propagation model. While deriving the feedforward transfer function, the output errors of the BPNN are back-propagated by using the way of the same connections applied in the feedforward mechanism. BPNN has the advantage that it can converge rapidly, especially with careful selection of the training parameters. A three-layer BPNN to solve an approximation problem and function fitting for optimization of grid-integrated ESSs is presented in Ref. [16].

4. *Engineering methods*: in engineering methods, energy consumption can be accounted on the basis of the ratings or characteristics of the end-users' appliances. Neglecting any historical information, the energy consumption of a sector can be thus fully developed using this method. Regarding the modeling of new technologies, engineering methods have the greater possible degree of capability and flexibility since the models are based on the physics of the end-uses. There are different techniques implemented in engineering methods such as distributions, archetypes, and sampling techniques.

   a. *Distributions technique*: this technique can be used in appliance distributions to determine the energy consumption in each consumer. The consumer consumptions are separately calculated. It does not account for interactions amongst consumers. The energy consumption is determined by multiplying appliance tenure, its energy use, rating, and the inverse of its efficiency. Thus the energy consumption is estimated by combining the consumption of appliances on the areal scale.

b. *Archetypes technique*: this estimation technique can be applied by classifying the housing stock as per its age, house type, size, etc. Archetype definitions for each major housing class are developed for the model to use as input data. The estimates of energy consumption in the developed model can be scaled up to denote the housing stock. The estimate is made by multiplying the results with the number of households.

c. *Sampling technique*: in this technique, actual samples of residence data are used as the input information to the model. It can be used in wide areas to identify regions with dense consumption. If the sample is representative of housing stock, the consumption can be approximated by incorporating relevant weightings to the outcomes. As there could be wide variations among housing, this technique demands a big database representative of the appliances.

9.4.5   Energy efficiency data statistics report

There are different techniques of evaluating the energy efficiency at different levels from households to a country or region. The generic energy efficiency indicator can be defined as:

$$\text{Energy efficiency indicator} = \frac{\text{Energy consumption}}{\text{Activity}} \qquad (9.42)$$

Their metrics are modified for different types of end-users.

The other index used for measuring energy efficiency at a household level is the home energy rating system (HERS) index. It is a measuring method deployed in housing to determine the energy efficiency. This method can tell us much about the house, for example, heating, cooling, and other factors which consume energy, and how it performs.

The report of this method will outline the energy features of the home and the expected cost of utility bills. It will also provide us with invaluable information about the existing home, like how efficiently it is operating and where one can make modifications for better energy savings.

The working principle of the HERS is that it assesses the energy efficiency of a home by considering reference performance scores. The lower the determined value, the more energy efficient the home and vice versa.

Variables considered in determining the HERS index include all exterior walls, floors, ceilings and roofs, attics, foundations, windows and doors, vents and ductwork, heating system, thermostat, home leakages, and other energy-consuming devices and spaces in the house. The statistics report in all categories can be available in two formats: comma delimited file (CSV) and SAS data file.

9.5 Application example: simulation of microgrid central controller for energy management of resilient low-carbon microgrid

9.5.1 The microgrid platform and simulation model

The simulation of the microgrid central controller (MGCC) is intended to address the microgrid platform shown in Fig. 9.2 which is inspired by an experimental microgrid in China.

The MGCC simulation was carried out and tested on a PSCAD simulation mode of the microgrid with a similar topology to Fig. 9.2. It comprises a 20 kW wind turbine, 40 kW PV system, 200 kW diesel genset, one battery electrical energy storage (BESS) system, and three loads. The BESS in the model is rated 100 kWh with 0.5 C charging/discharging rate hence supplying 50 kW for 2 hours. The first load (60 kW) is assumed to be a critical load which needs to be connected all the time while the other two (60 and 120 kW) are assumed to be deferrable loads which may be shed during periods of short power supply.

The microgrid platform is assumed to be operating in island mode all the time. The following general conditions and assumptions are considered in the simulation (Table 9.1).

9.5.2 Functions and overview of the MGCC

The MGCC designed in this simulation is expected to provide the following functions:

- Battery charge/discharge management.
- Decide whether to charge, discharge, or stall the battery based on the SOC and U_{dc} of the BESSs and operational conditions of the other DGs and loads in the microgrid.
- Ensure system stability and/or demand−supply equilibrium through:
  - Load management/load shedding.

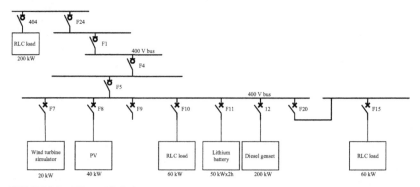

FIGURE 9.2 Microgrid platform.

- DG management/ unit commitment.
- Setting values adjustment (adjusting PQ setting values).
- Perform black starting for local service restoration during island operation.

The MGCC designed for energy management of the microgrid is shown in Fig. 9.3. The controller takes in measured and status signals from measuring instruments and sensors. It gives outputs commands and information for the generating units and the controllable loads.

TABLE 9.1 General parameters and assumptions in simulation.

No.	Parameter	Assigned value
1	Nominal bus line-to-line voltage	400 V)
2	System nominal frequency	50 Hz
3	Acceptable line-to-line voltage range	400 V + 10%
4	Acceptable frequency range	49.5−50.5 Hz
5	Time step of the MGCC	0.5 s (can be changed)

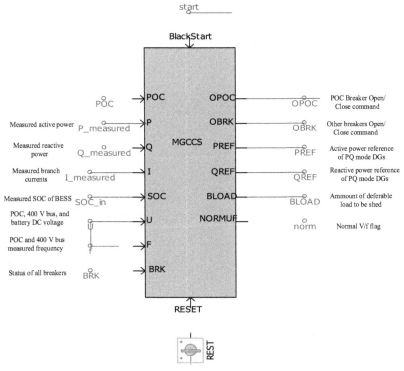

FIGURE 9.3 Overview and input/outputs of MGCC.

9.5.3 Operating algorithm of the MGCC

A diesel genset is the main DG operating at least with its rated minimum capacity and the two batteries will operate in active/reactive power (PQ) mode. The operating algorithms of the MGCC are presented in Figs. 9.4–9.6.

9.5.4 Testing procedure

The following procedure was followed to verify that the designed MGCC can achieve the intended functions.

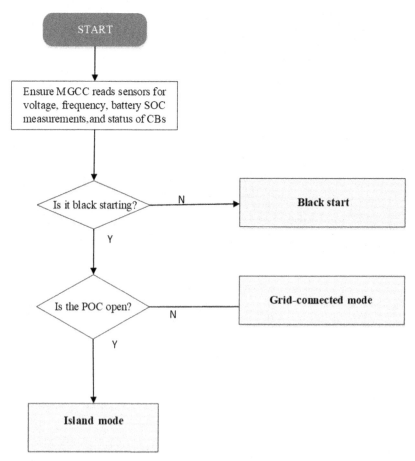

FIGURE 9.4 Summarized flowchart of the MGCC.

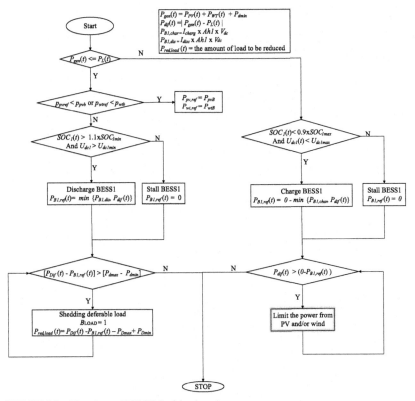

FIGURE 9.5   Flowchart of MGCC for island mode.

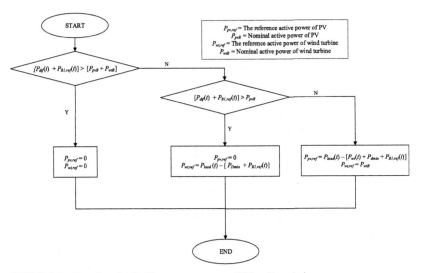

FIGURE 9.6   Flowchart for limiting power outputs of PV and/or wind.

9.5.5 Testing results

The following graphs show the results of the simulation experiment of the
procedure in Table 9.2. Fig. 9.7 shows the measured active power from

TABLE 9.2 Procedure for testing MGCC.

Step	Duration (s)	Description of status	MGCC expected output
	0	Start simulation with black starting	
1	0–10	**Black starting** The breakers and reference power settings are controlled by the black starting module $SOC_1 = 82 \rightarrow 22$ $P_{load} = 0.12$	$P_{wt,ref} = 0.02$ $P_{PV,ref} = 0.04$ $P_{ESS1,ref} = 0.05$ $P_{diesel} = 0.01$ (droop)
	10	Black starting ends and MGCC takes control	
2	10–12	$P_{gen,ren} < P_{load}$ $SOC_1 > 1.1 \times SOC_{min}$ ESS1 keeps discharging $SOC_1 = 82 \rightarrow 22$ $P_{load} = 0.12$	$P_{wt,ref} = 0.02$ $P_{PV,ref} = 0.04$ $P_{ESS1,ref} = 0.05$ $P_{diesel} = 0.01$ (droop)
	12	SOC_1 reaches its minimum	
3	12–14	BESS1 = fully discharged → stalled The diesel genset can supply the difference $SOC_1 = 1.1 \times SOC_{min} = 22$ $P_{load} = 0.12$	$P_{wt,ref} = 0.02$ $P_{PV,ref} = 0.04$ $P_{ESS1,ref} = 0.00$ $P_{diesel} = 0.07$ (droop)
	14	Add 90 kW load connect *breaker*	
4	14–16	Still the microgrid can carry the load $SOC_1 = 1.1 \times SOC_{min} = 22 \rightarrow$ stall $P_{load} = 0.21$	$P_{wt,ref} = 0.02$ $P_{PV,ref} = 0.04$ $P_{ESS1,ref} = 0.00$ $P_{diesel} = 0.15$ (droop)
	16	Set irrad → 0/no sunshine → shortage of generation → load shedding	
5	16–18	Irrad = 1000 → 0 (0 at 17 s) $P_{pv} = 0$ $SOC_1 = 1.1 \times SOC_{min} = 22 \rightarrow$ stall $P_{load} = 0.24$ $P_{gen}(0.18 + 0.2) < P_{load}(0.21)$ Shed deferrable load	$P_{wt,ref} = 0.02$ $P_{PV,ref} = 0.04$ $P_{ESS1,ref} = 0.00$ $P_{diesel} = $ droop $Db_{load} = 1$ (open breaker of deferrable load) $D_{load} = \Delta P$
	18	Irrad → 1000/sunshine comes back	

(Continued)

TABLE 9.2 (Continued)

Step	Duration (s)	Description of status	MGCC expected output
6	18–20	Irrad = 0 → 1000 (1000 at 19 s) P_{pv} = 0.04 Renewable generation is enough only to supply the load BESS = stall	$P_{wt,ref}$ = 0.02 $P_{PV,ref}$ = 0.04 $P_{ESS1,ref}$ = 0.00
	20	Decrease load by 60 kW → enough generation → BESS starts charging	
7	20–22	The renewable generation is enough only to charge BESS1 in addition to supplying the load BESS1 = charging SOC_1 = 22 → 82 P_{load} = 0.6	$P_{wt,ref}$ = 0.02 $P_{PV,ref}$ = 0.04 $P_{ESS1,ref}$ = −0.01
	22	SOC_1 reaches its maximum → Limit PV power	
8	22–24	BESS1 = fully charged → stalled The generation from renewable + P_{dmin} becomes higher than load SOC_1 > = 0.9 × SOC_{max} = 82 P_{load} = 0.06 Limit PV power	P_{pvref} = 0.03

Note: The SOC of the BESS is artificially controlled to match the steps in the procedure.

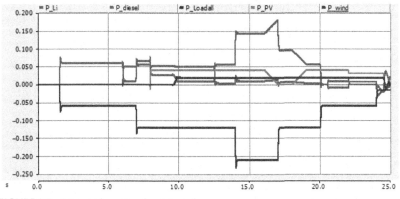

FIGURE 9.7 Measured powers of each branch.

each branch, which are the inputs for the MGCC. The set of changes applied in some of the input parameters for testing of the MGCC is shown in Fig. 9.8A−C. The output signals or commands from the MGCC according to the designed strategy corresponding to the changes made in Fig. 9.9 are displayed in Fig. 9.9A−D. The MGCC was then proved to effectively ensure stable operation of the microgrid under different conditions, balancing the generation with the load and hence keeping the voltage and frequency close to their respective standard values (1 p.u.) as shown in Fig. 9.10.

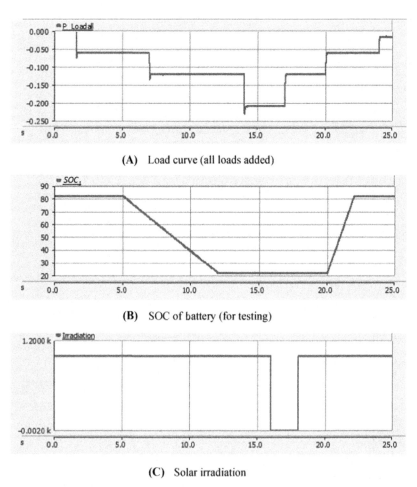

(A)   Load curve (all loads added)

(B)   SOC of battery (for testing)

(C)   Solar irradiation

FIGURE 9.8   Changes applied to input parameters for testing the MGCC: (A) load curve (all loads added); (B) SOC of battery (for testing); and (C) solar irradiation.

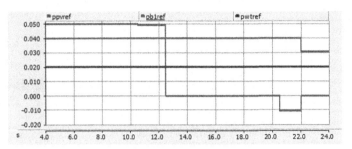

(A) Reference active power commands

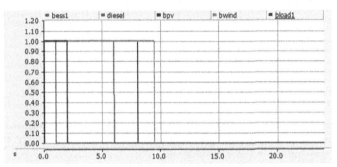

(B) Breaker commands

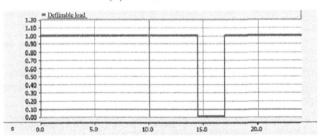

(C) Deferrable load breaker command (shed load command)

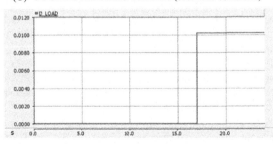

(D) Amount of load to be shaded

FIGURE 9.9 Outputs of MGCC for testing conditions: (A) reference active power commands; (B) breaker commands; (C) deferrable load breaker command (shed load command); and (D) amount of load to be shaded.

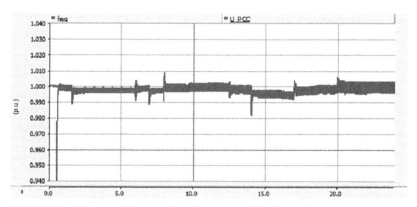

FIGURE 9.10 Voltage and frequency kept close to standard value.

References

[1] N. Liu, X. Yu, C. Wang, J. Wang, Energy sharing management for microgrids with PV prosumers: a Stackelberg game approach, IEEE Trans. Industr. Inform. (2017). Available from: https://doi.org/10.1109/TII.2017.2654302.

[2] C. Eyisi, A.S. Al-Sumaiti, K. Turitsyn, Q. Li, Mathematical models for optimization of grid-integrated energy storage systems: a review, in: 51st North American Power Symposium, NAPS 2019, 2019. Available from: https://doi.org/10.1109/NAPS46351.2019.9000190.

[3] M. Cain, R. O'Neill, A. Castillo, History of optimal power flow and formulations, Fed. Energy Regul. Comm. 1 (2012) 1−36.

[4] D.K. Molzahn, et al., A survey of distributed optimization and control algorithms for electric power systems, IEEE Trans. Smart Grid (2017). Available from: https://doi.org/10.1109/TSG.2017.2720471.

[5] Q. Li, S. Yu, A.S. Al-Sumaiti, K. Turitsyn, Micro water-energy nexus: optimal demand-side management and quasi-convex hull relaxation, IEEE Trans. Control. Netw. Syst. (2019). Available from: https://doi.org/10.1109/TCNS.2018.2889001.

[6] S.F. Santos, D.Z. Fitiwi, M. Shafie-Khah, A.W. Bizuayehu, C.M.P. Cabrita, J.P.S. Catalão, New multistage and stochastic mathematical model for maximizing RES hosting capacity—part I: problem formulation, IEEE Trans. Sustain. Energy (2017). Available from: https://doi.org/10.1109/TSTE.2016.2598400.

[7] S.F. Santos, D.Z. Fitiwi, M. Shafie-Khah, A.W. Bizuayehu, C.M.P. Cabrita, J.P.S. Catalão, New multistage and stochastic mathematical model for maximizing RES hosting capacity—part II: numerical results, IEEE Trans. Sustain. Energy (2017). Available from: https://doi.org/10.1109/TSTE.2016.2584122.

[8] F. Wang, et al., Multi-objective optimization model of source-load-storage synergetic dispatch for a building energy management system based on TOU price demand response, IEEE Trans. Ind. Appl, 2018. Available from: https://doi.org/10.1109/TIA.2017.2781639.

[9] R. Billinton, D. Huang, Effects of load forecast uncertainty on bulk electric system reliability evaluation, IEEE Trans. Power Syst. (2008). Available from: https://doi.org/10.1109/TPWRS.2008.920078.

[10] M. Bhoye, M.H. Pandya, S. Valvi, I.N. Trivedi, P. Jangir, S.A. Parmar, An emission constraint economic load dispatch problem solution with microgrid using JAYA algorithm, in: 2016 International Conference on Energy Efficient Technologies for Sustainability, ICEETS 2016, 2016. Available from: https://doi.org/10.1109/ICEETS.2016.7583805.

[11] I.N. Trivedi, M. Bhoye, R.H. Bhesdadiya, P. Jangir, N. Jangir, A. Kumar, An emission constraint environment dispatch problem solution with microgrid using Whale Optimization Algorithm, in: 2016 National Power Systems Conference, NPSC 2016, 2017. Available from: https://doi.org/10.1109/NPSC.2016.7858899.

[12] N. Nikmehr, S. Najafi-Ravadanegh, Probabilistic optimal power dispatch in multi-microgrids using heuristic algorithms, 2015. Available from: https://doi.org/10.1109/sgc.2014.7151035.

[13] R. Venkata Rao, Jaya: a simple and new optimization algorithm for solving constrained and unconstrained optimization problems, Int. J. Ind. Eng. Comput. (2016). Available from: https://doi.org/10.5267/j.ijiec.2015.8.004.

[14] S. Mirjalili, A. Lewis, The whale optimization algorithm, Adv. Eng. Softw. (2016). Available from: https://doi.org/10.1016/j.advengsoft.2016.01.008.

[15] Z.L. Lyu, X.Q. Wang, Z.X. Yang, Optimal dynamic dispatch of pareto frontier for microgrid based on MOIBBO algorithm, in: IEEE Power and Energy Society General Meeting, 2018. Available from: https://doi.org/10.1109/PESGM.2017.8274155.

[16] X. Xiong, Y. Ji, M. Wu, Y. Hui, Microgrid power optimal control with Markov decision process by using the specific policy of wavelet packet-fuzzy control, in: Proceedings of the Second IEEE Conference on Energy Internet and Energy System Integration, EI2 2018, 2018. Available from: https://doi.org/10.1109/EI2.2018.8582216.

[17] T. Başar, G.J. Olsder, Dynamic Noncooperative Game Theory, second ed., 1998.

[18] R. Wang, Q. Li, B. Zhang, L. Wang, Distributed consensus based algorithm for economic dispatch in a microgrid, IEEE Trans. Smart Grid (2019). Available from: https://doi.org/10.1109/TSG.2018.2833108.

[19] R. Eberhart, J. Kennedy, New optimizer using particle swarm theory, in: Proceedings of the International Symposium on Micro Machine and Human Science, 1995. Available from: https://doi.org/10.1109/mhs.1995.494215.

[20] E. Atashpaz-Gargari, C. Lucas, Imperialist competitive algorithm: an algorithm for optimization inspired by imperialistic competition, in: 2007 IEEE Congress on Evolutionary Computation, CEC 2007, 2007. Available from: https://doi.org/10.1109/CEC.2007.4425083.

[21] M. Wang, G. Pan, Y. Liu, A novel imperialist competitive algorithm for multithreshold image segmentation, Math. Probl. Eng. (2019). Available from: https://doi.org/10.1155/2019/5982410.

[22] IEEE, Systems: man and cybernetics, in: 2018 IEEE International Conference on Systems, Man, and Cybernetics, SMC2018, 2018.

[23] K.Y. Wong, M. Tanaka, K. Tanaka, T.T. Lee, A new nonconvex design algorithm for optimal polynomial fuzzy control, in: Proceedings of the 2018 IEEE International Conference on Systems, Man, and Cybernetics, SMC2018, 2019. Available from: https://doi.org/10.1109/SMC.2018.00500.

[24] Y.J. Chen, M. Tanaka, K. Tanaka, H.O. Wang, Nonconvex stabilization criterion for polynomial fuzzy systems, in: Proceedings of the IEEE Conference on Decision and Control, 2013. Available from: https://doi.org/10.1109/CDC.2013.6761066.

[25] K. Tanaka, M. Tanaka, Y.J. Chen, H.O. Wang, A new sum-of-squares design framework for robust control of polynomial fuzzy systems with uncertainties, IEEE Trans. Fuzzy Syst. (2016). Available from: https://doi.org/10.1109/TFUZZ.2015.2426719.

[26] M.H. Khooban, T. Niknam, A new intelligent online fuzzy tuning approach for multi-area load frequency control: self adaptive modified bat algorithm, Int. J. Electr. Power Energy Syst. (2015). Available from: https://doi.org/10.1016/j.ijepes.2015.03.017.

[27] C. Caraveo, F. Valdez, O. Castillo, Optimization of fuzzy controller design using a new bee colony algorithm with fuzzy dynamic parameter adaptation, Appl. Soft Comput. J. (2016). Available from: https://doi.org/10.1016/j.asoc.2016.02.033.

[28] D. Li, W. Y. Chiu, H. Sun, Demand side management in microgrid control systems, in: M.S. Mahmoud, (Ed.), Microgrid: Advanced Control Methods and Renewable Energy System Integration, Elsevier Ltd, 2017, pp. 203−230.

[29] M. Gheydi, A. Nouri, N. Ghadimi, Planning in microgrids with conservation of voltage reduction, IEEE Syst. J. (2018). Available from: https://doi.org/10.1109/JSYST.2016.2633512.

[30] IEEE, IEEE Application Guide for IEEE Std 1547(TM), IEEE standard for interconnecting distributed resources with electric power systems, IEEE Std 1547.2-2008, 2009. Available from: https://doi.org/10.1109/IEEESTD.2008.4816078.

[31] C. Concordia, S. Ihara, Load representation in power system stability studies, IEEE Trans. Power Appar. Syst. (1982). Available from: https://doi.org/10.1109/TPAS.1982.317163.

[32] Z. Wang, J. Wang, Review on implementation and assessment of conservation voltage reduction, IEEE Trans. Power Syst. (2014). Available from: https://doi.org/10.1109/TPWRS.2013.2288518.

[33] R. Singh, F. Tuffner, J. Fuller, K. Schneider, Effects of distributed energy resources on conservation voltage reduction (CVR), in: IEEE Power and Energy Society General Meeting, 2011. Available from: https://doi.org/10.1109/PES.2011.6039702.

[34] S. Mohajeryami, I.N. Moghaddam, M. Doostan, B. Vatani, P. Schwarz, A novel economic model for price-based demand response, Electr. Power Syst. Res. (2016). Available from: https://doi.org/10.1016/j.epsr.2016.03.026.

[35] D.P. Bertsekas, The auction algorithm: a distributed relaxation method for the assignment problem, Ann. Oper. Res. (1988). Available from: https://doi.org/10.1007/BF02186476.

[36] C.L. Chang, J.C.H. Peng, A decision-making auction algorithm for demand response in microgrids, IEEE Trans. Smart Grid (2018). Available from: https://doi.org/10.1109/TSG.2016.2634583.

[37] T. Logenthiran, D. Srinivasan, T.Z. Shun, Demand side management in smart grid using heuristic optimization, IEEE Trans. Smart Grid (2012). Available from: https://doi.org/10.1109/TSG.2012.2195686.

[38] A. Karapetyan, M. Khonji, C.K. Chau, K. Elbassioni, H.H. Zeineldin, Efficient algorithm for scalable event-based demand response management in microgrids, IEEE Trans. Smart Grid (2018). Available from: https://doi.org/10.1109/TSG.2016.2616945.

[39] W. Kou, K. Bisson, S.Y. Park, A distributed demand response algorithm and its application to campus microgrid, in: Conference Record of the Third IEEE International Workshop on Electronic Power Grid, eGrid 2018, 2018. Available from: https://doi.org/10.1109/eGRID.2018.8598668.

[40] Z. Fan, A distributed demand response algorithm and its application to PHEV charging in smart grids, IEEE Trans. Smart Grid (2012). Available from: https://doi.org/10.1109/TSG.2012.2185075.

[41] International Energy Agency, Energy Efficiency and Digitalisation, International Energy Agency, 2019.

[42] M. Parti, C. Parti, The total and appliance-specific conditional demand for electricity in the household sector, Bell J. Econ. (1980). Available from: https://doi.org/10.2307/3003415.

[43] W. Yu, B. Li, Y. Lei, M. Liu, Analysis of a residential building energy consumption demand model, Energies (2011). Available from: https://doi.org/10.3390/en4030475.

Chapter 10

Communication requirements of microgrids

10.1 Introduction

A fundamental requirement in a microgrid system is its modes of operation, including grid-connected and island modes. In island mode there is minimal load shedding and disruption during operation mode of transition, and optimal power flow should be ensured with the implementation of appropriate frequency and voltage stability control system. Moreover, during a transition from island to grid-connected operating mode, the system should provide the capability to allow resynchronization with a minimum impact to sensitive loads [1]. Microgrids are also equipped with higher-level monitoring and controlling systems, such as energy management system or Supervisory Control and Data Acquisition (SCADA). The special requirements for the protection of microgrids also mean there are advanced protection systems that need more measurements and sensing. For those control and protection systems to function properly, a robust, deterministic, and interoperable communication system should be deployed [2]. As shown in Fig. 10.1, communication links appear between grid and distributed energy resources (DERs) of microgrid, among DERs, and even between control elements within a DER. The communications among the control elements are time critical in that efficient algorithms should minimize the delay and computational resource requirements. Furthermore, the communication and security architectures must be versatile enough to support various communication patterns among control schemes such as unicast, multicast, and broadcast communications.

The communication system demands a medium to interconnect the units' operation with appropriate command flows and executions. There are three types of communicating media in use, copper-wired, fiber optics, and wireless technologies.

The communication networks may also be supported with different communication protocols, of which the common ones are Internet Protocol (IP), Modbus, Distributed Network Protocol (DNP), and International Electrotechnical Commission (IEC) 61850. These all are discussed in the following sections.

Microgrid Protection and Control. DOI: https://doi.org/10.1016/B978-0-12-821189-2.00010-3

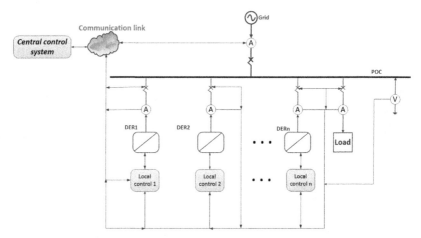

FIGURE 10.1   The general framework of microgrid topology with communication links.

10.2   Role of communication in microgrids

In order for microgrids to operate, coexist, and connect to the grid, the communication network plays a great role to provide the means for the control elements of microgrids to transfer control signals among the components. The basic requirements of a communication network to support control signaling in the microgrid include real-time performance guarantees evaluated via worst case delay performance analysis; security providing confidentiality and integrity guarantees while respecting the real-time delay boundaries; high availability [1]; reliability; robustness; scalability; and quality-of-service as needed in the application of smart grids [3].

With regard to delay performance, according to the study in [4], the maximum latency requirements in a microgrid communication system should be less than 4 ms, seconds, minutes, hours, and days/weeks/months, respectively, for protective relaying system, wide area situational awareness monitoring system, substation and feeder SCADA system, monitoring noncritical equipment and marketing pricing information system, meter reading and longer-term pricing information system, and collecting long-term usage data system.

The communication system also plays a key role in coordination of the control system. For instance, a very limited sharing of nonlinear loads and load-dependent frequency and voltage control in droop control is challenging. Such problems can be tackled by adding a layer of hierarchical control. Hierarchical control configuration in turn depends on a communication system which helps to coordinate the local control units and central control system. The protection system is also supported by a robust communication system. Whether the system is centralized or decentralized, the protection system can be implemented with the communication tools. It plays an important role to help the relays to identify the fault location and isolate it from the system. It can also be used to restore the power where some generators are disconnected after isolating the fault, but it is important to utilize the communication networks to isolate the fault as fast as possible [5].

As the are of microgrid development grows, the method of synchronization with the grid is probably one of the biggest challenges. For instance, providing an automatic grid transfer mechanism represents one essential feature of the microgrid, but ensuring the system's stability during and after the self-transfer between the microgrid operating modes (i.e., grid-connected and isolated) implies significant challenges. Moreover, the seamless transition becomes more difficult as the number of units involved increases. Traditional synchronization methods are not generally suitable for such complex systems [6]. Hence, a communication system can best tackle best challenge.

The role of a communication system in a microgrid basically depends on the microgrid control system design and the component count and types. The deterministic nature of real-time control demands signal delivery without delays. Signals carried in the communication system vary based on the device capability and the role of that device in the microgrid, but could be characterized as feedback from the DERs to the microgrid controller, and control signal traffic in the opposite direction [7].

Secure information storage and transportation are extremely vital for power utilities, especially for billing purposes and grid control. Providing system reliability has become one of the most prioritized requirements for power utilities. The quality-of-service in smart-MGs is defined by the accuracy and effectiveness with which different information such as the equipment's state, load information, and power pricing are delivered in a timely manner to the respective parties. Thus communication supports those functions and services in the microgrid to be carried out fast and efficiently.

End-to-end communication with low latency and sufficient bandwidth along with advanced controls should be robust enough to prevent cyber-attacks and provide system stability and reliability. The communication structure should be based on the preferred communication technology.

The communication between the power supplier and power customers is a key issue of the smart grid. Performance degradation like delay or outage may compromise stability. Therefore, a quality of service mechanism must be provided to satisfy the communications requirements (e.g., high-speed routing) and a quality of service (QoS) routing protocol must be applied in the communications network.

Many smart meters, smart sensor nodes, smart data collectors, and renewable energy resources are joining the communications network. Hence, the smart grid should handle the scalability with the integration of advanced web services and reliable protocols with advanced functionalities, such as self-configuration and security aspects.

10.3 Communication media for application in microgrid

Signals need a channel to follow, so that they can move from place to place. Fiber optic and copper cables are examples of these communication mediums or channels. In addition to such wired media, wireless media can also function for signal communications to move from place to place. To provide system reliability, robustness, and availability at the same time with appropriate installation costs, a hybrid communication technology mixed with wired and wireless solutions can be used. These three types of communication media in use for microgrid system application are briefly presented in this section.

10.3.1 Copper cable

There are two typical types of copper wire cables used for communication, twisted pair and coaxial cables. The wires may be shielded or unshielded depending on the design types and needs. The so-called wired networks establish a connection between various devices through cables and routers. Copper cables as mean of a transmission medium are susceptible to interference, noise, and signal attenuation, mostly attributed to the electromagnetic coupling fields. The physical constraints of the environment bring in unfavorable influence on the choice of the copper cable. However, the dual conductor dependency and the material (copper) characterize the physical components of copper-based cables used for transmission lines. Previously used to transport voices, the technology has been expanded to serve high-speed data services (ADSL, VDSL, and so on...) [8]. A wired transmission medium provides high-speed connectivity but poses constraints like immobility and extensive cabling [9].

The characteristics of copper cables create a level of limitation on the signal flowing in the channel. Considering the metal components of copper cables, with the induced resistance, the electrical conductivity characteristic of the medium affects the speed of data transfer during transmission. The signal in the copper medium can generate harmonics and degrades the signal over time. Copper conductor cables are also sensitive to noise, cross-talk, and electromagnetic interference, unlike fiber optic links where interference is minimal. A couple of techniques have been implemented to improve performance in the copper transmission medium, such as twisting of copper cables or wire transposition [6]. The copper medium can induce a high amount of attenuation, thus limiting the coverage footprint.

10.3.2 Fiber optics

An optical fiber is a thin, flexible, transparent fiber that acts as a waveguide, or "light pipe," to transmit light between the two ends of the fiber. Optical fibers are widely used in fiber-optic communication, which permits transmission over longer distances and at higher bandwidths than other forms of communication. This technology is a precise means to supply power to sensors, actuators, and other electrical consumers with high isolation demands [10]. In Ref. [10], a smart power management concept is introduced in the application of a 6 W optical power link with integrated optical data transmission which controls the laser output power with respect to changing electrical load, and optimizes the operating point of the photovoltaic (PV) cell, and thus increases system efficiency while varying load operation. As a result, it minimizes operating temperatures of laser and PV cell, and eventually prolongs the lifetime of the system. Fiber optics has many advantages like with its wider wavelength, signal security, and small size and weight. It also has drawbacks with its high investment cost, its need for more expensive optical transmitters and receivers, and it is more difficult and expensive to splice than wires, as well as being fragile. Two cables are needed for two-way communication as the propagation of signals in fiber optics is unidirectional. Taking the optimal points of its pros and cons, the fiber optics communication can be implemented in the application of microgrid system.

Unlike the wired medium, the fiber-based transmission medium relies on the transfer of light pulses in the transmitting process of digital data. Due to the characteristics of fiber optics, less attenuation can be observed in the network over long distances. However, limitations are also highlighted in the fiber-optics medium to signal composition, especially when traveling over long distances [8].

10.3.3   Wireless communication

Wireless communication can include many types based on its operation principle, function, and structure. Satellite communication, infrared communication, radiowave communication, microwave communication, Bluetooth, global position system (GPS), Wi-Fi_33, WiMAX, Zigbee, and 5G are the main communication technologies used with varying data rates. The WiMAX stands for Worldwide Interoperability for Microwave Access and it is a technology for a point to multipoint wireless network. WiMAX have wireless network similar to Wi-Fi_33 but it has higher speed over a wide area and higher accuracy compared with the Wi-Fi_33 [11]. Every WiMAX network has a Base Station (BS). The data packets traverse through the BS. The frequency range of the WiMAX is up to 11 GHz and it has speed of 75 mbps over a distance of 50 km.

To meet the cost and technical requirements in the application of a microgrid system, the appropriate communication technology and system should be chosen. Wireless technologies with constrained bandwidth and security and reduced installation costs can be a good choice for large-scale smart grid deployments. On the other hand, wired technologies with increased capacity, reliability, and security can be costly.

Wi-Fi_33 is a form of low-power wireless communication used by many electronic devices such as laptops, systems, smartphones. In a Wi-Fi_33 setup, a wireless router serves as the communication hub. These networks are extremely limited in range due to the low power of transmissions allowing users to connect only within close proximity to a router or signal repeater. Wi-Fi_33 is common in home-networking applications which provide portability without any need of cables. Wi-Fi_33 networks need to be secured with passwords for security purposes in order not to be accessed by others.

The main advantages of Wi-Fi_33 communication are its ease of integration and convenience, mobility, and expandability. However, it has cons in being subject to interference, lower range for large structure, security problems, and slower speed.

Advantages of wireless communication:

- Any data or information can be transmitted faster and with a high speed.
- Maintenance and installation are less costly for these networks.
- The internet can be accessed from anywhere wirelessly.
- It is very helpful for operators as they can be in touch with the centers.

The main drawback of wireless technology is that an unauthorized person can easily capture the signals which spread through the air. Hence, it is very important to secure the wireless network so that the information cannot be misused by unauthorized users.

10.4 Communication protocols for application in microgrid

With respect to the communications, the technologies may consist of dedicated duplex channels between a generation utility and the system operator for telemetry and telecontrol with additional voice/data channels over the public switched telephone network as a backup. The energy distribution network does not only allow for the physical transfer of energy but also supports information communication technologies interfaces enabling real-time information exchange related to the scheduling, monitoring, control, and protection of the interconnected DERs [12]. Communication system protocols are essential in microgrid implementation to ensure stable, reliable, and optimal operation. Harnessing the modern and secure communication protocols will significantly strengthen the system reliability and robustness.

One important communication capability is to deliver data on time to support the related functions. Thus several works have proposed different communication architectures which aim to achieve timely data delivery. Several Ethernet-based approaches have been implemented to provide a real-time access for control networks, fieldbuses, sensors, or actuators [6]. In the same way, many proprietary communication protocols are already available for information exchange and monitoring. Nevertheless, the standardization of communication for the grid is decisive for future smart grids. The implementation of standardized protocols and technologies is a reliable way to accomplish the functions for DER integration [7]. There are several guidelines and recommendations for nonproprietary protocols and for interoperation capabilities in electric utility applications [8−11].

Some efficient and reliable standardized protocols have been implemented for the management of large data amounts. The very popular protocols used for microgrid control and protection implementations, including IP, Modbus, DNP, and IEC 61850, are presented and discussed in the following subsections.

10.4.1   Internet Protocol

IP is generally configured on several protocol layers. It identifies lost information and errors and triggers retransmissions. It offers to support the connectivity, routing, and transport services that are required for microgrid applications. However, it increases the needed layers in the protocol definition and consequently the data processing time. IP can be used for nonrealtime applications.

The general procedures of the IP working principles (Fig. 10.2):

Step 1: data are broken into packets.

Step 2: as per IP protocols, the packets travel from router to router.

Step 3: the packets are reassembled into the original whole.

10.4.2   Modbus

Modbus is an application layer messaging protocol, positioned at level 7 of the open systems interconnection (OSI) model, which provides client/server communication between devices connected on different types of buses or networks. The industry's serial de facto standard since 1979, it is truly open and the most widely used network protocol in the industrial manufacturing environment. The Modbus protocol provides an industry standard method that Modbus devices use for parsing messages. The Internet community can access Modbus at a reserved system port 502 on the Transmission Control Protocol/Internet Protocol (TCP/IP) stack. Modbus is used to monitor and

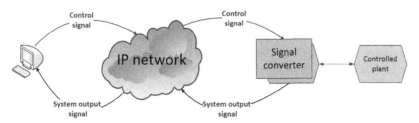

FIGURE 10.2   The block diagram of the IP network-based application.

program devices, to communicate intelligent devices with sensors and instruments, and to monitor field devices using personal computers (PCs) and human–machine interfaces [13]. It is widely used over a variety of physical layers. The primary advantage of Modbus is its simplicity for small devices and the very large range of devices that have some sort of Modbus interface. It is widely used in process control and SCADA systems.

10.4.3 Distributed Network Protocol

DNP was developed to achieve the requirements including high data integrity, flexible structure, multiple applications, minimized overhead, and open standards. Like Modbus, DNP is a byte-oriented protocol. DNP is an Ethernet standard communication protocol developed on the basis of three-layer enhanced performance architecture. DNP is widely used over a variety of physical layers. The DNP defines the information exchange by means of a comprehensive data-objects specification and by the transmission of small data-packets. For the data transaction, the messages are processed in a deterministic sequence [14].

The three layers include a data-link layer describing the required services to be provided by the DNP physical layer, a pseudotransport layer segmenting application layer messages into multiple data-link frames, and an application layer describing the message format, services, and procedures [13]. DNP plays important roles in the application of smart grid topologies to accomplish the functions of information exchange, monitoring, and control processes for integrating DERs with the grid [14].

DNP is specifically designed for use in SCADA applications. It is highly standardized, with relatively high compatibility and interoperability between devices from different manufacturers.

10.4.4 IEC 61850

IEC 61850 is an international standard defining communication protocols for intelligent electronic devices at electrical substations. The abstract data models defined in IEC 61850 can be mapped to a number of protocols. Current mappings in the standard are to MMS (Manufacturing Message Specification), GOOSE (Generic Object-Oriented Substation Event), SMV (Sampled Measured Values), and soon to web services. These protocols can run over TCP/IP networks or substation local area networks (LANs) using high-speed switched Ethernet to obtain the necessary response times below 4 ms for protective relaying.

Self-describing devices and object-oriented peer-to-peer data exchange capabilities are the most significant superiorities of IEC 61850 over the other common standards. Logical nodes (abstract data objects) are the main elements of the IEC 61850 object oriented virtual model, which consists of standardized data and data attributes. The virtual model expresses a physical (logical) device and number of logical nodes. Each logical node contains data elements, which are standard and related to logical node functions. Most of the data objects are composed of common data classes, involving basic data objects, status, control, and measurement.

Each data element consists of a number of data attributes with a data attribute type which belongs to functional constraints. This standard provides an efficient and reliable interoperability framework between different field devices in protection applications. For that, the GOOSE messaging protocol will be utilized to send control messages from microcontroller unit (MCU) to the system's intelligent electronic devices (IEDs) in order to adjust their settings accordingly.

IEC 61850 GOOSE messages are broadcast layer 2 messages of the OSI model that enable fast publish/subscribe message exchange between end devices within 3−4 ms during the fault condition. It is also important to realize the presence of merging units (MUs) at both ends of a transmission line. These MUs will digitize current measurements at the source and broadcast them over the designated network in the form of IEC 61850 SMV messages. Generally, IEC 61850 for microgrid application is the emerging protocol and taken to be a promising solution [2].

10.4.5   Implementation of IEC 61850-based microgrid communication

10.4.5.1   FPGA-based communication for the 10 kV microgrid

10.4.5.1.1   Background

At present, the mainstream energy storage, load, wind power, and PV devices are basically reserved with serial communication interfaces such as controller area network (CAN), RS485, universal asynchronous receiver/transmitter (UART), and I2C. If the reserved communication interface is directly used, the system will be extremely complicated, and the transmission distance and data transmission time will be limited. The communication interface reserved directly by the devices cannot reach the communication requirements of a 10 kV microgrid system containing wind, solar, storage, loads, and a dynamic disturbance control system. The traditional solution that the microgrid local control devices use is based on the Advanced RISC Machines (ARM) method. Another solution the microgrid local control devices is made based on the Field Programmable Gate Array (FPGA) method. With regard to the FPGA method, in order to solve the problem of communication delay, the serial status information of the microgrid devices is composed of the Ethernet frame through the local device of the microgrid to realize the Ethernet communication between the local device and the central controller.

These two solutions can significantly improve the communication time delay between the central controller and the local control system and devices in the microgrid. For the ARM-based solution, due to the serial execution mechanism, multiple tasks cannot be executed at the same time. So, when the system's task is too burdened, it will be difficult to meet the real-time communication requirements of the microgrid. For the latter method, if only FPGA is used without ARM, the flexibility of the system will be limited, and this solution is rarely used in practical projects.

According to the analysis above and the actual test environment, the 10 kV microgrid system mainly connects the master–slave controllers, local devices, and acquisition devices through optical fiber. The communication mode between different devices is mainly based on the Ethernet frame of self-defined message format formulated by SV/GOOSE messages to form a high-speed Ethernet control transmission. It adopts a ring network and gateway mode instead of point-to-point communication architecture between the controller and various equipment, mainly considering the complexity of system architecture and the real-time data interaction between various devices in microgrid. The reason that the SV/GOOSE message format is referred to is that IEC 61850-9-2SV (Sampled Value) and object-oriented substation event GOOSE are two communication protocols with high real-time requirements in the international IEC 61850 standard.

10.4.5.1.2 Brief introduction of SV/GOOSE protocol-based communication

This section mainly introduces the general communication network structure diagram in a microgrid, as shown in Fig. 10.3. It includes the communication between the controllers and the local acquisition devices and the local devices as well as the communication between the local devices and the optical communication ports of the local acquisition devices. The optical communication ports between the controller and the local devices are the communication mechanism of FPGA external expansion network ports. The function it needs to realize is the network communication structure mode of the IEC 61850 standard Ethernet communication protocol.

This communication method is mainly performed by using the analog values, switch values, messages, and other key data in the SV/GOOSE message data that the devices of the microgrid control system send. They are analyzed and processed by the FPGA part based on Xilinx Zynq-7000SoC and Altera IV chips, and then sent to the ARM part and the Digital Signal Processing (DSP) part of Xilinx Zynq-7000SoC for processing.

This kind of communication mainly uses the parallel execution mechanism of FPGA. For the part suitable for parallel logic processing in the decoding process, Hardware Description Language (HDL) description is adopted to improve the time delay of the process. In addition, the sequential process of field by field parsing in SV/GOOSE message is abstracted into Finite State Machine by HDL, which can not only describe the process well, but also improve the real-time performance and reliability of the SV/GOOSE message analyzing system.

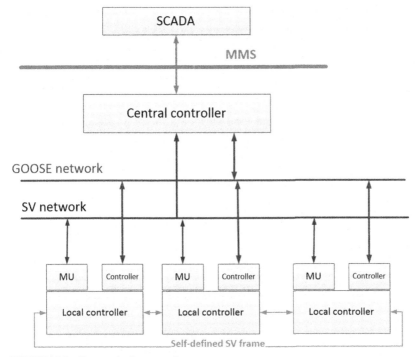

FIGURE 10.3   Communication network structure.

10.4.5.1.3   Realization of SV/GOOSE message communication

Analyzing the microgrid's real situation, the communication mode is mainly divided into the ring network and gateway SV/GOOSE message communication transmission mode.

According to the transmission mechanism and format of SV/GOOSE in the IEC standard, the self-defined message and TCP/IP message are mixed together to transmit. In this case, the real-time transmission of the self-defined message is ensured by different processing methods. The FPGA part realizes packet identification and classification. Through the FPGA part, the Ethernet message is decoded and the packet type is identified. If it is a TCP/IP packet, it is transmitted to the operation system for processing. If it is a self-defined message, it will be transmitted to the real-time chip for processing.

When the FPGA part is used to encode and decode the hardware message, because of its parallel processing characteristics, the data can still be effectively processed when the network transmission is large, but if it is ARM, it may lose some data when the network transmission is large.

Based on network transmission in the form of SV/GOOSE protocols in IEC 61850 standard, the data (such as switch signals) collected in a microgrid control system by the local acquisition device are transmitted to the FPGA to buffer.

Then FPGA quickly packages and sends the data based on the source address, destination address, and other configuration information of the SV/GOOSE protocol as long as the requirements to controllers and local devices in the microgrid (the SV condition of 80 points per cycle or 256 points per cycle and the transmission interval mode of GOOSE message—heartbeat message) are satisfied. Then, the FPGA module will send the SV/GOOSE Ethernet message to the gateway, respective devices, and the controller. The gateway is connected to the respective controller and local devices in the microgrid. So, in order to meet the needs of the system, FPGA at the receiving end decides the SV and GOOSE frames and decodes them, then buffers and uploads the data required by each user to the respective controller and local devices in the system.

10.4.5.1.4   Description of FPGA based SV/GOOSE communication

SV/GOOSE Ethernet protocol under the IEC 61850 standard is only used in FPGA communication of microgrid systems. Compared with the ordinary Ethernet TCP/IP protocol, SV/GOOSE only carries out data transmission processing at the physical layer and data-link layer on the communication protocol stack. It is implemented in the physical layer, which can reflect the timeliness of its processing, as shown in Fig. 10.4.

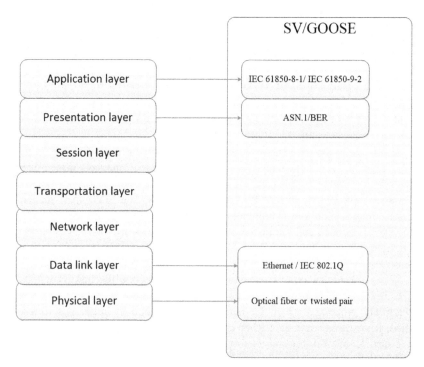

FIGURE 10.4   SV/GOOSE communication protocol stack.

The Ethernet frame structure of the sample value (SV)/GOOSE message is shown in Table 10.1 along with the meaning of each field. The first field is the frame header with a length of 8 bytes, including 7 bytes of presynchronization code and 1 byte of frame definer. The second field is the media access control (MAC) address, including the MAC destination address with a length of 6 bytes and the MAC source address with a length of 6 bytes. The third field is VLAN with a length of 4 bytes, which is optional. The fourth field is Ethernet type with a length of 2 bytes, which is fixed for the GOOSE message value set as 0x88_B8. The value of the SV message is fixed as 0x88_BA. The fifth field is the application identity (APPID) header including APPID, length, reserved bit-1, and reserved bit-2 with 2 bytes each, in which the APPID application ID is 2 bytes length, which is the message length from the beginning of APPID to the frame check sequence. The sixth field is an application protocol data unit, which is APDU (GOOSE-PDU/sav-PDU), of less than 1480 bytes. The seventh field is a frame check and the length of the test sequence is 4 bytes.

The SV message of the self-defined protocol of the microgrid adopts the communication structure of publisher/subscriber. SV message is a time-driven communication mode, that is, it sends a sample value every fixed time period (analog digital sampling is 80 points per cycle or 256 points per cycle). The main transmission requirement of the SV message is real-time and fast. When the message transmission is lost due to network reasons, the

TABLE 10.1 Ethernet frame structure of SV/GOOSE message.

Name of field	The field's function	Length
Frame header	Leading code	7 bytes
	Frame qualifier	1 byte
MAC address	Destination address	6 bytes
	Source address	6 bytes
VLAN (optional)	TPID	2 bytes
	TCI	2 bytes
Ethernet type		2 bytes
APDU header	APPID	2 bytes
	Length	2 bytes
	Reserved Bit 1	2 bytes
	Reserved Bit 2	2 bytes
APDU		<1493 bytes
Frame check sequence		4 bytes

publisher (voltage and current sensors) will continue to collect the latest voltage and current information. And the subscribers, such as the protection system, must be able to detect it. This can be reflected by the sample counter parameter (SmpCnt).

In the SV message, the protocol data unit sav-PDU adopts abstract syntax notation (ASN). 1 syntax rules definition, as shown in Fig. 10.5. The first field, 0x60, is the Tag value of sav-PDU, with a length of 1 byte. The second field is the length value of sav-PDU, with two choices: 1-byte format and at least 2-byte long format. noASDU indicates the number of ASDU contained in sav-PDU. For each ASDU, the following information is included: *svID* representing the sample value identifier; *SmpCnt* representing the sampling counter; *confRev* indicating the configuration version number; *smpSynch*, Boolean quantity, indicating whether the sampling value is synchronized

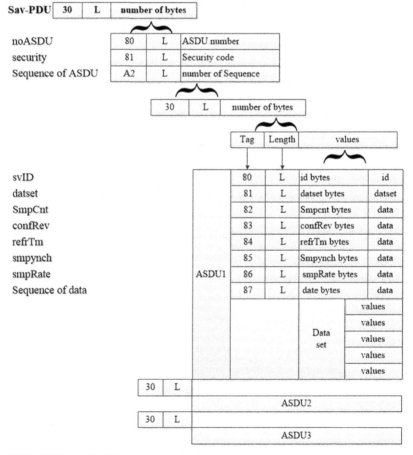

FIGURE 10.5   sav-PDU of SV frame protocol.

with the clock signal or not; and *Sequence of Data*, such as current and voltage values of devices collected by integrated development environment (IDE).
The sending process of the GOOSE message of the self-defined protocol of the microgrid is executed according to the sending rule of the GOOSE message, as shown in Fig. 10.6, where T0 is the heartbeat time. The device normally sends the current status every T0 time, so the message at this time is called the heartbeat message. When ARM finds that the value of any member of the GOOSE data changes, the local acquisition devices will immediately send all the set data, and then send the second frame and the third frame at T1 interval, the fourth frame at T2 interval, and the fifth frame at T3 interval. The sending time interval of subsequent messages gradually increases until the last message interval is restored to heartbeat time.

The heartbeat interval of the GOOSE message is T0 in Fig. 10.5. According to the 61850-implementation specification, the time allowed to keep the message displayed is 2T0. If the acceptance time exceeds 2T0, the message is judged to be lost; if the next frame GOOSE message is not received within 2 times of the time allowed to keep the message displayed, it is judged that the communication is interrupted. After judging the interruption, the device sends the GOOSE chain breaking alarm. Therefore in the process of communication, GOOSE realizes the intelligent detection of the circuit connection and disconnection between devices by continuous self-test, and overcomes the defect that a traditional cable circuit fault cannot be automatically detected.

According to engineering practice, T0 is set to 5000 ms, T1 to 2 ms, T2 to twice T1, and T3 to 2 × T2. Therefore, the time interval of four retransmissions of bit change message is as follows: the first retransmission interval is 2 ms, the second retransmission interval is 2 ms, the third retransmission interval is 4 ms, and the fourth retransmission interval is 8 ms. After four retransmissions, the GOOSE message is forced to return to heartbeat time in

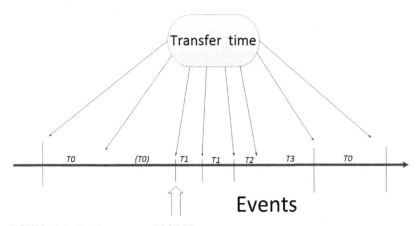

FIGURE 10.6 Sending process of GOOSE message.

order to reduce the system burden.where $T0$ is retransmission time under stable condition (no event for a long time); $(T0)$ is retransmission time under stable conditions may be shortened by events; $T1$ is the shortest transmission time after the event occurs; and $T2$, $T3$ is the retransmission time until stable conditions are obtained;

10.4.5.2 The FPGA communication module for the 10 kV microgrid system

In the 10 kV microgrid system of FPGA communication system framework, it is mainly divided into three modules: central controller and local acquisition devices, central controller and local controllers, and local controllers and local controllers. The local acquisition devices are mainly used to collect, and process the local information (voltage, current analog quantity, and switch value) of the microgrid system, and send it to the central controller. The local controllers mainly process and analyze the information that is collected by local acquisition devices and uploaded by local controllers, and then issue instructions and status information to local controllers for control and adjustment. The local controllers mainly realize the data acquisition, processing, adjustment, and state control of the local devices (such as lead—carbon and lithium batteries, ultracapacitors, active harmonics, etc.) in the microgrid system. It is uploaded to the central controller for decision-making. At the same time, it receives the status information from the central controller and processes and realizes it in time.

10.4.5.2.1 A mechanism between central controller and local acquisition devices

The dual CPU structure can be adopted between the central controller and the local acquisition devices, as shown in Fig. 10.7. The low-voltage differential signaling (LVDS) bus is used between the backplane to realize high-speed data interaction between the two CPU boards. The reason why the dual CPU board used is that the data requirements of the decision data processing center and the data processing center for displaying are different. For example, when making decisions, it is not necessary to convert a binary system into a decimal system, but it converts a binary system into a decimal system when it displays the data. Dual CPU can avoid occupying the computing resources of the decision data processing center due to the display and improve the time delay of the decision response. In addition, when the burden of a single CPU is

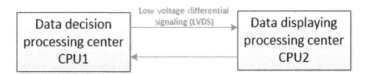

FIGURE 10.7 Dual CPU structure of local devices.

too large, the computing resources of the CPU display can be used for decision data processing, which expands the adaptability of the system.

The schematic diagram of each local acquisition device in the SV/GOOSE network architecture of IEC 61850 communication protocol is shown in Fig. 10.8. The device is mainly composed of two parts. One is the SV/GOOSE message generator, which is the source of various IEDs in the microgrid system, which mainly generates SV/GOOSE message data to be decoded. The other part is the SV/GOOSE message encoding and decoding the processing devices.

Fig. 10.9 is the schematic diagram of a local acquisition device encoding and decoding processing device, which is the embodiment of the SV/GOOSE frame protocol coding and the decoding structure diagram under IEC 61850 communication standard. Gigabit optical fiber communication mode is adopted, and a photoelectric conversion module is added between the physical (PHY) network ports to realize real-time conversion between photoelectric devices. The analog data, such as voltage sampling value, current sampling value, state information, switching value, etc., are collected and processed by local devices. The SV/GOOSE protocol under the IEC 61850 standard is used to encode and decode, and SV/GOOSE format Ethernet transmission and reception is realized.

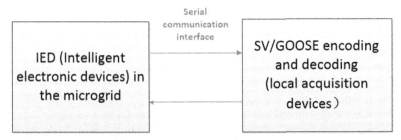

FIGURE 10.8 Schematic diagram of local acquisition device.

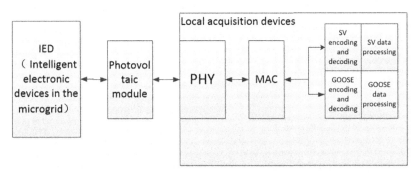

FIGURE 10.9 Schematic diagram of local acquisition coding and decoding.

10.4.5.2.2 A mechanism between the central controller and local controllers

For the 10 kV microgrid, the controller and local acquisition devices of ZYNQ module include ARM and FPGA. According to the data flow direction, the message from the central controller, acquisition devices, and local controllers on the spot of the Ethernet self-defined IEC 61850 SV/GOOSE enter into the Ethernet receiving module after the optical fiber module, the PHY chip, which is the RX module. It mainly completes the Ethernet frame in the receiving module, and decides if the destination address of the receiving frame is consistent with the source address of the local device. If it is consistent, the data in the frame will be delivered to the ARM section of the CPU chip. After carrying out the correlation calculation in the ARM, operation instructions will be directly sent to the local control devices through a serial interface. If it is inconsistent, the data will be directly sent to the Ethernet frame binarized by an Ethernet transmission module through the PHY chip and optical module (Fig. 10.10).

The running data of the microgrid, according to the data flow, is sent to the ARM section of the CPU chip for relevant numerical calculations through the FPGA serial port receiving module in the CPU chip, such as effective value and maximum value. Then, the data will be dealt with in the ARM and sent to the Ethernet module of the FPGA part, namely the transmission (TX) module.

According to the self-defined IEC 61850 SV/GOOSE message format of Ethernet frames, finally the data are sent out through the PHY chip and optical module. The receiving and sending data flow direction of the above microgrid central controller is shown in Fig. 10.11.

According to Fig. 10.11, as per IEC 61850 standard, the encoding/decoding process of the central controller is embodied through the SV/GOOSE frame protocol coding structure. In order to realize real-time photoelectric conversion, a gigabit optical fiber communication mode is adopted and a photoelectric conversion module between the front-end ports of PHY chip is added. The analog quantities (voltage and current sampling values, state

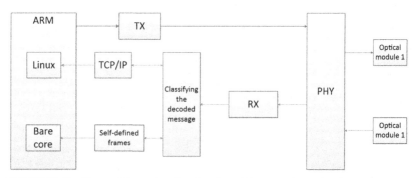

FIGURE 10.10 The receiving and sending data flow of the central controller device.

information of the switch, etc.) collected by the local acquisition devices and the Ethernet in the SV/GOOSE format are received and dealt with by the local controllers.

10.4.5.2.3   A mechanism between local control devices

The SV/GOOSE frame format of IEC 61850 is adopted in the ring network mode. In this mode, as shown in Fig. 10.12, the local control device sends and receives frames to/from the other local control devices. FPGA determines whether the sending frame is valid or not. When the data frame is valid, the valid data are buffered and DSP is informed to read it and vice versa (Fig. 10.13).

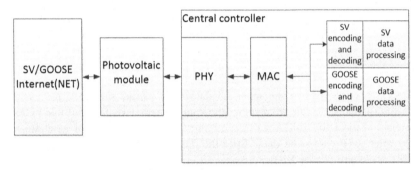

FIGURE 10.11   Schematic diagram of the central controller.

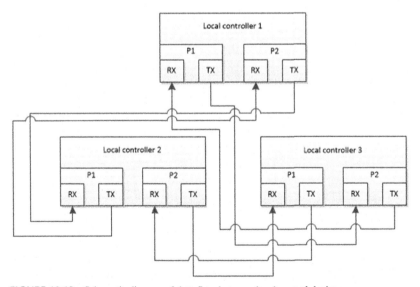

FIGURE 10.12   Schematic diagram of data flow between local control devices.

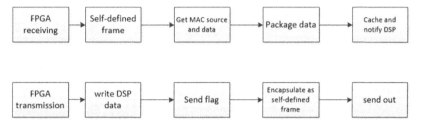

FIGURE 10.13 The process of FPGA sending and receiving data. Note: Receiving frame decision process: MAC decision → APPID decision → frame number → memory read (DSP). Sending frame decision process: DSP to memory write data → send mark to decide → FPGA encapsulation → send (optical fiber).

10.4.5.3 Interactive data format of FPGA and DSP

The data are divided into related data zones, which can be used by the FPGA for further decisions (Table 10.2).

10.4.5.3.1 DSP writes to data zone

The DSP writes data to the specified data zone (to the same address), and the FPGA is responsible for sending the data immediately. For multiple bytes writing of the DSP, FPGA will cache and set them to FPGA or generate a flag after DSP writes the data. After the data are sent by FPGA, the flag is reset or cleared (Table 10.3).

According to Table 10.2, the DSP writes the packet of the data, adds the local MAC address, and combines the data length. FPGA takes out the effective data and sends it in SV frame format.

10.4.5.3.2 DSP receives from data zone

After each optical port receives the data sent by another device, FPGA will receive it if the MAC address is accessible, otherwise, the data will not be received. For each frame, the valid data transmitted are extracted in the format of Table 10.2, and then are encapsulated in the format of Table 10.4. The data are stored in the DSP shared memory area, informing the DSP to update in the form of interruption. DSP clears the shared buffer and flags after reading and writing. The maximum size of a data zone is 2kbytes. This operation is used for each FPGA optical fiber port.

FPGA decodes the frames according to the data format in Table 10.2, and stores the valid data into the buffer zone according to the data format in Table 10.4. DSP reads the address in the buffer zone and extracts the transmitted valid data.

As described above, the communication transmission method can meet the accuracy and speed of the experimental requirements of communications for the microgrid system. In the process of the fast transmission data will be

TABLE 10.2 Frame of DATA format interaction between FPGA and DSP.

Destination address MAC	Source address MAC	Data length	Data zone	CRC
6 bytes	6 bytes	2 bytes	46–1500	4 bytes

TABLE 10.3 The list of cache data.

Register address	Name	Data length	Note
	Writing sign	2 bytes	DSP read-only, FPGA write-only
	Sending sign	2 bytes	DSP write-only, FPGA read-only
	DSP Writing operation counter	2 bytes	DSP plus 1 every writing
	FPGA Reading operation counter	2 bytes	FPGA plus 1 every reading
	Buffer data	2K bytes	Write in Table 10.2 format

TABLE 10.4 Buffer list of accepted data.

Name	Data length	Note
Writing sign	2	DSP read-only, FPGA write-only
Sending sign	2	DSP write-only, FPGA read-only
DSP Writing operation counter	2	DSP plus 1 every writing
FPGA Reading operation counter	2	FPGA plus 1 every reading
Data length	2	Parse the received data and extract the valid data in Table 10.1 format
Valid data	46–1500	Parse the received data and extract the valid data in Table 10.1 format

encoded and decoded between devices in the 10 kV microgrid system, the FPGA module encodes and decodes the SmpCnt of a data frame between the local devices. The data written by DSP module in the periodic operation of dynamic disturbance control system are accumulatively coded by FPGA. When the receiver decodes SmpCnt, it judges whether the SmpCnt is continuous or not, and then it judges whether the data transmission is qualified or not, which also intuitively reflects the time delay.

Thus this method can realize the periodic signal transmission of the dynamic disturbance system of the 10 kV microgrid with the communication time of 330 μs. The optical communication is transmitted by a dynamic disturbance control system, and the delay is only less than 8 μs, which meets the design requirements of the experimental and practical microgrid systems.

The energy storage devices mentioned above include lithium batteries, supercapacitors, and lead–carbon batteries. The load includes domestic and industrial electrical equipment and other inductive, capacitive, and resistive loads.

Based on the above framework, the acquisition devices, control equipment, and the central controller in the 10 kV microgrid combined are able to set up high-speed, real-time data interactions between each piece of equipment in the microgrid, which guarantees the rapid response between the central controller and the local controllers working for all condition of disturbances.

References

[1] V. Kounev, D. Tipper, A.A. Yavuz, B.M. Grainger, G.F. Reed, A secure communication architecture for distributed microgrid control, IEEE Trans. Smart Grid (2015). Available from: https://doi.org/10.1109/TSG.2015.2424160.

[2] I. Ali, S.M. Suhail Hussain, Communication design for energy management automation in microgrid, IEEE Trans. Smart Grid (2018). Available from: https://doi.org/10.1109/TSG.2016.2606131.

[3] V.C. Güngör, et al., Smart grid technologies: communication technologies and standards, IEEE Trans. Ind. Inform. (2011). Available from: https://doi.org/10.1109/TII.2011.2166794.

[4] R. Bikmetov, M.Y.A. Raja, K. Kazi, B. Chowdhury, S.M.H. Zaidi, Role of PONs with the conventional communication systems used in today's smart- and MicroGrids, 2014. Available from: https://doi.org/10.1109/HONET.2014.7029374.

[5] F.H. Hany, F. Nevin, M.E. Mohammad, M. Osama, B. Sukumar, An enhancement of protection strategy for distribution network using the communication protocols, IEEE Trans. Ind. Appl. (2020). Available from: https://doi.org/10.1109/TIA.2020.2964638.

[6] I. Serban, S. Cespedes, C. Marinescu, C.A. Azurdia-Meza, J.S. Gomez, D.S. Hueichapan, Communication requirements in microgrids: a practical survey, IEEE Access. (2020). Available from: https://doi.org/10.1109/ACCESS.2020.2977928.

[7] A. Bani-Ahmed, L. Weber, A. Nasiri, H. Hosseini, Microgrid communications: state of the art and future trends, 2014. Available from: https://doi.org/10.1109/ICRERA.2014.7016491.

[8] K.A. Ogudo, M.H. Mthethwa, D.M.J. Nestor, Comparative analysis of fibre optic and copper cables for high-speed communication: South African context, 2019. Available from: https://doi.org/10.1109/ICABCD.2019.8851021.

[9]   U. Okpeki, J. Egwaile, F. Edeko, Performance and comparative analysis of wired and wireless communication systems using local area network based on IEEE 802.3 and IEEE 802.11, J. Appl. Sci. Environ. Manag. (2019). Available from: https://doi.org/10.4314/jasem.v22i11.3.

[10]   H. Helmers, A. Cornelius, M. von Ravenstein, D. Derix, C. Schöner, 6W optical power link with integrated optical data transmission, EEE POWER Electron. Regul. Pap. 35 (8) (2020) 7904–7909.

[11]   IEEE, IEEE standard for local and metropolitan area networks part 16: air interface for fixed broadband wireless access systems, 2004. Available from: https://doi.org/10.1109/ieeestd.2004.226664.

[12]   M. Short, M. Dawood, C. Insaurralde, Fault-tolerant generator telecontrol over a microgrid IP network, 2015. Available from: https://doi.org/10.1109/ETFA.2015.7301470.

[13]   J. Makhija, Comparison of protocols used in remote monitoring: DNP 3.0, IEC 870-5-101 & Modbus, Electronics Systems Group, 2003.

[14]   E. Padilla, K. Agbossou, A. Cardenas, Towards smart integration of distributed energy resources using distributed network protocol over ethernet, IEEE Trans. Smart Grid (2014). Available from: https://doi.org/10.1109/TSG.2014.2303857.

Chapter 11

Application cases of industrial park microgrids' protection and control

11.1 Background

The microgrid is a concept relative to the traditional power grid. It refers to a small power generation and distribution system that is composed of distributed power sources, energy storage devices, energy conversion devices, loads, and the related controlling, monitoring, and protection systems. It is an autonomous system that can operate in conjunction with the external power grid or in isolation. From a micro perspective, a microgrid can be regarded as a small power system with a complete set of power generation, transmission, and distribution systems that can achieve local power balance and energy optimization. From a macro perspective, a microgrid can be considered as a "virtual" power source or load in the distribution network. Compared with the external power grid, it appears as a single controlled unit and can satisfy users' requirements for power quality and power supply reliability.

Issues such as growing interest on renewable energy, clean utilization of coal, and reduction of carbon emission have been forcing the world to vigorously consider the option of deploying microgrids. Microgrids create opportunity to integrate renewable energy resources. The cost of solar and wind power generation has dropped dramatically in recent years making them cost competitive with traditional fossil fuel power generation technologies. This has enabled the microgrid to achieve the clean energy goal in an economically feasible way while completing the main task of ensuring reliable and stable power supply.

Reliable and secure power supply has become a basic need for economic development and people's daily life in any country. Temporary power interruption in residential buildings will only bring a little inconvenience to family life. However, when it happens to hospitals, military bases, and other key facilities, it may endanger the safety of human life or a country's security. One of the effective ways to avoid this risk is through using microgrids.

The microgrid integrates distributed power generation to the existing physical network. It uses energy storage devices and control devices to

Microgrid Protection and Control. DOI: https://doi.org/10.1016/B978-0-12-821189-2.00011-5

smooth out the system fluctuation, maintain the balance of power generation and load in the network, and ensure stability of voltage and frequency. Local administrations develop power supply systems according to their own resources and local conditions, so as to maximize the local consumption of renewable energy. The microgrid can also be used as a system to participate in peak load regulation, voltage regulation, and frequency regulation of the power grid to effectively ensure the stability of the grid. This has been encouraged by the opening of the power auxiliary service market.

During disastrous events in the power grid, the grid-connected microgrid can operate independently by separating itself from the grid to form an isolated network so as to ensure an uninterrupted power supply for critical loads. In disaster-prone areas, a microgrid of different forms and scales can quickly restore power supply to critical loads after disasters, and has the ability of black start.

Microgrid can supply power to areas beyond the reach of the power grid. In remote areas, the population density is low, the construction cost of traditional power grid is high, and those areas are sometimes rich in scenic resources. The microgrid is an economical, effective, and environment-friendly choice for such places.

This chapter presents a detail discussion on two microgrids in China. Following the recent international trend, China has given huge attention toward green and low-carbon microgrids. Microgrid represents the future development trend of energy, and is considered as an important means to implement China's revolution on energy production and consumption. It is an important scheme for promoting the transformation of energy development and management through innovative application of "Internet plus" in the energy field, which is of great significance for promoting energy conservation and emission reduction and achieving sustainable energy development. At the same time, the new energy microgrid is an alternative for the power grid expansion and addressing the community in need of electricity, which is in line with the country's direction of power system reform.

The current power grid of China is characterized by high voltage, long-distance transmission, and large capacity. The six regional power grids have been interconnected, and the grid structure has become increasingly complex. In theory, it can play the role of interregional accident support and backup, and realize the optimal allocation of power resources. However, the large-scale alternating current (AC) synchronous grid has a large range of lower-frequency oscillation and instability, and its dynamic stability occurrence is difficult to control, likely resulting in large-scale blackout.

On the other hand, after the separation of management of power plants and the grid, the market stakeholders have become diversified, the contradiction between the power plants and the grid has increased, and the coordination between the power plants and the grid have become more difficult. The safety management of the grid equipment is not in place, which poses a threat to the safe and stable operation of the power system. Hence, compared with conventional centralized power stations, the microgrid can combine with the existing power system to form an efficient and flexible new system.

The new round of power system reform proposes incremental distribution business, encourages the orderly investment of social capital and operation of incremental distribution network, promotes the construction of distribution network, and improves the operation efficiency of the distribution network. In November 2018 the first microgrid type distribution network pilot project of Hainan Mei'an Science and Technology New City was connected to the grid, which promoted the progress of the microgrid in power system reform.

Remote and island areas have been the best market for microgrid development in China. Most of such type of microgrids in China are mainly concentrated in the northwest region, such as Shiquanhe microgrid in Ngari region of Tibet, Dingebengongsi microgrid in Dinge County of Naqu region of Tibet, and Chenbaerhuqi microgrid in Hulunbeir City of Inner Mongolia. China has a large number of islands, many of which are inhabited by residents. Most of them rely on diesel for power generation, but the supply time of diesel power is limited and the transportation cost is high. These islands also have a lot of renewable energy which allows for construction of renewable energy—based microgrids that can solve the problem of power shortage. Most of these demonstration projects are concentrated on the Southeast coast, such as the smart microgrid of Dongao Island in Zhuhai City, Guangdong Province, the Dongfushan Island microgrid in Zhejiang Province, and the Yongxing Island microgrid in Sansha City, Hainan Province. In August 2015 the Yantai Changdao distributed generation and microgrid access control project passed the acceptance of the National Development and Reform Commission, and was officially completed and put into operation. This is the first island microgrid project in northern China, which can realize isolated grid operation, ensure continuous power supply to important users, and greatly improve the power supply capacity and reliability of the long island power grid when the external large power grid collapses.

Following China's power system policy reform, the distribution network business has gradually become market-oriented, which opened a great potential for the investment, construction, and operation of smart microgrids. In 2016 China's NDRC and the National Energy Bureau issued the policy on promoting the development of "Internet plus" smart energy, and the notice on implementing the "Internet plus" smart energy demonstration project, which provided a strong promise for the development of smart microgrids. It can be said that China's smart microgrid policy conditions have gradually opened up enormous opportunities.

China's microgrid market space is already huge and expected to grow rapidly even more with the development of China's economy and cities expanding. Many places that are not connected to the power grid, such as the urban—rural fringe, small towns, marginal rural areas, and mountainous areas, create opportunities for the deployment of microgrids. It is roughly estimated that China's smart microgrid holds hundreds of billion Yuan of investment potential in 2020 [1]. At present, China's smart microgrid demonstration projects have been blooming everywhere, and made great breakthroughs in the technical field.

This chapter reviews two projects in China, located in two provinces of the country and intended for two different main purposes. The demonstrational microgrid is mainly planned and utilized to serve as a base for research and development works in the area. There is a brief discussion on the technology products based on the microgrid and the achieved technical advantages through the use of those technologies are discussed in 11.2.5. The industrial microgrid is a commercial microgrid installed in an industry park. The project involves the cooperation of three parties (i.e., a funder, a developer and operator, and user/consumer). The special business model and technical aspects of the microgrid are discussed later.

11.2 Demonstrational microgrid testbed

11.2.1 Introduction

The microgrid testbed is located in an industry park area. The area has large energy consumption and high grid power tariff which results in industrial institutions paying high electricity bills. To address this issue and other reasons, a microgrid that integrates renewable energy sources [wind and solar photovoltaic (PV)] and other distributed energy resources (DERs) is developed for an industry park in the area. In addition to the financial aspect, the microgrid is ultimately intended to serve as a demonstrational and testing platform for researching, developing, standardizing, and promoting the microgrid technology and different technological products.

The testbed has been used as a platform for research and development, testing and application of a series of microgrid technologies such as dynamic stability control system, transient stability control system, microgrid fault detection and protection system, microgrid intelligent load monitoring system, and so on. The products passed multiple evaluation and certification stages and have been credited to be of the internationally accepted level.

The microgrid has also served as an experimental base for development of two International Electrotechnical Commission (IEC) standards and one national standard on microgrids, namely:

- IEC TS 62898-3-1:2020, "Microgrids—Part 3-1: Technical requirements—Protection and dynamic control" [2];
- IEC TS 62898-2:2018, "Microgrids—Part 2: Guidelines for operation" [3]; and
- GB/T38953-2020 Technical requirements for relay protection of microgrid [4].

Extensive research and development works have been carried out based on this microgrid, resulting in publications (Refs. [5−14]), patents ([15−19]), and copyrights ([20]). The microgrid has also been serving as a platform for developing, testing, and verifying the satisfaction of expected standard performance of the following new products:

1. Microgrid load monitoring and control system
2. Microgrid transient disturbance control system
3. Microgrid dynamic disturbance control system
4. Microgrid optimal energy efficiency management system
5. Microgrid protection systems
6. Renewable energy and load forecasting models

11.2.2 Architecture and components

The microgrid has various types of distributed generation, energy storage, and load as shown in Fig. 11.1. The power generation is composed by two wind turbines (2.0 MW each), 280 kW PV, 1 MW microturbine, 500 kW diesel generator. The wind turbines are connected to 10 kV busbars, while the PV system is connected to 400 V bus. The load of the industrial park is mainly (about 70%) composed of electric motor loads in the production site of the park. In the year 2019 the microgrid has an annual energy yield of about 5 million MW contributing to about 70% of the consumption of the industry park. An energy-saving rate of about 14.3% was achieved through the smart

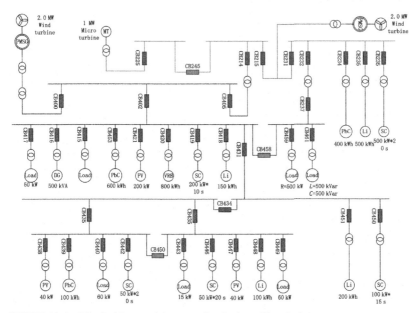

FIGURE 11.1 Electrical layout of demonstrational microgrid testbed.

energy efficiency system applied in the microgrid that consists of 5.8% saving from lighting loads and 8.5% saving from air-conditioning loads.

11.2.2.1 Wind turbines

The microgrid consists of two permanent magnet direct driven (PMDD)-type wind turbine units rated 2.0 MW each. The turbines have a hub height of 80 m, cut-in wind speed of 2.5−3 m/s (depending on air density), cut out speed of 25 m/s, and rated speed of 10.6 m/s. The rotor diameter of the turbine is 103 m. The output of the wind turbines is connected to the 10 kV bus through a transformer.

11.2.2.2 Photovoltaic system

The total rated capacity of the PV system in the microgrid is 280 kWp. The PV system is made of 240 kW polycrystalline silicon, 20 kW monocrystalline silicon and 20 kW cadmium telluride modules installed on the rooftop of the three buildings and a greenhouse in the park. The modules are grouped into three PV arrays (200 kWp, 40 kWp, and 40 kWp) connected to the 400 V busbar through central inverters.

11.2.2.3 Microturbines

The microgrid consists of a 1 MW micro gas turbine, which is a combined cooling heating and power (CCHP) system. The basic operation principle of microturbine is in such a way that compressed air is burnt in the combustion chamber which generates high-temperature and high-pressure air that ultimately drives the turbine. It contains only one moving part and adopts the air-cooled technique, achieving a compact structure and requiring no liquid lubricants or coolants to maintain. Moreover, it can operate on a variety of fuels and exhaust residual heat which can be recovered for additional energy.

11.2.2.4 Diesel generators

In addition to the renewable DER units, the microgrid consists of a 500 kW diesel generator. The generator is equipped with a special smoke filtering device and an enclosure to reduce the sound and air pollution significantly.

11.2.2.5 Energy storage systems

There are two kinds of energy storage systems (ESSs) in the microgrid, energy-intensive and power-intensive ESSs, serving different purposes. The energy-intensive ESS components are the lithium-ion (Li-ion) battery (950 kWh), vanadium redox flow battery (VRB; 800 kWh), and lead carbon (PbC) battery (1100 kWh). The supercapacitor makes the power-intensive ESS. Lithium battery is under replacement. The energy-intensive ESS are used for energy arbitraging, renewable energy smoothing, peak load shifting, and dynamic disturbance rejection, while the power-intensive ESS primary serves the function of transient stability and improves the power quality of a microgrid system.

- *VRB*: The construction of a VRB cell consists of a positive and negative half-cell separated by a proton exchange membrane. Each half-cell has an inert electrode with an electrolyte flowing over it. During discharging, a reduction reaction of $V_5^+ + e^- \to V_4^+$ occurs at the positive electrode (cathode) which accepts electrons. An oxidation reaction, on the other hand, involves the chemical reaction of $V_2^+ \to V_3^+ + e^-$ taking place at the negative electrode (anode) which emits electrons as a result of the reaction. These reactions are reversed when the battery is charged. VRB is a kind of chemical battery which has the advantages of flexible design, fast charging/discharging rate, large charging/discharging power, safety, stable operation, easy maintenance, low self-discharging, and long life cycle, etc. One disadvantage with VRB is the large space requirement. The VRB in this microgrid, for instance, covers an area of 90 m^2.
- *Li-ion battery*: Li-ion batteries are becoming a dominant component in microgrid projects across the world. They are being preferred for their high cycle efficiency, low self-discharge, low maintenance requirements, and applicability for both short-term high-power and long-term low-power applications [21]. The relatively high price used to be the main issue hindering wider deployment of Li-ion batteries for microgrid applications. However, recent years have seen a declining price and hence better market in microgrid projects. The microgrid also consists of four sets of Li-ion battery−based ESSs (150 kWh, 500 kWh, 200 kWh, and 100 kWh).
- *PbC battery*: It is a relatively new type of battery which is a modification from the traditional lead−acid battery with the addition of carbon to the negative electrode plate. They are reported to be less costly compared to lithium-ion and flow batteries [22]. With the intention of making technical comparisons through extensive experimentation for the selection of batteries to future projects and investigating application-specific characteristics of the batteries, the microgrid includes all the three types of batteries. The PbC battery−based ESS in the microgrid has a total capacity of 1100 kWh constituting three individual sets rated 600 kWh, 400 kWh, and 100 kWh.
- *Supercapacitor*: a supercapacitor is a technology whose characteristic lies in the middle of a conventional capacitor and a battery. They have super high power density (kilowatt per kilogram) compared to batteries, and have moderate energy density and extremely high charging and discharge rate. In addition, supercapacitors have many advantages such as convenience for transient control, high conversion efficiency, and wide working temperature range. With the capability of supplying both active and

reactive power, it can be used to address transient events such as voltage sag and rise as well as frequency fluctuation.

There are five sets of supercapacitor in the microgrid each of which are rated 200 kW*10 s, 500 kW*20 s, 100 kW*15 s, 50 kW*20 s, and 50 kW*20 s. The microgrid is based on an integrated control strategy and platform that is designed to comprehensively control the energy flow between the generation and load ends in the industry park so as to:

- maximize the incorporation and utilization of renewable energy;
- provide power to users with optimal economy, safety, and reliability;
- ensure stable operation and reliable protection system;
- achieve grid friendly and user-friendly grid interconnection; and
- improve the comprehensive efficiency of energy utilization.

The main component of the monitoring and control system of the microgrid and the functional architecture of the microgrid control system are shown in Fig. 11.2 and Fig. 11.3, respectively. The integrated energy management platform considers various factors such as different energy supply end characteristics, different load types, different energy-consuming equipment, and power quality needs to optimize the management of energy flow between the power grid, the load, and the ESS. It provides diversified services for the grid side and user side.

The overall architecture of the monitoring and control system of the microgrid includes the following major software and hardware components:

- data acquisition and control terminal devices;
- communication system;
- cloud and data storage system; and
- application (software) platform.

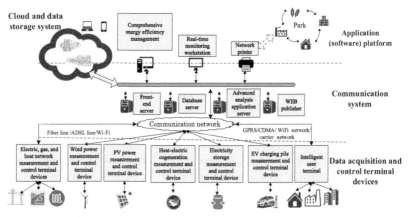

FIGURE 11.2 Main component of the monitoring and control system of the demonstrational microgrid.

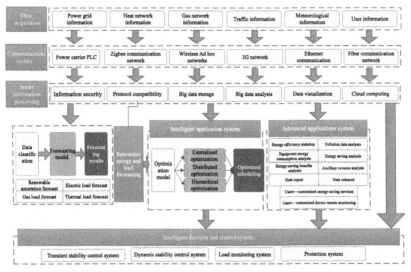

FIGURE 11.3   Functional architecture of the microgrid control system.

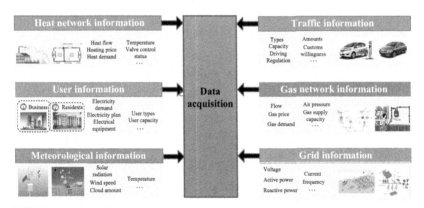

FIGURE 11.4   Smart data acquisition system.

11.2.3   Functional components

11.2.3.1   Data acquisition and control terminal devices

The data acquisition and control terminal device is mainly responsible for the data collection, classification, and processing of information of the microgrid's network which constitutes the electric network, natural gas network, and heat network. This includes the information of the power generation equipment, energy storage equipment, and the various types of loads in the microgrid (refer to Fig. 11.4). The smart data acquisition terminal devices refer to the smart measuring instruments and sensors that collect those data. The collected information is uploaded to the local server through the communication network. The integrated energy management platform processes the information, makes decisions, and interacts with network components in real time again through the communication network. The smart data acquisition and control terminal devices can also switch and/or continuously adjust the energy consumption by the user-end devices based on commands from the integrated energy management platform to achieve energy quality and energy-saving services.

IEC 61970/61968/61850, IEC 1888 and other standards can be used as reference standard for data acquisition and transmission. The open multiservice network can support a seamless end-to-end service, realize the interaction between users and the power grid, and allow the "plug and play" of various intelligent devices.

11.2.3.1.1 Electrical data acquisition terminal device

An electrical data acquisition terminal device is a terminal equipment integrating intelligent meters and various kinds of intelligent sensors, which is used to collect and preprocess the data on operational status and other parameters of the electrical equipment and electrical network of the microgrid. The remotely communicated module can collect the real-time data of each data acquisition device and respond to the command of the control system to control the local device. Its basic functions are:

- displaying the instantaneous power generation and consumption, control information, meter reading data, terminal parameters, maintenance, and other information;
- sampled measurement of voltage, current, power, power factor, active power, reactive power, etc.;
- alarming and warning functions of self-diagnostic alarm, factory settings and operation parameter change alarm, alarm of operational status of measurement unit, warning sign of measuring device abnormality, alarm of terminal electrical power failure;
- interfacing for RS485, RS232, infrared, general packet radio service (GPRS), code division multiple access (CDMA), public switched telephone network (PSTN), optical modules to achieve communication with electric energy meters, meter reading devices, sensors and main stations, and supports domestic and foreign metering protocols;
- storing historical daily data for over 60 days and historical monthly data for over 12 months.

11.2.3.1.2 Thermal data acquisition terminal device

Thermal quantity acquisition terminal device collects and preprocesses the real-time data of each equipment in the heating system in the microgrid. The remotely communicated module can collect real-time data from each data acquisition device, and receive and implement control instructions to control the local devices. It provides the following main functions:

- *Data collection function*: collect information such as temperature, pressure, flow rate, thermal power, etc.;
- *Monitoring functions*: display real-time collected heat-related information and equipment status;
- *Alarm function*: temperature and pressure over limit alarm, self-diagnostic alarm, factory settings and operation parameter change alarm, alarm for operational status of metering units, warning sign of metering device abnormality, alarm for terminal thermal power outage;
- *Data storage function*: capable of storing historical data for more than 60 days.

11.2.3.1.3   Gas data acquisition terminal device

The gas data acquisition terminal device can collect and preprocess the gas information of the devices in the gas supply network in real time, and upload related data through the communication network. Its basic functions are the following:

- *Metering function*: collect, analyze, and process the amount of gas;
- *Data transmission function*: information can be uploaded and received through the data transmission module;
- *Query function*: provide query function for real-time and historical data;
- *Alarm function*: alarm for pressure over limit, gas leakage, and other emergencies.

11.2.3.1.4   End-user smart terminal devices

End-user smart terminal devices integrate information collection and automatic control, are installed in the target buildings of various types (residential, commercial, industrial, agricultural, and other users) in the microgrid. The devices are used to monitor and collect data regarding energy-consuming equipment through smart meters and sensors, and upload the collected information to the integrated energy management platform. At the same time, the smart user terminal control device can be directly connected to various energy-using equipment in the building electrical lines, heating pipelines, and natural gas pipelines. Through the control instructions issued by the integrated energy management platform, the energy-consuming equipment in the building can be efficiently controlled, which can provide users with energy-saving services, improved quality of energy use and other functions. The human−machine interface of the smart terminal device provides functions such as display and query of the remaining energy, payment balance, supply interruption notification, payment notification, and liquidated damages reminder notification issued by the integrated energy management platform. It also allows querying of historical energy records, historical payment records, historical data statistics, energy prices, prepaid fees, etc. Users can also flexibly connect with smart terminal devices using personal computers and mobile phone through wireless networks to view and manage the status of energy-consuming devices anytime anywhere.

11.2.3.2 Communication system

The communication system that transfers data and commands between the local data acquisition and control units, the energy management system, the local server and cloud server is made of optical fiber, wireless, and other communication methods.

The microgrid is based on an integrated and high-speed two-way communication network (shown in Fig. 11.5) and involves advanced sensor and

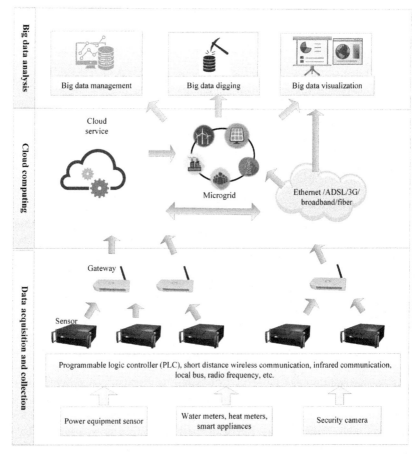

FIGURE 11.5 Communication system of the microgrid.

measurement technology, advanced equipment technology, advanced control method, and advanced decision support system technology. The broadband communication ensures real-time market transactions and seamless connection and real-time interaction between members of the power grid, natural gas network, and thermal network.

At present, there are various communication protocols applied in the fields of distribution system automation, smart power, and smart home. Fieldbus, Distributed Network Protocol 3 (DNP3), and other industrial communication protocols are still widely used in distribution network communication systems. However, in different smart power distribution demonstration projects, different protocols such as Ethernet, GPRS, Long Term Evolution (LTE), Wi_Fi_33, Zigbee, etc. have been applied to achieve reliable and fast communication. In addition, the next generation 5G wireless network technology is also actively developing. 5G is expected to be a new technology to provide all-round high-performance multidevice communication services. In future microgrids, all kinds of distributed devices may be connected to the system either through an industrial communication network or through an open network such as the Internet. To adapt to the coexistence of multiple communication protocols, the microgrid must be able to support the mutual conversion of the abovementioned various communication protocols to ensure system compatibility.

Considering the large number of distributed devices in the microgrid, especially the large number of controllable loads, it is obviously not realistic for the developer to plan and design the communication system manually. It is also required for the microgrid to be scalable with the possibility to undertake upgrading and expansion of the power distribution system. Thus user-side energy management system and the dispatching control arrangement shall have sufficient communication capabilities, and support the ability to plug and play distributed devices.

The prerequisite for realizing the coordination, optimization, and control of distributed devices is that various distributed devices are "visible" to the microgrid. Therefore the information interface needs to be able to identify the identity and type of distributed devices after they are connected, and this requires the support of standard communication protocols. At present, there are not that many standard communication protocols that support the access of distributed devices. Considering the large number of distributed devices, Internet Protocol version 6 (IPv6) can be considered as the network layer protocol. Since IPv6 can support the addressing of up to 3.4×10^{38} devices, it can provide a large enough address space for the microgrid. The use of the transmission control protocol (TCP) and internet protocol (IP) (TCP/IP) protocol suite also facilitates the interconnection with third-party systems such as smart homes. The choice of the protocol should also be based on the data communication requirements of the coordination and control of distributed equipment (what data need to be collected and transmitted, the amount of data, etc.), and the microgrid's requirements for data transmission reliability.

11.2.3.3 Central monitoring and control system

The SCADA (supervisory control and data acquisition) system with data collection, monitoring, and control functions is particularly important in microgrids. The traditional SCADA technology of the power grid has some limitations, and can no longer meet the requirements of the microgrid. More intelligent means must be provided to assist the operation and control of the smart microgrid.

The conferred microgrid is an integrated energy system that includes several energy systems, including electric power, natural gas, and heat which are independent and, at the same time, coupled to each other through various energy conversion units. The "smartness" of the system is based on the traditional SCADA concept that syndicates additional customized features that reflect the special characteristics and requirements of the network. The main purposes of the monitoring and control system of the microgrid are to provide the online information of the equipment and system status, to improve the system operation efficiency and ensure operation safety, and to assist the analysis of incidents and their causes. According to the direction of energy flow, the system consists of two parts: control of energy generation and control of energy consumption.

On the energy supply side, renewable distributed generation systems and combined heat and power systems are integrated and controlled in a centralized manner. A single device will have a significant impact on the operation of the system. The measurement accuracy will directly determine the control decisions. Thus reliability and accuracy of measurement and communication should be emphasized. The smart control system collects information from the smart meters at the bottom layer, selects essential parameters to monitor the operation status of DG sources such as wind turbines and PV systems, receives weather information from the weather station, collects parameters of the ESS, and considers the gas and heating networks' operational status. The collection and monitoring of the microgrid operation details is achieved through a three-tier structure of *terminal equipment—concentrator—data center*. The energy efficiency management platform uses appropriate functions to accurately determine the operating status of the equipment based on the preset information. At the same time, the energy management platform will make overall calculations based on the parameters of multiple elements of the microgrid to ensure the normal operation of a single device while also taking into account the cooperation between the individual units to achieve efficient system operation.

On the load side, SCADA's monitoring and control targets are the energy storage devices, electrical loads such as lighting, air-conditioning, electric vehicles, and other loads. It is characterized by the large amount of data, scattering of constituting units, small dimension of single parameters, and diverse data types. The concentrator partitions the data and assigns addresses to the individual equipment data while collecting equipment parameters to achieve decentralized monitoring, centralized management, and coordination. Because of the large number of devices and the need for sufficient address space, IPv6 or a similar communications protocol can be considered for addressing and grouping. Small-dimension data require that the measuring devices have sufficient accuracy and reliable information communication system.

The status information collected at both supply and consumption ends of the network will be merged into the real-time data and stored in the integrated control strategy platform and the cloud through a reliable communication network to be used by the system for monitoring, prediction, and optimization function modules. The preset information and related programs will be used to determine the operating status of the equipment and obtain appropriate scheduling and operating instructions to realize the monitoring and control of the equipment and ensure the safe and efficient operation of the system.

Advanced cloud computing and other methods are used to process the large-scale data collected from the different components of the microgrid. The top level of the integrated control strategy and platform for the microgrid is the user interface application software, which realizes the functions of real-time monitoring of system status, generation and load forecasting, optimal scheduling, demand side management, energy efficiency analysis, etc.

The microgrid uses an open source management software platform built by using smart communication, cloud computing, and big data analysis technologies. It strives to achieve the data collection, management, analysis, and user interaction functions of the microgrid, and supports the power trading, carbon trading, distributed power supply and electric vehicle charging facilities monitoring and maintenance, energy-saving services, interactive electricity consumption, demand response, and other new types of business.

11.2.3.4   Cloud computing technology

The monitoring and control in microgrids involve a large amount of information (such as equipment status, voltage, current, frequency, etc.) from customers, smart buildings, smart homes, and smart appliances as well as the distributed generation systems in the microgrid. Cloud computing is a good option for the processing, computing, and storage of such a large amount of data for monitoring and controlling activities [23].

Cloud computing is a mode of accessing and improving the availability of computing resources (including networks, servers, storage, applications, services, etc.) anytime, anywhere, on demand, and conveniently over the network. The energy management system of a microgrid can be hosted in the cloud which allows to gather the advantages of cloud computing in terms of higher flexibility, storage capacity, and processing capacity [23]. The cloud computing environment enables a more efficient utilization of the available instrumentation and communication infrastructure. It is especially relevant in case of a group of microgrids being operated by one operator. Such a cloud framework based on the concept of Internet of Things is applied for the distributed optimization of networked microgrids in Ref. [24], and the configuration is labeled as "Internet of Microgrids." It allows economical and efficient implementation of different power management strategies on microgrids that are deployed over a wide area [25].

There are modified versions of the cloud computing concept named *fog computing* and *edge computing* that push the computation to the lower layers in the communication hierarchy. In edge computing, the data analytics and computation tasks happen directly on the devices to which the sensors are connected, while in fog computing, the data processing and execution of applications are performed at the fog nodes which are a collection of devices and gateways. Cognitive computing that represents self-learning systems that work on deep learning algorithms and big data analytics to be applied in automated information technology models and end-user devices is an interesting option of edge computing for application in control and protection of microgrids.

Cloud computing faces constraint related to communication delay due to servers being located far from clients, and the large amount of data traffic in the core network. Real-time applications are mostly time sensitive and fog computing and edge computing can better achieve their delay requirements. Fog computing and edge computing follow a distributed approach and have advantages of lower delay, lower noise or interference, and higher security. Thus fog computing and edge computing environments are better suited for decentralized microgrid control.

In the demonstrational microgrid, some functions have been combined with cloud computing, such as computing related to forecasting of wind and PV output power and other applications by leasing the Alibaba cloud platform to store large amounts of data using distributed storage.

In the future, it is planned to incorporate more features of the Internet of Energy concept, especially on exploring the marketing aspect through cloud computing devices, including e-commerce platform technology, and corresponding marketing models for Internet marketing. Internet of Energy supports B2B (business to business), B2C (business to consumer), C2C (customer to consumer), and other business models that make use of the strong interconnection and interoperability of the Internet to support the trading activities of various market entities at anytime and anywhere, such as power generators (including distributed power and microgrid operators), network operators, users, wholesale or retail power selling companies, etc.

11.2.3.5 Renewable generation and load forecasting

The module for generation and load forecasting is used to forecast the demand for supply of electricity, heat, cold, and gas in the microgrid based on meteorological information, historical data, and real-time data. Short-term forecasting is of special importance for optimal operation of microgrids. The forecasting module is a component of the central energy management system of the microgrid. It consists of the following functions:

- Short-term wind power forecasting (STWF)
- Short-term PV power forecasting (STPF)
- Short-term load forecasting (STLF)

The forecasting models cover a time horizon of 24 hours (a day) with an hourly time step and are updated four times a day.

The forecasting models are statistical models that are trained using an extensive amount of historical data. Once the models are trained and validated, they may also need real-time data related to weather condition and status of the network or network components to do the forecasting. The main input data for the forecasting modules are summarized in Table 11.1.

Different methods are applied to forecast the generation of PV and wind systems and the load demand by the microgrid. Most of the techniques are based on a variety of machine learning or artificial intelligence techniques. The methods in Table 11.2 have been tried and applied in the microgrid.

11.2.3.6 Optimal dispatching

The dispatching module is mainly responsible for comprehensive optimal dispatching of the DER in a multienergy microgrid. This module takes generation and load forecasting information and equipment operational status as input data, constructs a corresponding objective function according to user

TABLE 11.1 Input data for renewable generation and load forecasting.

	STWF	STPF	STLF
Input data	• Wind speed • Air density • Air pressure • Air temperature • Humidity • Historical wind power[a]	• Shortwave solar radiation • Surface/air temperature • Cloud cover (low, middle, and high) • Humidity • Surface pressure • Hour of day • Day of year • Historical PV power[a]	• Air temperature • Hour of day • Type of day • Power tariff • Historical load data

[a]Historical data are used for training the forecasting models.

TABLE 11.2 Forecasting methods applied for renewable generation and load forecasting.

	STWF	STPF	STLF
Forecasting methods	• ANFIS [6] • GA-PSO-ANFIS [26] • ANNs [9] • HHT-GA-ANN [5] • ANN-ANFIS-SVM [27]	• Wavelet-PSO-SVM [8] • ANN [28] • GA-PSO-ANFIS [12]	• Wavelet-NARX [29] • EMD-PSO-ANFIS [30]

ANFIS, Adaptive network–based fuzzy inference system; *ANN*, artificial neural network; *EMD*, empirical mode decomposition; *GA*, genetic algorithm; *HHT*, Hilbert–Huang transform; *NARX*, Nonlinear AutoRegressive with eXogenous-based neural network; *PSO*, particle swarm optimization; *SVM*, support vector machine.

demand or the purpose of system optimization of the microgrid, chooses an appropriate optimization algorithm to solve, generates optimal control strategy of multiple energy resources, and completes the optimal control of generation, storage, and load equipment in the microgrid.

The following machine learning–based methods have been applied for effective use in optimal dispatching of the microgrid testbed.

• Regrouping particle swarm optimization [10]
• Modified particle swarm optimization [14]
• Genetic algorithm optimization [7]
• Mixed-integer linear programming [13]

11.2.3.7 Dynamic and transient disturbance control systems

Microgrids are susceptible to instability due to their low system inertia, especially when operating in island mode. To make up for this issue dynamic disturbance control systems and transient control systems are devised and applied for the respective types of disturbances.

According to IEC TS 628989-3-1, dynamic disturbances include disconnection of a DG unit, generation from DGs exceeding power limits, poor load sharing between DGs, incapability of the grid-forming generator, and unplanned switching of loads [2]. There could also be dynamic disturbances caused by poorly designed or poorly tuned control systems. The demonstrational microgrid, similar to most of the microgrids, involves a large proportion of renewable energy sources which are usually fluctuating and intermittent in nature. When the generation from such DG units reaches a high proportion, the system may approach its stability limit so that further increases in generation may lead to the loss of system stability. Nonlinear loads and planned islanding are also other causes of dynamic disturbance. To suppress disturbances caused by such events in the microgrid, a dynamic disturbance control system is developed. This control system operates in cooperation with energy-intensive ESSs in the form of VRB, PbC, and Li-ion batteries. Detailed discussion of dynamic disturbance control system design is provided in Chapter 7, Dynamic Control of Microgrids, of this book.

Transient disturbances are disturbances of large magnitude and short duration, taking a time step of 50 ms or less [2]. The common causes of transient disturbances are transfer between grid-connected and island modes, switching of large load and big DG units, and fault clearing. A transient disturbance control system that makes use of a power-intensive ESS (a supercapacitor) is employed in the microgrid testbed to handle the impact of those types of events. The supercapacitor effectively operated by the transient control system provides a very fast power injection or absorption capability to cover temporary active or reactive power mismatches. More detail on the control system design and frequency and voltage stabilization is provided in Chapter 8, Transient Control of Microgrids.

11.2.3.8 Protection system

Protection of a microgrid is significantly different from that of the conventional grid. As discussed earlier in this chapter, the demonstrational microgrid is dominantly composed of DG units and ESSs interfaced through electronic converters to the microgrid network. Both wind turbines and the microturbines are naturally AC sources. However, the turbines are direct drive type which means their output is of variable speed/frequency. The microturbines, on the other hand have output power frequency much higher than the system nominal frequency of 50 Hz. Thus back to back AC−DC−AC converters are used to interface the wind turbines and the microturbines to the electrical network of the microgrid. The PV and ESSs are naturally DC, which makes the need of converters inevitable. The existence of the converter interface is of a great deal in respect to the fault detection and protection system of the microgrid. That is due to the fault current magnitude being low (not more than twice the rated current) due to the current handling capacity of the electronic switches in the converters.

The existence of multiple DG sources and the complicated architecture of the microgrid, shown in Fig. 11.1, also make the conventional protection system, such as overcurrent protection, less effective. Thus advanced protection schemes are developed and implemented in this microgrid. This includes total harmonic distortion (THD)−based fault detection and location, machine learning−based technique (augmented with wavelet and Fourier transforms) [11], communication-supported (Park-transformed disturbance voltage-based) strategy [31], and others.

As there are similarities in the initial characteristics of faults and the other types of disturbances, the protection system is coordinated with the above-discussed control systems to avoid false tripping of protection devices.

11.2.4 Advanced functions

11.2.4.1 Energy conservation and efficiency services

In the smart microgrid testbed, an advanced energy efficiency service is provided by the comprehensive control strategy and platform of the microgrid. The energy efficiency function is implemented through the application function of comprehensive energy dispatching, using real-time collection and monitoring of various energy consumption and energy efficiency data of the energy generating and consuming units. The procedure involves the closed-loop control of "detection−control−execution−feedback." The approach used in the microgrid makes full use of the existing advanced energy efficiency management theory and the matured industry experience.

The following procedure is followed in devising the energy conservation and efficiency service:

1. collecting extensive data on energy consumption and emission;
2. summarizing the energy rules by analyzing the key performance indicators and clustering of energy consumption characteristics of the different loads; and
3. providing guidance and consultation on energy conservation and efficiency to ultimately achieve the purpose of energy conservation and emission reduction, and help enterprises harvest economic benefits.

The following data from the users and cluster of users in the microgrid are collected, classified, and analyzed for the energy efficiency and conservation scheme:

- electricity data for lighting equipment;
- electricity data of air-conditioning equipment;
- power data for office equipment; and
- data on electricity, gas, and heat consumption of industrial loads.

Environmental impact analysis of the microgrid is particularly important to cope with the increasingly serious environmental problems. The pollutant emission analysis function makes statistical analysis on the direct discharges of gases (CO_2, SO_2, etc.), sewage, and noise from the facilities in the microgrid and secondary or indirect discharges due to generation of electricity. It can also make horizontal comparisons between the emission of different departments, buildings, and business units, and generate environmental analysis reports.

The environmental emission analysis requires data about direct and indirect greenhouse gases, sewage, and noise emissions from electrical power generation equipment, heating equipment, and gas supply equipment.

Energy efficiency evaluation and analysis is an indispensable feature of establishing an efficient and clean microgrid. This function first establishes an energy efficiency evaluation index, then analyzes the comprehensive energy efficiency of the equipment, obtains the cost of multienergy comprehensive utilization, and predicts the future energy utilization cost. The latter step is evaluation of the energy security status. The results of energy efficiency analysis in this area where the microgrid is located and the industry average data are benchmarked to assess the potential of improving the energy efficiency of the microgrid. The steps are elaborated below.

1. *Establishing energy efficiency evaluation index*: With greenhouse gas emissions and energy consumption per unit output value as indicators, an index to measure the energy consumption within the microgrid is established.
2. *Comprehensive analysis of equipment efficiency*: Statistical analysis is made on the energy utilization of various devices, revealing high energy consumption and low efficiency devices, and generating efficiency analysis results.
3. *Comprehensive cost analysis and forecast*: This step classifies and analyzes the supply costs of various types of energy in the microgrid, considers the energy consumption and pollution emission, obtains the comprehensive power supply costs, and predicts the future supply costs according to the law of energy use.
4. *Energy safety analysis*: Through a comprehensive analysis of the reliability of energy supply and environmental emissions, the energy security status of the microgrid is obtained, which provides a basis for improving the conditions of energy supply.
5. *Industry comparison analysis*: By comparing the results of the analysis with the average level in the industry, the energy efficiency level of the microgrid can be obtained, and the potential to improve the overall energy utilization can be explored. Energy efficiency benchmarks and comparisons can be used to assess the microgrid's performance relative to industry participants or its previous performance, or to assess improvements in energy efficiency resulting from the implementation of energy efficiency measures. Benchmarking the energy efficiency of facilities enables energy auditors and managers to identify best practices that can be replicated.

The following can be used as benchmarks for energy efficiency analysis of the microgrid:

1. *Previous energy efficiency values*: comparing current and historical performance;

2. *Average industry performance*: compared to established performance measures, such as recognized average performance in the same type of organization;
3. *Best in the same industry*: the best in the industry, not the average;
4. *Best practices*: qualitatively compare with certain practices or technologies considered to be the best in the industry.

Energy efficiency diagnosis and the optimization module diagnoses and analyzes the unreasonable use of energy through a comprehensive assessment of the use of energy within the microgrid; optimizes the use of energy according to specific needs; and generates optimized energy utilization programs and equipment renovation plans.

The energy efficiency diagnosis service analyzes the following aspects:

1. *Energy demand*: Energy efficiency diagnostics first need to analyze the demand for various types of energy in the case of classified energy consumption statistics. That includes the electricity power demand, heat demand, and demand for natural gas.
2. *Opportunities for energy efficiency improvement*: Through the analysis of energy demand, the potential opportunities to improve energy utilization efficiency are found. In the microgrid, opportunities for energy efficiency improvement can be found from the following points to identify measures to improve the energy efficiency of lighting fixtures, air-conditioning systems, office equipment, and power equipment.
3. *Cost−benefit analysis of energy efficiency improvement*: After identifying the measures applicable to the improvement of the energy efficiency of the microgrid, the economic feasibility analysis of the improvement measures, the so-called "cost−benefit analysis," will be carried out to make specific recommendations for the implementation. It can be analyzed from the aspects of life cycle cost, net present value, internal rate of return, energy saving per investment ratio, and return period.

In the microgrid, the energy efficiency data statistics report can show the results of energy efficiency analysis in two ways: one is to display the analysis results in a fixed report format or to print directly, the other is to generate a customized statistics report according to user settings. The energy efficiency data statistics report includes the following:

- Operation of the equipment classified by energy sources: electricity, heat, gas, etc.
- Flow rate and flow distribution of various types of energy
- Efficiency of various types of equipment: lighting, air-conditioning, heating, power supply, air supply, office, power equipment, etc.
- Specific energy consumption
- Energy bills and analysis results
- Analysis results of energy use and generation models

- Comparing analysis results
- Optimized measures for energy conservation

11.2.4.2   Customized services

Customized services include two aspects, namely power customization and customized service. Power customization refers to the ability of users to obtain a specific quality of power. Customized service refers to the enrichment and improvement of various value-added high-quality service items. To provide unique customized services to better meet customer needs and improve customer satisfaction, the demonstrational microgrid uses the online-to-offline service mode, which enables users to make appointments to their homes, place orders online, and settle transactions through the cloud platform.

The customized services include the following:

- Maintenance services
- Deep data analysis
- Power quality service
- Equipment physical examination service
- Data copy service
- Special electricity services
- Fault diagnosis
- Energy-saving services
- Energy storage service
- Remote monitoring service
- Temperature customization service
- Interactive electricity sales service

11.2.4.3   Grid auxiliary services

While the primary objective is to meet the diverse energy needs of users in the microgrid, the microgrid is also capable of providing auxiliary services for the distribution network of the utility grid.

The following grid support functions can be provided by the microgrid:

- Frequency modulation service
- Peak shaving service
- Standby services
- Reactive power regulation
- Pressure regulation service (in nonelectric networks), etc.

11.2.5   Operational results

11.2.5.1   Fault protection

As discussed in Section 11.2.3.8, advanced protection schemes have been researched, developed, and implemented in this microgrid. Two of the

TABLE 11.3 Parameters after faults at different point in the microgrid.

Parameter	Sate	$THDU_a$ (%)	$THDU_b$ (%)	$THDU_c$ (%)	$THDI_a$ (%)	$THDI_b$ (%)	$THDI_c$ (%)	U_d	U_q	U_{a0}	U_{a2}	I_{a0}	I_{a2}
KM1[a]	1	1.340	1.200	1.210	1.070	1.070	1.080	269.23	29.810	0.095	0.828	0.123	0.147
KM2[b]	1	1.380	1.360	1.350	1.200	1.210	1.210	268.78	30.289	0.100	0.846	0.093	0.097
KM3[c]	1	1.360	1.230	1.210	2.570	2.600	2.650	269.26	0.419	0.108	0.754	0.099	0.276

[a]KM1, single line to ground fault at the branch connected to breaker 440 in Fig. 11.1.
[b]KM2, single line to ground fault at the branch connected to breaker 445 in Fig. 11.1.
[c]KM3, single line to ground fault at the branch connected to breaker 446 in Fig. 11.1.

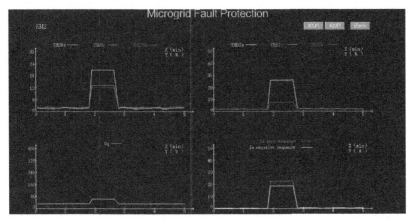

FIGURE 11.6 THD_U, THD_I, $U_{q.dist}$, and sequence currents for single line to ground fault.

techniques used for the detection and location of faults in the microgrid are the THD-based and Park transformation—based techniques, whose results are shown in Table 11.3 and Fig. 11.6. Findings from the use of other techniques such as machine learning— and communication-based methods are reported in Refs. [11] and [31], respectively.

11.2.5.2 Transient and dynamic disturbance stability

The transient disturbance control system was tested for its performance in handling a large motor starting (a typical cause of transient disturbance) in the microgrid. The results, shown in Fig. 11.7B (compared to Fig. 11.7A) proved that the control system was able to suppress the resulting disturbance within a cycle and keep the system stable.

The dynamic disturbance control system was also tested by increasing the PV generation in the microgrid. The system was operating stably without the dynamic disturbance control system until a third PV array was brought into the system. Then some disturbance started in the form of oscillation, as shown in Fig. 11.8. The control system was started a moment later and thus stabilized the system by suppressing the disturbance, as seen in the right half of the figure.

11.2.5.3 Renewable generation and load forecasting

The different artificial intelligence—based techniques listed in Table 11.2 were applied in the microgrid with varying results. Table 11.4 summarizes the forecasting accuracies achieved through the different methods. It is worth noting that the differences in accuracy measuring metrics, time period, and data used for training and testing mean it will not be correct to do a straight-forward comparison of the methods. Regardless of the slight variations in the

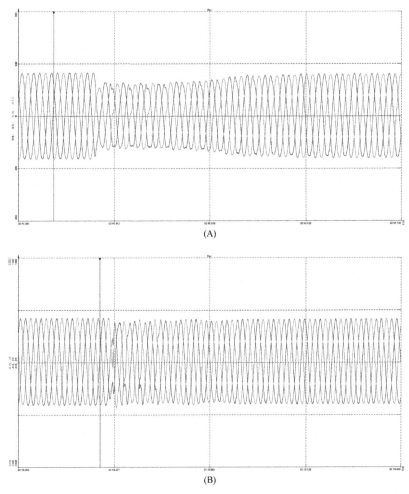

(A)

(B)

FIGURE 11.7 Voltage waveform during a transient disturbance event of large motor starting (A) without and (B) with transient disturbance control device.

outcomes of the methods, all of the applied techniques were good enough to achieve a significant level of accuracy.

11.3 Industrial microgrid

11.3.1 Introduction

The microgrid is built for an industry engaged in the production and sales of chemical products. The industry park has three production workshops, two office buildings, and two production lines that manufacture stretched nylon

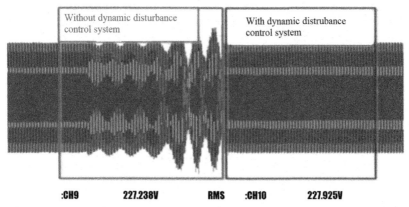

| :CH9 | 227.238V | RMS | :CH10 | 227.925V |

FIGURE 11.8   Dynamic disturbance control system field testing results (voltage waveform).

membranes simultaneously. The park requires approximately 11 MW of power supply out of which the microgrid accounts for about 82%. After the completion of the second phase of the project, there will be a total demand of about 20 MW.

The total microgrid project investment is 166 million Yuan, of which 80 million Yuan has been invested in the first phase (including the low-voltage energy storage and protection control systems reserved for the second phase of the project).

The project uses transient disturbance control, dynamic disturbance control and protection technologies along with energy analysis and prediction technology. The renewable energy penetration rate of the project is higher than 70%, and the reliability rate is higher than 99.99%. Two modes of grid-connected and island operation can be realized in the project. The switching time delay of island-to-grid-connected and grid-connected to island is less than 10 ms. The project has the ability of balancing the autonomous operation of power supply and demand and black start capability, and can guarantee continuous power supply to critical loads for a minimum of 2 hours.

In this microgrid project, many new technologies and equipment are adopted. Advanced microgrid control and protection technology, load and generation forecasting, power trading scheme, and optimal economic dispatching model are used in the project. New flexible PV support, double-sided PV panel, aluminum alloy cable, and other new technologies are also applied. This project is the first domestic innovation in accordance with IEC microgrid standards with state-owned intellectual property rights.

This microgrid is a bold move by a state-owned power company, a private investor, and a developer to jointly participate in microgrid business. It is a model of direct partaking of capital, technology, and market. Through joint efforts of all parties, the project explored the application of advanced control technology of distributed energy and smart microgrid according to

TABLE 11.4 Forecasting accuracies from different methods applied for short-term renewable generation and load forecasting.

Forecasting method	Accuracy metrics
STWF	
ANFIS [6]	MAE = 60.84 kW MAPE = 6.8%
GA-PSO-ANFIS [26]	MAE = 45.73 kW NMAE = 1.83 MAPE = 6.64%
ANNs [9]	MAE = 42.2 kW RMSE = 60.53 kW MAPE = 13.84%
HHT-GA-ANN [5]	NMAE = 0.33 MAPE = 7.62%
ANN-ANFIS-SVM [27]	MAE = 28.31 kW NMSE = 1.03
STPF	
Wavelet-PSO-SVM [8]	MAPE = 4.22% NMAE = 0.40%
ANN [28]	MAE = 1.393 MSE = 8.203
GA-PSO-ANFIS [12]	RMSE = 7.89 MAE = 3.98 NMAE = 5.31
STLF	
Wavelet-NARX [29]	MAE = 7.7834 GoF = 0.8310
EMD-PSO-ANFIS [30]	RMSE = 89.11 MAE = 60.97 MAPE = 9.86 NMAE = 5.32

ANFIS, Adaptive network–based fuzzy inference system; *ANN*, artificial neural network; *EMD*, empirical mode decomposition; *GA*, genetic algorithm; *GoF*, goodness of fit; *HHT*, Hilbert–Huang transform; *MAE*, mean absolute error; *MAPE*, mean absolute percentage error; *MSE*, mean square error; *NARX*, Nonlinear AutoRegressive with eXogenous-based neural network; *NMAE*, normalized mean absolute error; *NMSE*, normalized mean squared error; *PSO*, particle swarm optimization; *RMSE*, root mean square error; *SVM*, support vector machine.

local conditions, mixed ownership of microgrid business model, and microgrid power sales business. It formed a microgrid model with easy to copy characteristics, which can be gradually promoted to more places.

The main electric loads of the industrial park are production equipment loads. The main production equipment in the production line is electronic heating and motor equipment. Office loads are mostly computer and lighting systems which consume relatively less power. The starting current of motor loads is generally 5−10 times the rated current. Thus frequent startup and shut-down have a great impact on the grid. In addition, most of the electrical equipment in the power supply system are inductive motors, which consume a lot of reactive power. As a result, power factor correction will be reduced. The actual demand of the industry production line requires that the multienergy microgrid power system should quickly respond to power mutation, smooth output, and can also carry out reactive power compensation and harmonic control.

The industrial microgrid thus provides power supply for production and offices in the park by a multienergy microgrid system. All loads in the park are directly connected to a 10 kV bus through the 0.4/10 kV isolating trans-formers. PV, wind turbine, auxiliary power, and vanadium redox flow energy storage provide smooth and stable power supply for loads through advanced and intelligent microgrid control technology.

11.3.2 Architecture and components

The industrial microgrid is initially planned to be composed of 10 MW PV, 8 MW low-speed wind turbine, 1 MW CCHP power supply, 10 MWh VRB ESS, 700 kW*10 s supercapacitor, and 50 electric vehicle charging piles and other loads. The first phase of the project has been completed with 4.2 MW PV farm, 2 MW wind turbine, 4 MWh VRB storage, 200 kW*10 s superca-pacitor, and 6 electric vehicle charging piles. The second phase is planned to add a 6 MW wind turbine, a 5.8 MW PV, 1 MW CCHP, 6 MWh VRB type energy storage, and 500 kW*10 s supercapacitor battery. This would take the total capacity of the microgrid to be about 19 MW and will allow the micro-grid to provide backup power to the surrounding area. Advanced microgrid control technology is deployed in the system to ensure the power quality, harmonic distortion, and other performance parameters meet the standard requirements.

The main electrical layout of the microgrid is illustrated in Fig. 11.9 and it consists of the following main components.

11.3.2.1 Photovoltaic system

A total of 6 MWp PV system is installed in the project, in which 500 kWp PV panels are connected into the junction box and then connected to the 10 kV bus through the centralized inverter.

11.3.2.2 Wind turbine

The project is equipped with a 2.0 MW wind turbine with an output voltage of 690 V. The outgoing line is then connected to the 10 kV distribution bus with a 0.69/10 kV transformer, and networked with the ESS.

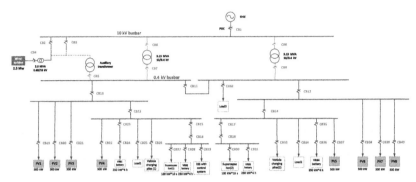

FIGURE 11.9 Electrical layout of the industrial microgrid.

11.3.2.3 Energy storage system

The ESS of the microgrid, at its current state, constitutes 4 MWh VRB-based energy-intensive energy storage and 200 kW*10 s supercapacitor-based power-intensive energy storage. The VRB and the supercapacitor are parallelly connected to the 10 kV bus through 0.4/10 kV isolation transformers and converters. The dynamic and the transient control systems are applied in the ESSs to regulate the power flow and suppress the disturbances.

11.3.3 Core technologies

To address the issues associated with the microgrid, core technologies are applied in the industrial microgrid project. These include protection system, transient and dynamic stability control technologies, renewable generation and load forecasting system, and smart communication technology.

The microgrid protection technology in the project deployed, through the functions of accurate fault location, fault type determination, and rapid fault isolation, provides testing and protection services for all components of the microgrid: distributed generation, point of connection (POC), lines, transformers, etc., to ensure the safe and stable operation of the microgrid.

The transient stability control systems can enhance the penetration of new energy, ensure that the microgrid will not suffer from system oscillation instability due to the excessive new energy, and ensure the reliable and high-quality power consumption of users under nonfault conditions, and also reflects the concept of energy conservation and carbon emission reduction. This control system can realize the smooth switching from grid-connected to island and vice versa, and the transient process of smooth switching can be controlled within 50 ms.

When the microgrid is required to switch to the island state, the energy storage with the dynamic stability control system responds in time to ensure the seamless switching of the system and ensure the continuous power supply of important loads in the island operation mode. It can not only ensure the continuous and reliable power supply to the load, but also realize the friendly coordination between microgrid and the grid. The dynamic process of smooth switching can be controlled within 2 s.

Renewable generation and load forecasting technology performs the generation and load predication in different time horizons so as to respond to the load monitoring system as per the user's need. It can also supply or store surplus in the storage systems while providing stable operation. The load side demand response is realized to achieve the economic and coordinated operation of the load.

Smart communication technology functions to integrate, cooperate, and coordinate the local controllers, the components and the central controller in the microgrid network to provide fast information flow in which the control requirements of the microgrid in an appropriate time range should be met.

11.3.4   Functional components

11.3.4.1   Renewable generation and load forecasting

The functions of generation and load forecasting are performing very well in the project. The accuracy of forecasting, including generation forecasting, load forecasting, faults, and other uncertain factors, directly affects dispatching scheme. It also affects evaluation of the impact of uncertainty on the optimal operation results of the microgrid, and take corresponding measures to improve the expected effect of system optimal operation.

Power generation prediction is mainly aimed at short-term and medium-term output energy prediction of intermittent renewable energy, which is realized by forecasting wind speed, solar intensity, and environmental temperature. The commonly used forecasting methods include a numerical weather forecasting model and historical data.

Load forecasting includes short-term forecasting of various types of loads, including cooling, heating, and electricity loads in the microgrid network. Considering the demand side response, electric vehicle charging, and the delay characteristics of cooling and heating load, the difficulty of load forecasting will be further increased.

11.3.4.2   Scheduling method

The microgrid scheduling system includes heuristic methods of optimization. The optimization strategy determines the system operation scheme as per the optimization objective, and the heuristic strategy generally determines the scheduling scheme as per the given scheduling logic for the limited system operation mode. Generally, the optimization strategy can obtain a better optimization effect than heuristic strategy, but in the actual project implementation, due to the existence of various uncertain factors, the optimization effect may be weakened. As the microgrid scheduling optimization problem is a complex practical engineering problem, the stochastic and multiobjective characteristics of the system should be taken into account when formulating the dispatching strategy, especially for the cooling, heating, and power cogeneration systems, the influence of the delay of the strategy implementation on the optimization results should be considered.

Hence, in the industrial microgrid, the monitoring strategy combining optimization and heuristic strategies is adopted. According to the load characteristics of the project, the optimal dispatching strategy is customized to make the dispatching strategy more reasonable.

The microgrid monitoring and control system contains an intelligent load management system. The main functions of load management system are real-time data acquisition, load prediction, power quality analysis, and data communication.

11.3.4.3 Transient and dynamic stability control systems

The transient stability control system mainly completes the functions including real-time data acquisition, load multitime scale prediction, fast fault protection of load, real-time power quality analysis, real-time power quality regulation, millisecond level accurate suppression of microgrid voltage and frequency disturbances, and fast power balance to ensure system stability when renewable energy generation gap fluctuations and load power change occur. Whereas the dynamic stability control system mainly provides the tasks of voltage and frequency dynamic stability regulation, voltage harmonic stability regulation, and reactive power compensation.

11.3.4.4 Demand side management

One of the main functions of the smart microgrid is to support the interaction between users and microgrid operator. Users can adjust their own energy usage strategies according to the current energy supply situation, and the microgrid operator can formulate the energy supply mode according to all users' energy strategies to meet users' needs. It can also support the interaction between users and microgrid operator. The smart microgrid will collect the energy use data of different users, calculate the corresponding energy use rules, formulate reasonable energy use strategies, and feed back to users as an optimal use strategy for users to choose. Therefore the smart microgrid control system should have the ability to receive and process user requests. And it can in a timely manner and accurately reflect the current energy supply situation with energy prices and other information feedback to users, involving interaction and demand management and other functional modules.

11.3.4.5 Communication network

The communication system of the microgrid adopts a variety of communication modes, such as optical fiber, broadband power line, ZigBee, Wi-Fi_33, Ethernet, and RS485, to realize data exchange and information sharing among microgrid devices in the park, and send the information to the micro-grid control center for monitoring.

A set of microgrid monitoring systems is built in the park to comprehensively monitor and control the generation side and load side equipment so as to achieve the purpose of energy saving, consumption reduction, and power sales. This all is supported by the communication system.

In addition, the monitoring system supports remote access and data interaction with microgrid control center which can be realized through a cloud platform.

In the implementation process of the microgrid, smart meters with multiple communication functions are installed for much equipment in the park. Many environmental monitoring instruments are installed in the production workshop such as temperature and humidity sensor, noise detection sensor, sunshine intensity detection instrument, etc. Through these communication function instruments or meters, the microgrid monitoring system can monitor the operation status, energy consumption, environmental conditions, and other information of the production line in real time. It is of great significance to reasonably arrange the production line, find the abnormal operation of the production line in time, and reduce the energy consumption.

The communication network is very important for a smart microgrid, and the goal is to maintain the availability, scalability, reliability, survivability, and security of the communication network in real time. The development of targeted management protocol is of great significance for the operation of smart microgrids. The management functions should include access identification of the communication network, remote deployment of management strategy, exception handling and repair, and setting and management of log files.

The following measures are taken for the practical application of the microgrid:

- The multifunctional ammeter with RS485 communication capability is installed on the load side.
- The multirate multifunctional meter and metering cabinet with RS485 communication capability are installed locally at the power generation side.
- Load switch and circuit breaker that can be controlled remotely.

11.3.4.6 Monitoring and control

The monitoring and control system of the microgrid project is the core control system for stable operation. The specific schemes include:

1. *Monitoring*: real-time acquisition of data (voltage, current, and operation status) of PV system, wind turbine, ESS, and the grid.
2. *Coordinated control*: the main control system coordinates the control of the PV system, wind turbine, ESS, and loads to achieve power balance and optimal operation.

3. *Grid-connected operation strategy*: the locally generated power will be primarily consumed by the industrial park and only the surplus power is sold to the grid. According to the real-time price of power grid, charge and discharge management of ESS is carried out, and the benefit of the price difference is brought by using peak−trough electricity price. During the off-peak period, the main control system operates the charging of the ESS, while it provides the discharge of the ESS during the peak period of the grid price.

When the microgrid monitoring and control system detects the grid fault, it will actively disconnect the connection with the grid, carry out the smooth transition from grid-connected to island, and improve the reliability of power supply. After the detection of grid fault recovery, the main control system will carry out the smooth transition from island to grid connection mode.

When the output active power of the new energy is greater than the active power required by the loads, the charging margin of the energy storage battery is detected. If the battery is not fully charged, the microgrid dynamic stability control system is controlled to charge to reduce the output active power of the grid connection point. If the battery has been fully charged, the residual power will be connected to the grid to realize the residual power consumption and effectively reduce the wind and PV power loss.

When the active power output of the new energy is less than the active power required by the loads, the ESS is detected if the capacity of the battery has discharge margin. If there is discharge margin, the dynamic stability control system of microgrid is controlled to discharge and maintain the power stability of the grid connection point. If the battery has no discharge margin, the relevant action will not be carried out, and the power difference will be supplemented from the grid.

The microgrid monitoring and control system realizes the power control of the whole microgrid system. So, it needs to obtain the real-time power status of each equipment in the microgrid, and at the same time accepts the power scheduling instructions, and adjusts the grid-connected power according to the power dispatching value. Therefore, to realize the above functions, it is necessary to collect the information of the relevant equipment and carry out the necessary communication control. The equipment to be measured and controlled is as follows:

- It communicates with the microgrid dynamic stability control system to obtain the current working state, charge and discharge action and power control.
- It communicates with battery management systems to obtain the working state of battery, and take necessary protection action when battery alarm occurs.
- It communicates with the wind turbine to obtain the current power data of the wind turbine. If necessary, the power limiting control should be carried out.

- It communicates with the measuring watt-hour meter to obtain the power value of the POC and the charging and discharging power value of power conversion system as the basis of grid-connected power control.
- It communicates with the intelligent load monitoring system to obtain the working state of the current loads, and to carry out switching action and power control.

11.3.5   Operational results

In this project, technical achievements were investigated while the microgrid was operating in different conditions. The aim was to demonstrate the performances of the microgrid associated with protection system, transient control system, and dynamic control system.

11.3.5.1   Fault and protection system

To verify the performance of the protection system, a single-phase short circuit test was carried out under island mode of the microgrid. First of all, a 0.7 Ω single-phase grounding resistance was connected in parallel at the lower terminal of load circuit breaker of CB26. To facilitate the test, the grounding branch was also connected with a circuit breaker. Then, the mode transfer from grid connected to island mode was carried out. The power quality and power data in island mode steady-state operation are shown in Table 11.5.

After the above operations were completed, the circuit breaker of the fault branch was closed remotely, resulting in the single-phase to ground fault. The fault was identified and successfully isolated by the microgrid fault protection system within 80 ms, and the isolated microgrid system returns to normal operation. The voltage, power, and current data of the fault process are as shown in Table 11.6.

TABLE 11.5 Voltage, current, and power quality in island mode steady-state operation.

Testing point	Parameters	Phase A	Phase B	Phase C
Lower terminal of switch CB26	Voltage (V)	230.20	230.08	229.24
	Current (A)	16.20	16.80	16.80
	Voltage THD (%)	400.00	350.00	350.00
	Active power (W)	3700.00	3850.00	3850.00
	Reactive power (Var)	3750.00	3850.00	3850.00
	Apparent power (VA)	49.96	16,000.00	15,500.00
	System frequency (Hz)		50.01	

As shown in Table 11.6, the current of phase A increases significantly, indicating that fault occurs at phase A, which is the single-phase current increase phenomenon during the fault process. After the fault lasts for about 80 ms, it is cleared. After clearance, the system returns to normal operation. The voltage, power, and current data under the recovery of normal operation are listed in Table 11.7.

The whole fault process and the real-time curve of voltage and current before, during, and after fault are shown in Fig. 11.10.

From the real-time curve analysis, the fault occurred at 04.02.345, the phase A current increases significantly, and the three-phase voltage imbalance appears. After the fault lasted for nearly 80 ms, it was identified and successfully isolated by the microgrid fault protection system, and the voltage returns to its normal operation after isolation.

TABLE 11.6 Voltage, current, and power data during fault.

Testing point	Parameters	Phase A	Phase B	Phase C
Lower terminal of switch CB26	Voltage (V)	227.02	229.54	229.68
	Current (A)	83.80	16.80	18.30
	Voltage THD (%)	6800.00	650.00	450.00
	Active power (W)	5650.00	3750.00	4150.00
	Reactive power (Var)	15,000.00	3850.00	4200.00
	Apparent power (VA)	49.94	16,000.00	15,500.00
	System frequency (Hz)		50.01	

TABLE 11.7 Voltage, current, and power measurements after fault clearance operating condition.

Testing point	Parameters	Phase A	Phase B	Phase C
Lower terminal of switch CB26	Voltage (V)	236.44	236.40	235.56
	Current (A)	0	0	0
	Active power (W)	0.00	0	0
	Reactive power (Var)	0	0	0
	Apparent power (VA)	0	0	0
	System frequency (Hz)		49.94	

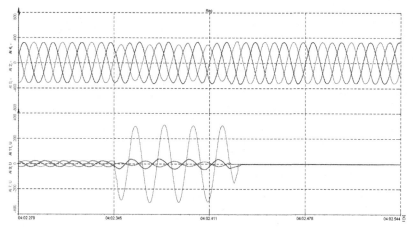

FIGURE 11.10   The real-time curve of voltage and current in the whole fault process.

11.3.5.2   Transient disturbance suppression

The suppression of transient disturbance of the microgrid under load startup was carried out in the project. The project electrical layout diagram is depicted as shown in Fig. 11.9. The project is composed of two submicrogrid systems (one section of microgrid bus is connected to the grid through the islanding circuit breaker of CB10, and the second section of microgrid bus is connected to the grid through the circuit breaker of CB12). The test was conducted on the first section of the microgrid bus. This test scheme was aimed to be carried out in island mode. At the beginning, the two sections were operating in grid-connected mode.

In this state, it was confirmed that the circuit breaker CB31 and CB36 of the vanadium redox flow energy storage were closed, and the circuit breaker CB25 of critical load was closed (note that this load is a critical load in the industrial park, no matter what kind of test is conducted, no power-off operation is allowed). The circuit breakers CB26, CB15, and CB16 of impulse motor branch were disconnected. The switch CB27 of supercapacitor branch was opened, and the electric branch was controlled by energy storage. The circuit breakers CB29, CB13, and CB10 were closed. The PV circuit breakers CB19, CB20, and CB22 were also closed, but the PV power generation in island mode was large, and the energy storage and load could not be completely absorbed after switching into island. So, the PV power should have been regulated before switching to island mode.

For this test, the results of voltage, current, and power quality of the microgrid were recorded before islanding and are listed in the Table 11.8.

When the above process was completed, seamless switching operation from grid-connected to island mode was carried out. The results during this islanding stage are listed in Table 11.9.

TABLE 11.8 Voltage, current, and power quality of the microgrid before switched to the island mode.

Testing point	Parameters	Phase A	Phase B	Phase C
Upper terminal of islanding switch CB10	Voltage (V)	229.39	230.3	230.7
	Current (A)	50.4	70.5	68.2
	Voltage THD (%)	2.57	2.55	2.63
	Active power (W)	5000.00	6000.00	9500.00
	Reactive power (Var)	9500.00	14,500.00	12,000.00
	Apparent power (VA)	11,500.00	16,500.00	15,500.00
	System frequency (Hz)		50.00	
Lower terminal of islanding switch CB10	Voltage (V)	229.5	230.4	230.7
	Current (A)	50.92	70.42	68.2
	Voltage THD (%)	2.55	2.55	2.62
	Active power (W)	5000.00	6000.00	9000.00
	Reactive power (Var)	9500.00	14,500.00	12,000.00
	Apparent power (VA)	11,500.00	16,000.00	15,500.00
	System frequency (Hz)		50.00	

The next process aimed to conduct the load step change. It was carried out when the motor load of 20 kW (switch on CB26) was connected to the microgrid network. The results of voltage and current waveforms are demonstrated in Fig. 11.11.

According to the curve in Fig. 11.11, the voltage is forced to drop at the moment of motor startup lasting nearly 200 ms. From the current curve, it can be seen that the instantaneous impulse current reaches about 10 times that of the normal current at the moment of motor starting, and it returns to normal after 200 ms. It shows that the capacity of power supply operating in voltage-frequency (U/f) mode is limited and the response speed to load step change is slow in island operation mode of the microgrid.

To compare and test the function of transient disturbance control strategy, the experiment first was operated in island mode after seamless switching. Then, the circuit breaker CB27 of supercapacitor branch was closed, and the voltage and current curves were recorded. After the state was stable, the motor load of 20 kW (switch on CB26) was connected into the operational network, and the voltage and current data were taken and the results are demonstrated in Fig. 11.12.

TABLE 11.9 Testing results of voltage, current, and power quality during islanding stage.

Testing point	Parameters	Phase A	Phase B	Phase C
Upper terminal of island main switch CB10	Voltage (V)	229.75	231.2	231.29
	Current (A)	0	0	0
	Voltage THD (%)	2.61	2.52	2.57
	System frequency (Hz)		50.01	
Lower terminal of island main switch CB10	Voltage (V)	229.52	229.24	229.66
	Current (A)	0	0	0
	Voltage THD (%)	1.43	1.36	1.52
	System frequency (Hz)		49.98	

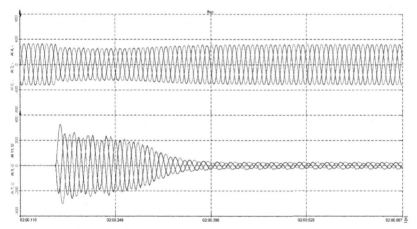

FIGURE 11.11   Instantaneous voltage–current curve without transient disturbance under motor load startup.

According to the voltage curve in Fig. 11.12, the voltage is forced to drop at the moment of motor startup, but the time of voltage drop lasts only about 40 ms. From the current curve in the figure, it can be seen that the instantaneous impulse current is only about three times that of the normal current at the moment of starting of the motor load, and it only lasts for 60 ms. After that, it gradually recovers to its normal value. It shows that the capacity of U/f power supply was limited and the response speed to impulse load was slow under island operation mode of the microgrid with

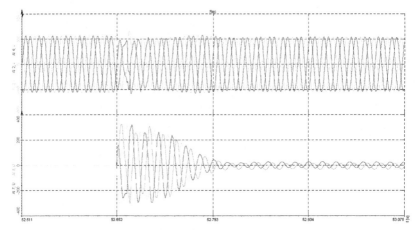

FIGURE 11.12 Instantaneous voltage and current curve with transient disturbance control strategy under motor load startup.

the absence of the transient disturbance control strategy. However, with the addition of this strategy, the system gains the ability to recover quickly from the transient disturbance to the normal voltage level.

11.3.5.3 Dynamic disturbance suppression

Tests were carried out to verify the capability of the dynamic disturbance control system to suppress voltage and frequency oscillation in island mode of operation due to high penetration of renewable energy. It was conducted on one section of the microgrid bus. The two sections of the microgrid were basically operating in grid-connected and switched to island mode.

After switching from grid-connected to island, the circuit breaker CB34 of load branch was closed, and then all branch circuit breakers CB19, CB20, CB21, and CB22 of PV power generation system were closed. During this time, the voltage, current, and power quality were measured and the results are listed in Table 11.10.

When the PV power in four branches was gradually being increased, the microgrid dynamic disturbance control system was activated only to start the converter. When the PV power increases to a certain level, the system begins to show a resonance phenomenon, and the system voltage and current oscillate. The voltage and current show small amplitude oscillation. The more PV power increases, the greater is the oscillation amplitude that occurs. The oscillation voltage curve is illustrated in Fig. 11.13.

With the further increase of PV power capacity, the system voltage still tends to be stable. The voltage, current, and power quality under this condition are shown in Table 11.11 below.

TABLE 11.10 Voltage, current, and power quality of the network.

Testing point	Parameters	Phase A	Phase B	Phase C
Circuit breaker of 1-1-F5	Voltage (V)	222.92	223.43	224.2
	Current (A)	280.5	281.53	270.2
	Voltage THD (%)	2.02	1.98	2.01
	System frequency (Hz)		50.12	

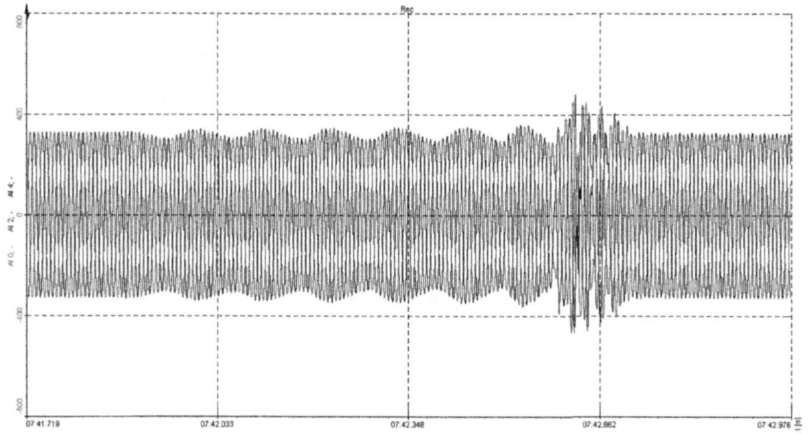

FIGURE 11.13   Real-time curve of voltage oscillation under the variation of PV power generation.

TABLE 11.11 Voltage and current data after dynamic stability control.

Testing point	Parameters	Phase A	Phase B	Phase C
Circuit breaker of 1-1-F5	Voltage (V)	218.06	217.53	218.27
	Current (A)	311.40	309.40	299.5
	Voltage THD (%)	2.23	2.07	2.32
	System frequency (Hz)		50.12	

11.3.5.4   Seamless switching between grid-connected and island modes

Seamless switching was undertaken to demonstrate the performance of the transient and dynamic control strategies. Firstly, the microgrid was switched to island mode. The voltage, current, and power quality of the microgrid before switching to island mode are shown in Table 11.12.

TABLE 11.12 Voltage, current, and power quality of grid-connected mode.

Testing point	Parameters	Phase A	Phase B	Phase C
Upper terminal of islanding switch CB10	Voltage (V)	229.34	230.32	230.68
	Current (A)	50.4	70.5	68.2
	Voltage THD (%)	2.57	2.57	2.63
	Active power (W)	5000.00	6000.00	9500.00
	Reactive power (Var)	9500.00	14,500.00	12,000.00
	Apparent power (VA)	11,500.00	16,500.00	15,500.00
	System frequency (Hz)		50.02	
Lower terminal of islanding switch CB10	Voltage (V)	229.46	230.36	230.74
	Current (A)	50.92	70.38	68.23
	Voltage THD (%)	2.57	2.54	2.65
	Active power (W)	5000.00	6000.00	9000.00
	Reactive power (Var)	9500.00	14,500.00	12,000.00
	Apparent power (VA)	11,500.00	16,000.00	15,500.00
	System frequency (Hz)		50.02	

Then, all PV generation power, ESS power, and load power at the lower terminal of the switch CB10 were imposed to imbalance (convenient for unintentional seamless switching). In this state, the islanding main breaker CB10 was tripped off, and the microgrid was operating in the island mode. At the switching moment, the transient disturbance control system detects the power imbalance of the island operation system, and starts the supercapacitor battery instantly to balance the power in the island microgrid. Due to its fast-discharging rate, the supercapacitor cannot support the power balance of the system for a long time, but is suitable for a transient disturbance support. Hence, the transient disturbance control system did quickly adjust the power balance among PV generation, ESS, and loads. The voltage and current power quality in the island mode of operation is as in Table 11.13.

Comparing the grid-connected and island modes of operation of the microgrid, the highest power quality voltage THD of 1.51 can be seen in island mode operation, while the minimum voltage THD of 2.56 for grid-connected mode was recorded.

As illustrated in the real-time curve of the voltage of Fig. 11.14, the results show that the power quality of the island microgrid system is better

TABLE 11.13 Voltage, current, and power quality of island mode.

Testing point	Parameters	Phase A	Phase B	Phase C
Upper terminal of islanding switch CB10	Voltage (V)	229.74	231.17	231.26
	Current (A)	0	0	0
	Voltage THD (%)	2.63	2.57	2.54
	System frequency (Hz)		50.05	
Lower terminal of islanding switch CB10	Voltage (V)	229.47	229.31	229.59
	Current (A)	0	0	0
	Voltage THD (%)	1.47	1.35	1.54
	System frequency (Hz)		49.97	

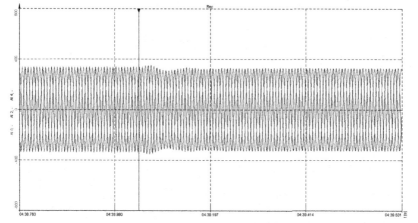

FIGURE 11.14 Waveforms of voltage with the implementation of transient disturbance control strategy under seamless switching from grid-connected to island mode.

than that of power grid due to the implementation of microgrid transient disturbance control systems under the condition of seamless switching from grid-connected to island mode.

It can be seen from the real-time curve that there is no voltage sag in the seamless switching process. The switching process only takes 60 ms to complete. There is no equipment shutdown and power interruption during the switching process. The system verifies unplanned seamless switching.

The test of switching from island to grid connection was also carried out on the above setup. While the system started to operate in the island mode, the grid-connected circuit breaker CB10 was closed. At the moment of closing the circuit breaker CB10, the transient disturbance control system detected the closing command of the islanding circuit breaker, and quickly adjusted the converter side of the equipment in the closed-loop control system so that the island power system can complete the synchronization task in a very short time. When the difference between the amplitude, phase, and frequency of the island voltage and the amplitude, phase, and frequency of the grid voltage reaches the synchronization standard, closing of the breaker was carried out. The voltage, current, and power quality before the synchronization of unplanned island to grid connection are shown in Table 11.14.

The voltage, current, and power quality curve after unintentional island mode to grid connection is described in Table 11.15.

It can be seen from the power quality (Table 11.15) that the voltage THD of the power grid-connected mode of operation of the microgrid is 1.67% before the unplanned island to grid connection, while voltage THD in island mode is 1.37%. After successful seamless switching, the system was in the grid-connected mode, and the power quality of voltage THD recorded is 1.67%. The results show the capability of the transient disturbance control system in the islanding state of the microgrid, and the power quality is better than that of the grid-connected one.

In Fig. 11.15, the period before 03.52.068 was the synchronization process between the grid voltage and the island microgrid carried out. At 03.52.068 the transient disturbance system sent the closing instruction to the main breaker CB10, and the island microgrid system was successfully closed. As shown on the curves from the figure, the amplitude, phase, and frequency of the island microgrid voltage are almost the same as that after the synchronization was adjustment made. There was no equipment failure

TABLE 11.14 Voltage, current, and power quality before unintentional island transfer to grid connection.

Testing point	Parameters	Phase A	Phase B	Phase C
Upper terminal of islanding switch CB10	Voltage (V)	229.79	230.5	230.61
	Current (A)	0	0	0
	Voltage THD (%)	1.72	1.69	1.63
	System frequency (Hz)		50.02	
Lower terminal of islanding switch CB10	Voltage (V)	229.66	229.46	229.88
	Current (A)	0	0	0
	Voltage THD (%)	1.48	1.37	1.54
	System frequency (Hz)		50.02	

TABLE 11.15 Voltage, current, and power quality after unintentional island transfer to grid connection.

Testing point	Parameters	Phase A	Phase B	Phase C
Upper terminal of islanding switch CB10	Voltage (V)	229.14	229.66	229.79
	Current (A)	48.82	69.90	68.03
	Voltage THD (%)	1.65	1.64	1.65
	Active power (W)	5000.00	5000.00	9500.00
	Reactive power (Var)	9000.00	15,000.00	12,000.00
	Apparent power (VA)	11,000.00	16,000.00	15,500.00
	System frequency (Hz)		50.02	
Lower terminal of islanding switch CB10	Voltage (V)	229.29	229.78	230.00
	Current (A)	48.90	69.50	67.90
	Voltage THD (%)	1.67	1.65	1.67
	Active power (W)	5000.00	5000.00	9500.00
	Reactive power (Var)	11,000.00	16,000.00	15,500.00
	Apparent power (VA)	9000.00	14,500.00	12,000.00
	System frequency (Hz)		50.03	

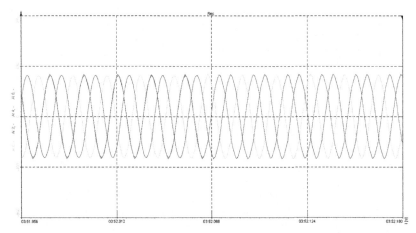

FIGURE 11.15 Voltage waveform during switching from island to grid-connected mode.

during the closing process, which realizes the seamless unplanned switching from island to grid-connected mode.

References

[1] FORWARD Business Information Co. Ltd, Report of industry strategic planning and enterprise strategic planning on China electrical power industry (2019-2024), 2019.

[2] IEC, IEC TS 62898-3-1:2020, Microgrids—Part 3-1: Technical requirements—Protection and dynamic control, 2020.

[3] IEC, IEC TS 62898-2:2018, Microgrids—Part 2: Guidelines for operation, 2018.

[4] Standardization Administration of China, GB/T38953-2020: Technical requirements for relay protection of microgrid, 2020.

[5] D. Zheng, M. Shi, Y. Wang, A.T. Eseye, J. Zhang, Day-ahead wind power forecasting using a two-stage hybrid modeling approach based on SCADA and meteorological information, and evaluating the impact of input-data dependency on forecasting accuracy, Energies 10 (12) (2017). Available from: https://doi.org/10.3390/en10121988.

[6] D. Zheng, A.T. Eseye, J. Zhang, H. Li, Short-term wind power forecasting using a double-stage hierarchical ANFIS approach for energy management in microgrids, Prot. Control. Mod. Power Syst. 2 (13) (2017). Available from: https://doi.org/10.1186/s41601-017-0041-5.

[7] L. Wang, L. Zhang, C. Xu, A. Tesfaye Eseye, J. Zhang, D. Zheng, Dynamic economic scheduling strategy for a stand-alone microgrid system containing wind, PV solar, diesel generator, fuel cell and energy storage: a case study, in: IOP Conference Series: Earth and Environmental Science, 2018, Available from: https://doi.org/10.1088/1755-1315/168/1/012006.

[8] A.T. Eseye, J. Zhang, D. Zheng, Short-term photovoltaic solar power forecasting using a hybrid Wavelet-PSO-SVM model based on SCADA and mteorological information, Renew. Energy 118 (2018) 357–367. Available from: https://doi.org/10.1016/j.renene.2017.11.011.

[9] A. Tesfaye, J.H. Zhang, D.H. Zheng, D. Shiferaw, Short-term wind power forecasting using artificial neural networks for resource scheduling in microgrids, Int. J. Sci. Eng. Appl. 5 (3) (2016). Available from: https://doi.org/10.7753/ijsea0503.1005.

[10] H. Li, A.T. Eseye, J. Zhang, D. Zheng, Optimal energy management for industrial microgrids with high-penetration renewables, Prot. Control. Mod. Power Syst. 2 (12) (2017). Available from: https://doi.org/10.1186/s41601-017-0040-6.

[11] S. Netsanet, J. Zhang, D. Zheng, Bagged decision trees based scheme of microgrid protection using windowed fast Fourier and wavelet transforms, Electron. 7 (5) (2018). Available from: https://doi.org/10.3390/electronics7050061.

[12] Y.K. Semero, J. Zhang, D. Zheng, PV power forecasting using an integrated GA-PSO-ANFIS approach and Gaussian process regression based feature selection strategy, CSEE J. Power Energy Syst. 4 (2) (2018). Available from: https://doi.org/10.17775/cseejpes.2016.01920.

[13] Y.K. Semero, J. Zhang, D. Zheng, Optimal energy management strategy in microgrids with mixed energy resources and energy storage system, IET Cyber-Physical Syst. Theor. Appl. 5 (1) (2020) 80–84. Available from: https://doi.org/10.1049/iet-cps.2019.0035.

[14] A.T. Eseye, J. Zhang, D. Zheng, D. Wei, Optimal energy management strategy for an isolated industrial microgrid using a modified particle swarm optimization, in 2016 IEEE International Conference on Power and Renewable Energy, ICPRE 2016, 2016, pp. 494–498, Available from: https://doi.org/10.1109/ICPRE.2016.7871126.

[15] A.T. Eseye, Z. Dehua, Wind power generation field wind energy prediction method and equipment, CN107124003A;CN107124003B, 2017.

[16] A.T. Eseye, Z. Dehua, Method, relay and system for microgrid protection, CN108110738A;CN108110738B, 2018.

[17] L. Gangju, Z. Dehua, Z. Yunzhi, Compensation device and method for transmission power angle between micro-grid and large power grid, CN104167728A;CN104167728B, 2014.

[18] Z. Dehua, Z. Wei, W. Dan, C. Xiaohai, Control method and equipment for super-capacitor in micro-grid, CN106549407A; CN106549407B, 2017.

[19] D. Zheng, W. Zhang, D. Wei, F. Qiu, Microgrid dynamic stability control system and microgrid dynamic stability control method, US 10,637,242 B2, 2020.

[20] S.N. Alemu, Goldwind PV forecaster-software copyright, 2017SR467382, 2017.

[21] A. Berrueta, I. San Martín, P. Sanchis, A. Ursúa, Lithium-ion batteries as distributed energy storage systems for microgrids, Distributed Energy Resources in Microgrids: Integration, Challenges and Optimization, Rajeev K. Chauhan; Kalpana Chauhan, Ed., 2019, pp. 143−183.

[22] Y. Chen, Z. Yang, Y. Wang, SOC estimation of lead carbon batteries based on the operating conditions of an energy storage system in a microgrid system, Energies (2019). Available from: https://doi.org/10.3390/en13010033.

[23] K. Monteiro, M. Marot, H. Ibn-Khedher, Review on microgrid communications solutions: a named data networking—fog approach, in 2017 16th Annual Mediterranean Ad Hoc Networking Workshop, Med-Hoc-Net 2017, 2017, Available from: https://doi.org/10.1109/MedHocNet.2017.8001656.

[24] E. Harmon, U. Ozgur, M.H. Cintuglu, R. De Azevedo, K. Akkaya, O.A. Mohammed, The internet of microgrids: a cloud-based framework for wide area networked microgrids, IEEE Trans. Ind. Inform. 14 (3) (2018) 1262−1274. Available from: https://doi.org/10.1109/TII.2017.2785317.

[25] T. Rajeev, S. Ashok, A cloud computing approach for power management of microgrids, in: 2011 IEEE PES International Conference on Innovative Smart Grid Technologies-India, ISGT India 2011, 2011, Available from: https://doi.org/10.1109/ISET-India.2011.6145354.

[26] D. Zheng, Y.K. Semero, J. Zhang, D. Wei, Short-term wind power prediction in micro-grids using a hybrid approach integrating genetic algorithm, particle swarm optimization, and adaptive neuro-fuzzy inference systems, IEEJ Trans. Electr. Electron. Eng. 13 (11) (2018) 1561−1567. Available from: https://doi.org/10.1002/tee.22720.

[27] S. Netsanet, J. Zhang, D. Zheng, R.K. Agrawal, F. Muchahary, An aggregative machine learning approach for output power prediction of wind turbines, in: 2018 IEEE Texas Power and Energy Conference, TPEC 2018, 2018, Available from: https://doi.org/10.1109/TPEC.2018.8312085.

[28] S. Netsanet, D. Zheng, L. Zhang, M. Hui, Input parameters selection and accuracy enhancement techniques in PV forecasting using artificial neural network, in: 2016 IEEE International Conference on Power and Renewable Energy, ICPRE 2016, 2017, Available from: https://doi.org/10.1109/ICPRE.2016.7871139.

[29] S. Netsanet, J. Zhang, D. Zheng, Short term load forecasting using wavelet augmented non-linear autoregressive neural networks: a single customer level perspective, in: 2018 IEEE 3rd International Conference on Big Data Analysis, ICBDA 2018, 2018, Available from: https://doi.org/10.1109/ICBDA.2018.8367717.

[30] Y.K. Semero, J. Zhang, D. Zheng, EMD$-$PSO$-$ANFIS-based hybrid approach for short-term load forecasting in microgrids, IET Gener. Transm. Distrib. (2020). Available from: https://doi.org/10.1049/iet-gtd.2019.0869.

[31] D. Zheng, A.T. Eseye, J. Zhang, A communication-supported comprehensive protection strategy for converter-interfaced islanded microgrids, Sustainability (2018). Available from: https://doi.org/10.3390/su10051335.

Index

Printed in the United States
by Baker & Taylor Publisher Services